Grassland ecosystems of the world: analysis of grasslands and their uses

THE INTERNATIONAL BIOLOGICAL PROGRAMME

The International Biological Programme was established by the International Council of Scientific Unions in 1964 as a counterpart of the International Geophysical Year (IGY). The subject of the IBP was defined as 'The Biological Basis of Productivity and Human Welfare', and the reason for its establishment was recognition that the rapidly increasing human population called for a better understanding of the environment as a basis for the rational management of natural resources. This could be achieved only on the basis of scientific knowledge, which in many fields of biology and in many parts of the world was felt to be inadequate. At the same time it was recognised that human activities were creating rapid and comprehensive changes in the environment. Thus, in terms of human welfare, the reason for the IBP lay in its promotion of basic knowledge relevant to the needs of man.

The IBP provided the first occasion on which biologists throughout the world were challenged to work together for a common cause. It involved an integrated and concerted examination of a wide range of problems. The Programme was co-ordinated through a series of seven sections representing the major subject areas of research. Four of these sections were concerned with the study of biological productivity on land, in freshwater, and in the seas, together with the processes of photosynthesis and nitrogen-fixation. Three sections were concerned with adaptability of human populations, conservation of ecosystems and the use of biological resources.

After a decade of work, the Programme terminated in June 1974 and this series of volumes brings together, in the form of syntheses, the results of national and international activities.

INTERNATIONAL BIOLOGICAL PROGRAMME 18

Grassland ecosystems of the world: analysis of grasslands and their uses

EDITED BY

R. T. Coupland

Department of Plant Ecology
University of Saskatchewan
Saskatoon, Canada

CAMBRIDGE UNIVERSITY PRESS

CAMBRIDGE
LONDON · NEW YORK · MELBOURNE

Published by the Syndics of the Cambridge University Press
The Pitt Building, Trumpington Street, Cambridge CB2 1RP
Bentley House, 200 Euston Road, London NW1 2DB
32 East 57th Street, New York, NY 10022, USA
296 Beaconsfield Parade, Middle Park, Melbourne 3206, Australia

First published 1979

Printed in Great Britain at the
University Press, Cambridge

ISBN 0 521 21867 5

Contents

Table des matières

Содержание

Contenido

List of Contributors

Chief editor

R. T. Coupland	Department of Plant Ecology, University of Saskatchewan, Saskatoon, Saskatchewan S7N 0WO, Canada

Subeditors

R. Misra	Department of Botany, Banaras Hindu University, Varanasi–221005, India
M. Rychnovská	Ecological Department, Institute of Botany, Czechoslovak Academy of Sciences, Stará 18, Brno 66261, Czechoslovakia
L. Ryszkowski	Department of Agrobiology, Polish Academy of Sciences, Swierczewskiego 19, 60–809 Poznań, Poland
W. M. Willoughby	Division of Animal Production, Pastoral Research Laboratory, CSIRO, Armidale, New South Wales 2350, Australia

Authors

Ambasht, R. S.	Department of Botany, Banaras Hindu University, Varanasi–221005, India
Balátová-Tuláčková, E.	Ecological Department, Institute of Botany, Czechoslovak Academy of Sciences, Stará 18, 66261 Brno, Czechoslovakia
Bazilevich, N. I.	Dokuchaev Soil Institute, 7 Pyzhevski per. Moscow 17, USSR
Biederbeck, V. O.	Canada Agriculture Research Station, Swift Current, Saskatchewan, Canada
Clark, F. E.	Soil and Water Conservation Research Division, Agricultural Research Service, United States Department of Agriculture, P.O. Box E, Fort Collins, Colorado 80523, USA
Dash, M. C.	Postgraduate Department of Biological Science, Sambalpur University, Sambalpur, Orissa, India
Davidson, R. L.	Division of Entomology, Pastoral Research Laboratory, CSIRO, Armidale, New South Wales 2350, Australia

Dwivedi, R. S.	Department of Botany, Banaras Hindu University, Varanasi–221005, India
Dyer, M. I.	College of Natural Sciences, Natural Resource Ecology Laboratory, Colorado State University, Fort Collins, Colorado 80523, USA
French, N. R.	Natural Resource Ecology Laboratory, Colorado State University, Fort Collins, Colorado 80523, USA
Gołębiowska, J.	Department of Agricultural Microbiology, Institute of Soil Science, Agricultural Academy, Wotyńska 35, Poznań, Poland
Gupta, R. K.	Division of Plant Science, Soil Conservation Research Centre, 218 Kaulagarh Road, Dehra Dun, India
Haas, H. II,	Ritterstr. 16, D2957, Westoverledingen, West Germany
Hutchinson, K. J.	Division of Animal Production, Pastoral Research Laboratory, CSIRO, Armidale, New South Wales 2350, Australia
Joshi, M. C.	Department of Botany, Birla Institute of Technology and Science, Pilani, India
King, K. L.	Division of Animal Production, Pastoral Research Laboratory, CSIRO, Armidale, New South Wales 2350, Australia
Misra, K. C.	Department of Botany, Banaras Hindu University, Varanasi–221005, India
Numata, M.	Laboratory of Ecology, Faculty of Science, Chiba University, Chiba, Japan
Paul, E. A.	Department of Soil Science, University of Saskatchewan, Saskatoon, Saskatchewan S7N 0WO, Canada
Ricou, G. A. E.	Laboratoire de Zoologie Agricole, Institut National de la Recherche Agronomique, 16 Rue de Dufay, Rouen 76, France
Sims, P. L.	Department of Range Science, Colorado State University, Fort Collins, Colorado 80523, USA
Singh, J. S.	Department of Botany, Kumaun University, Nainital 263002, India
Singh, K. P.	Department of Botany, Banaras Hindu University, Varanasi–221005, India
Speidel, B.	Hessische Lehr- und Forschungsanstalt für Grünlandwirtschaft, Eichhof in Bad Hersfeld, West Germany

Till, A. R. — Division of Animal Production, Pastoral Research Laboratory, CSIRO, Armidale, New South Wales 2350, Australia

Titlyanova, A. A. — Institute of Soil and Agrochemistry, USSR Academy of Sciences, 18 Sovetskaya, Novosibirsk 99, USSR

Úlehlová, B. — Ecological Department, Institute of Botany, Czechoslovak Academy of Sciences, Stará 18, 66261 Brno, Czechoslovakia

Van Dyne, G. M. — Department of Range Science, Colorado State University, Fort Collins, Colorado 80523, USA

Vickery, P. J. — Division of Animal Production, Pastoral Research Laboratory, CSIRO, Armidale, New South Wales 2350, Australia

Wójcik, Z. — Institute of Geography, Nowy Swiat 72, Warsaw, Poland

Yadava, P. S. — Life Sciences Division, Jawaharlal Nehru University, Centre of Post-Graduate Studies, Imphal–795001 (Manipur), India

Foreword

The first official mention of the Grasslands Biome studies in IBP is in a report of a meeting of the Productivity of Terrestrial Communities section – IBP(PT) – held in Warsaw 30 August to 6 September 1966. The main purpose of the technical sessions at Warsaw was to consider 'the principles and methodology of secondary productivity of terrestrial ecosystems' but among the working groups which met during that meeting was one devoted to the study of grassland ecosystems. As IBP News (SCIBP, 1966) records: 'The working team concerned with grassland ecosystems discussed the project of a fully coordinated research programme. Five specialists from different countries were concerned with this during the meeting, and it was decided to elaborate the project subsequently by mail.'

The grasslands studies became Theme 1 of IBP(PT). Later, when the distinction was sharpened between major studies on ecosystems and investigations on special topics such as certain groups of consumers, the grasslands projects became collectively known as the Grasslands Biome Studies. By 1972, when the majority of IBP field studies were terminated, some thirty nations had shared in the work of the grasslands group, and several hundreds of scientists had been involved along with a much greater number of support staff.

The general aims of the grasslands theme were summarized in IBP News (SCIBP, 1969) as follows: 'Grasslands are one of the most important of terrestrial ecosystem types. Large areas occupy the interior of the principal continents, and provide, when managed for crop or meat production, a major source of man's food. The object of the IBP programme is to learn more about the organic production and overall energy flow of the world's grassland areas. Wherever feasible, concurrent studies with those on natural grasslands will be made on managed areas (pastures and croplands) and successional grasslands in the same region. This will make possible the provision of ecologically sound recommendations for the future utilization of the world's grassland areas.'

The considerable task of drawing up detailed plans and later of ensuring a high degree of coordination among the various projects was performed by a small Working Group. It met for the first time in 1967 under the chairmanship of Dr R. G. Weigert and was attended by scientists from six countries. It was held, appropriately, at Saskatoon, Canada, where, under the direction of Dr R. T Coupland, a Canadian IBP Grasslands team was already at work. Over the period 1967–74 the Biome Coordinating Committee, with various changes in chairmanship, met on fifteen occasions.

In 1972 the Scientific Committee of IBP (SCIBP) appointed Chief Editors who, in association with subeditors or coordinators, were to be responsible for the preparation and completion of the synthesis volumes.

IBP brought together scientists from countries with different types of government; with very different attitudes towards the place of science in their culture; and with very different levels of scientific development. In addition, the acceptance of English as the language of communication within IBP placed an added burden on many scientists throughout the world. These strains were compounded because those of us with ready access to tape recorders, typewriters, secretaries, copying machines, a superabundance of paper and large, if not unlimited, postage accounts, produced a mass of voluminous reports and correspondence that must have added many hours of labour to scientists beyond the English-speaking world. Now, having studied once again many of the reports received during the course of the grasslands studies, I want to place on record the appreciation of the English-speaking contingent to those many workers who had to struggle with a language which lacks the logic of a formalized grammar. Not least among their problems was that of appreciating those subtleties of meaning which, although not clearly defined in a mother tongue, are, through custom and practice, understood by those born to it.

Whilst the basic approaches to scientific discovery may appear universal, the degree of universality tends to be exaggerated and this is certainly the case in a diverse and very often diffuse subject such as ecology. Ecology was, and to a considerable extent remains, a highly personalized science. In *The evolution of IBP* (Worthington, 1975), I quoted a statement of Margalef: 'Ecosystems reflect the physical environment in which they have developed, and ecologists reflect the properties of the ecosystems in which they have grown and matured. All schools of ecology are strongly influenced by a *genius loci* that goes back to the local landscape.'

Nowhere else in IBP(PT) was the 'ecologist–locale' link more in evidence than in the Grasslands Biome Studies. On many occasions the energy expended in arguments on aims, methods and significance of results appeared to be approaching an explosion level but compensating control systems, usually in the form of good humour, succeeded in keeping the 'reactors' below the critical point of disintegration. It is not the first time in the history of biology that heated discussions have paved the way for scientific clarification. IBP productivity studies in particular have been fraught with arguments arising from differences in philosophy, definitions, and the complexity of the organic systems being studied. It is worth recalling a comment by Professors K. Petrusewicz and A. Macfadyen (1970) in their preface to *Productivity of terrestrial animals*: 'It has not always been easy to agree on what to say or how to say it. Our discussions have been many and prolonged and the final outcome has usually been based, we believe, on understanding rather than on compromise.' It can certainly be said that discussions within the grasslands theme were many and prolonged.

The USA and Australian IBP grasslands teams were composed largely of scientists whose background and interests centred on domestic animal production. The European, Japanese and Canadian teams on the other hand were less concerned with 'large' consumers and devoted considerable attention to the invertebrate and small vertebrate components of grassland ecosystems. In the early stages of IBP many European ecologists viewed with some scepticism the wholehearted devotion of their North American confreres to the 'systems approach'. They pointed to many unresolved taxonomic problems; to the paucity of knowledge on the feeding habits of key organisms; to the fragmentary information available on the seasonal changes in the contribution of known species to different trophic levels; and, above all, to the need for a thoroughgoing analysis of production within species or groups of organisms as distinct from obtaining knowledge on yields or standing crops.

I believe that it is important to recall that the operations phase of IBP(PT) reached its maximum in 1971, the year when ecologists should have been celebrating the centenary of the birth of A. G. Tansley who in 1935 added the term *ecosystem* to the ecologist's vocabulary. Until the advent of the computer and the creative imagination of a new generation of ecologists with the ability to use mathematical techniques to the full, the ecosystem as conceived by Tansley remained a mental construct, valuable and certainly genuine, unlike the Philosopher's Stone, but seemingly beyond total analysis.

Thirty years after Tansley's publication, IBP provided the opportunity for a thorough examination of the interaction of organisms and environment which together, to use Tansley's phrase, formed one physical system. First, the Canadian IBP(PT) Committee, and later the USA National Committee, decided on a systems approach in their programmes, using large multidisciplinary teams. In both cases grasslands were selected as the first ecosystems to be studied intensively, using the still embryonic ideas of systems ecology.

At the time that IBP(PT) embarked upon a systems approach, it should be remembered that the systems which were to be studied were largely unknown. In engineering and economics, where systems dynamics had become a way of life, the systems could be simulated with a high degree of precision and the relationship between individul components and particular outputs could be explored. The systems ecologist, at the beginning of IBP, was faced with a very different situation. Knowledge of the component parts and processes was fragmentary even in the most thoroughly explored ecosystems. Thus, the systems ecologist was forced to look at his systems in a manner very different from that followed by his colleagues in aeronautics or in business management.

Natural and cultivated grasslands provide man with the major part of his food supply. They cover at least 23 % of the land surface of the globe. Whilst some parts are becoming desert, elsewhere forests are being felled and, on balance, the grasslands biome is on the increase. Because of its extensive nature and diverse vegetational form, the grasslands biome has straddled three biomes in IBP – grasslands, tundra and arid lands. Scientists from all three biomes met at various workshops to discuss common problems, especially those arising from the utilization of the biomes for grazing. In 1969, largely as a result of the successful application of the systems approach in the IBP Grasslands Biome Studies, it was proposed that grasslands specialists from IBP and FAO should meet to explore how far the knowledge gained from IBP studies could be applied and developed further for use on a world scale. The discussions took place and other international organizations showed interest, particularly UNESCO. A proposal for an International Grazing Lands Programme was approved by SCIBP on behalf of the IBP(PT) biome investigators and it received the support of FAO and UNESCO, but it failed to obtain adequate financial support. Nevertheless the exploratory studies made by the tundra, arid lands and grassland biomes were not completely discarded. The MAB Project No. 3 on the *Impact of human activities and land use practices on grazing lands: savanna, grassland (from temperate to arid areas), tundra* owes much to these earlier discussions initiated by IBP.

The two volumes produced on the IBP grassland investigations can be regarded as independent approaches to the problems of studying grassland ecosystems. This volume, titled *Grassland ecosystems of the world* (ed. R. T. Coupland), concentrates on the structure, development and utilisation of the world's grasslands, largely by producing extensive word models of the types of grassland which exist. The volume ends with a summary of the major components of the world's grassland ecosystems. The other volume, *Grasslands, systems analysis and man* (eds. A. Breymeyer & G. M. Van Dyne), is a synthesis of the massive amount of data collected during IBP grasslands studies. Its main purpose is to emphasize the dynamic aspects of the grassland ecosystem by giving emphasis to those processes related to productivity. It provides models and submodels which permit the assessment of the effects of changes or stresses within a grassland ecosystem.

All who have contributed to these volumes are aware of gaps in their attempt to produce a synthesis. The end result is a distillation of numerous points of view and the volumes reflect differences of emphasis and of approach and, above all, disagreements – may it be remembered that disagreements in science are a way of exploring weaknesses and, when accepted in a constructive manner, are the lifeblood of scientific development.

It is always easy to say 'If only it had been possible to...'. All of the major biome studies were inadequately funded during the synthesis phase. This limited the opportunities to bring editors, chapter authors and contributors together for discussion. Furthermore, some contributors were unable to give enough time to the synthesis operation because they had to become involved full-time in other occupations. Nevertheless, in spite of these and other difficulties the grassland biome studies have advanced our knowledge of grassland ecosystems and they are a major contribution towards the formulation of a theory of ecosystems. These are major achievements. Taken together, they should ensure that the management of the world's grasslands can be placed on a more rational basis than has been possible in the past. The results are especially timely because the world may be facing changes in weather patterns which could well reduce food yields below today's levels. Thus, IBP Theme 1 has made a contribution towards the fulfilment of the overall aim of IBP to investigate *The biological basis of productivity and human welfare*.

J. B. Cragg
Killam Memorial Professor
Faculty of Environmental Design
University of Calgary, Canada

Contents of IBP 19

Part III. System utilization

Part I. Introduction

1. Background

R. T. COUPLAND

The objective of research in the grassland habitat theme of IBP was to learn more about the organic production and overall energy flow and nutrient circulation through this important type of ecosystem. While emphasis was to be given to natural grasslands, wherever resources permitted concurrent studies were made in managed areas (pastures and cropland) under comparable conditions of soil and climate. Early in the development of IBP 30 nations started grassland studies (SCIBP, 1969).

This is one of two volumes in which the results of these studies are discussed. Here we will consider trophic structure and function in each of several kinds of grassland (and some related herbaceous ecosystems). In the companion volume, *Grasslands, systems analysis and man* (Breymeyer & Van Dyne, 1979), the processes of production and flow of energy and matter are emphasised by bringing together for each of them information from all studies. The processes are then combined into subsystem and system models in both diagrammatic and analytic presentations. Thus, one wishing to know about annual production of herbage in tropical grasslands, for example, will find a detailed treatment in Part IV of this volume based principally on interpretations by the biomass (harvest) method. However, for a consideration of the dynamics of the photosynthetic process in grasslands as a whole, the reader is referred to the companion volume.

Because of the increasing degree of intensity of management for crop and meat production, natural grasslands are being changed at an accelerating rate. It is urgent to obtain more information about grasslands in their present condition before they are further modified. This is essential to an understanding of mechanisms and processes that control various components of grassland ecosystems. This basic information is required to manage these resources effectively and efficiently. One of the major goals of IBP was to further an understanding of how ecosystems work and the extent to which they can be manipulated by man (SCIBP, 1969). This problem also involves the relationship between the structure and function of communities of organisms within ecosystems. Multidisciplinary studies were established in several countries to provide the necessary information. Each of these intensive studies provides data within one grassland ecosystem, for an analysis of trophic structure, interspecies relationships, primary and secondary production, decomposition, energy flow, nutrient cycling, and the mechanisms by which these observed characteristics are achieved and maintained. Resources were available in only a few study sites for all these undertakings. Many of the studies are concerned solely with an evaluation of organic production or

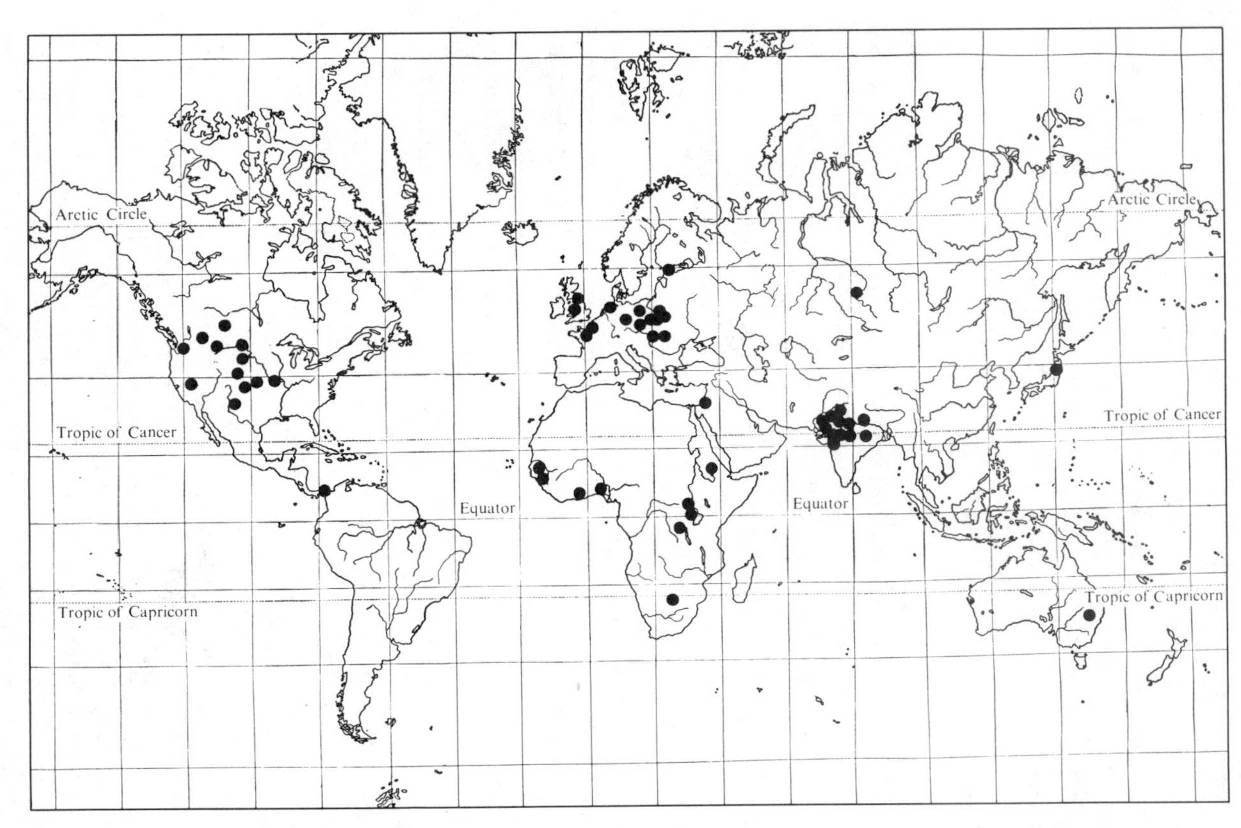

Fig. 1.1. Location of the study sites from which the data discussed in this volume were collected.

energy flow of a plant community of a given level of consumers. Nevertheless, they constitute supporting studies which permit the application of knowledge gained from the intensively studied sites to these less intensive ones. The distribution of study sites that are discussed in this volume is indicated in Fig. 1.1.

Activities of the IBP Grassland Working Group

Because of the vast combined undertaking that was implied by the large number of grassland studies that were undertaken in IBP, it was necessary to develop various means of coordination. The first step in this direction was the development of a manual of methods. A draft was discussed at an international meeting held in Saskatoon, Canada, in 1967 in which 27 participants from five countries took part. The resulting manual (Milner & Hughes, 1968) was publicized among all projects. As studies proceeded many advances were made in methodology, and the original IBP guide on methods rapidly became outdated. The need for continued exchanges among the participants became evident as more sophisticated methods were developed.

The format for five meetings to review progress of IBP grassland studies was set by a meeting in Canada in 1969 (Coupland & Van Dyne, 1970). Subsequent meetings were held in 1971–2 in the Ivory Coast (IBP/PT Grassland Working Group, 1973), in 1972 in the USA (IBP/PT Grassland and Tundra Working Groups, 1972), in 1973 in Poland, and in 1974 in India (IBP/PT Grassland Working Group, 1974) (Appendix 1*a*, *b*). In each instance papers were presented from each study site and the local study areas were visited. Two meetings were held on study sites, at Matador in Canada and at Lamto in the Ivory Coast. The duration of each meeting was from 5 to 12 days and the participants numbered from 31 to 86, representing 8 to 19 countries.

At the first of these meetings a Coordinating Committee (Appendix 1*c*) was set up to act as an executive to the group. Continued contact between meetings was arranged through a Grasslands Newsletter which was edited and distributed from the Canadian study site, designated at that time by SCIBP as the 'International Centre for IBP Grasslands Studies'. This newsletter appeared in a series of eight numbers from May, 1969 to April, 1973. The Coordinating Committee arranged for the review meetings and studied the problem of achieving worthwhile synthesis of the data. It was soon realised that traditional means of data analysis, based on summaries of averages, were not sufficient to cope with the necessity of developing dynamic models of ecosystem function.

Data synthesis was developed by means of 'workshops'. The first such meeting was arranged to be held in 1970 at the time of the XI International Grassland Congress in Australia. Researchers were encouraged to submit data summaries well in advance of the meeting, but too few were received to provide an adequate basis for workshop sessions, and the meeting, therefore, concentrated on discussing problems associated with international synthesis. The outcome was a proposal to set up an international centre for synthesis of grassland data. This plan was discussed at the Australian Arid Zone Conference in 1970 and was presented to FAO at the time of the IBP General Assembly in Rome later that year. The plan was then modified into the 'International Grazing Lands Project', to include tundra, and was presented again to FAO and UNESCO officials in 1971. It was received sympathetically, but the magnitude of the funding required was such that it could not be developed in time to satisfy IBP data-synthesis needs. Accordingly, the Coordinating Committee turned its attention to arranging further workshops.

The second workshop was held in conjunction with the Tundra Working Group in 1972 in the headquarters of the USA Grassland Biome Study (IBP Grassland and Tundra Working Groups, 1972). Much effort was expended on data in advance and they were processed into computer memory for use during the workshop. Data sets were supplied from 51 grassland study sites in 20 countries. This meeting was very successful as an educational exercise, in developing some preliminary submodels, and in acquiring a data base. Preliminary plans were made for a two- to three-month workshop in 1974 which would permit combined data-synthesis activity by authors of the planned volumes, but this workshop never reached fruition.

Concurrently with these data-synthesis activities considerable effort was given by the Coordinating Committee to the development of a data bank. This was first discussed at the 1969 general meeting, and a decision was reached to make a modest beginning by filing data sets in the International Centre. The first real progress in acquiring such data sets was achieved at the 1972 workshop. As a result, it was decided to concentrate data-bank activities at the headquarters (in Fort Collins) of the USA/IBP Grassland Biome Study. This decision was subsequently given general support at a general meeting in Warsaw in 1973 as the method to be followed for acquiring data sets for the preparation of this and the companion grassland volume. Arrangements were made for transmission of data sets to authors. Thus, the plans for joint data-synthesis activities at one data centre never materialized. Instead, data analysis was mostly done at project level and the authors used, for the most part, data summaries supplied to them. Researchers were not equally effective in transmitting data sets to the data bank. Some failed to comply with the Coordinating Committee's requests, even though they were advised that failure to do so could result in lack of full recognition of their contribution to IBP. Nevertheless, authors were asked to solicit data directly from project participants, to supplement data from the data bank, and to review published records of IBP and IBP-related studies. Consequently, the intention here is to report on the state of knowledge available in the mid

1970s concerning structure and function in grassland ecosystems. Failure to recognise IBP research in some sites results from failure of workers to make their data available to the data bank, in some cases no doubt because progress in data collection and analysis was not sufficiently advanced.

An Editorial Committee (Appendix 1*c*) took over the coordinating role from the Coordinating Committee late in 1972. This change was associated with the official conclusion of the data-gathering phase of IBP and the emphasis on publication. This group met on three occasions and is responsible for the general structure of this and the companion volume. It also continued the coordinating role in connection with the 1973 and 1974 general meetings, which became a basis for acquiring data and information for the synthesis volumes. From 1974 onwards coordination between the volumes was by exchanges between editors under the supervision of the PT representative on the IBP Publications Committee and through the Executive Secretary of that Committee.

The activities of the Grassland Working Group were synchronised to some extent with those of the Tundra and Aridlands groups. This was possible through the keen interest of the PT Convenor in the activities of all groups, by the attendance of members of one group at the meetings of others and by exchanges of newsletters. As has been mentioned already, the three groups cooperated in the presentation of the International Grazing Lands Project to United Nations agencies. As well, the Grassland and Tundra groups worked together in the major data-synthesis workshop (1972) at Fort Collins. An 'input workbook' was prepared in advance of this meeting to provide each of the 62 participants with access to available data. The 'output workbook' (IBP/PT Grassland and Tundra Working Groups, 1972) was distributed more widely.

The Grassland Editorial Committee was conscious not only of the large amount of information which had been collected on grassland ecosystems but of the differences in approach which had been followed at different centres. Considerable thought was given to the problem of how best to weld these differences into a worthwhile synthesis. It was finally agreed that the work should be presented in two complementary parts, each part requiring a volume to itself. This has allowed the editors to make use of a large number of contributors familiar with particular kinds of grassland (this volume) and to allow others to make an assessment of processes applicable to grasslands in general.

Thirty-five authors from 10 countries spread over four continents have collaborated in the production of the present volume. Each author has been associated wih an IBP grassland study in his own country and has compared, as far as possible, the results of one major component of the ecosystem with studies elsewhere in the kind of grassland with which he is familiar. These contributions from each kind of grassland have been assembled by subeditors who, for the most part, have been leaders in IBP grassland studies within their own countries. The numbers of scientists involved in these studies were too large for all to be authors. To those who contributed information and data to this synthesis we are thankful. An attempt has been made in Appendix 2 to acknowledge the cooperation of some of these participants by indicating the names of those who should be contacted for information about a particular ecosystem component in a particular site.

A great variety of both published and unpublished information resulted from IBP grassland studies. At the time the contributions to this volume were being prepared, most of this information had not yet appeared in the 'open' literature. Much of it was in the form of technical reports, the theses of post-graduate students, local publications, and in manuscripts of scientific papers and other synthesis volumes. Authors did not always have access to this information.

Types of grasslands

For the purpose of interpreting the results of IBP research in a large number of sites of varying nature, it was considered desirable to distinguish groups of sites, within each of which similarities are greater than between groups. The similarities of importance here are in respect of attributes that might be expected to account for differences in trophic structure and ecosystem function.

Various approaches have been used to classify the grasslands of the world (Shantz & Marbut, 1923; Bews, 1929; Clements & Shelford, 1939; Roseveare, 1948; Moore, 1970). While local and regional classifications are typically and justifiably based on botanical composition, this is an inadequate basis for an inter-continental classification, because of discontinuities in the distribution of plant species. Instead it is necessary to consider similarities in conditions of climate and soil in grouping of sites.

The classification of grasslands followed in this volume was developed to facilitate the presentation of IBP results. Although arbitrary, it permitted the regional grouping of study sites and so, to a large extent, it simplified problems of communication. Six kinds of grassland are recognised on the basis of climate, successional status and land use. These are discussed in five parts of this volume, as indicated by Fig. 1.2. Tropical grasslands are separated from temperate grasslands. Within the temperate zone the natural grasslands are distinguished from semi-natural types. The semi-natural types have been divided further into those used primarily for hay and those that are grazed by domesticated livestock. Finally, two types of arable grasslands are recognised that may be present in any of the other grassland regions, but were primarily studied in the temperate zone, namely, those seeded to perennial forages and those planted to annual crops.

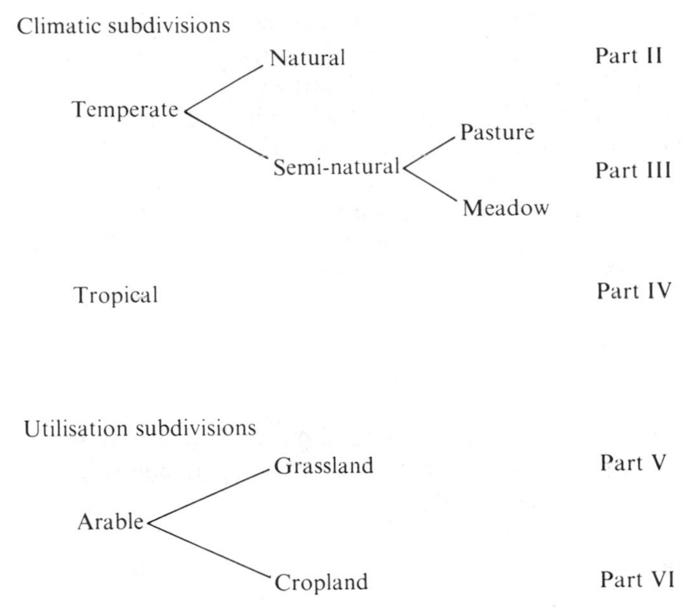

Fig. 1.2. Arrangement of sections in this volume.

The degree of human interference is reflected in this classification. Natural grasslands are those in which climate is the prime controlling factor under light to moderate grazing pressure by ungulates. However, even in this group, some sites have undergone considerable modification because of grazing by domesticated herbivores. Semi-natural grasslands are deforested areas, in regions of forest climate, that are held in a relatively stable condition by various natural and man-induced means, such as flooding, mowing, grazing and other treatments that prevent reinvasion of natural forest. They are dominated by perennial graminoids (grasses and grass-like plants) mostly native to the region. Arable grasslands are generally maintained with a cover of perennial exotic species (usually dominated by grasses, but sometimes with associated legumes) usually by periodic tillage and reseeding. Cropland implies annual tillage and seeding of annual herbaceous species. Both croplands and arable grasslands occur in all zones. Grasslands of the tropics are considered as one group because of the difficulty of interpreting their successional status in many instances and because their nature is often quite different and less understood from those of the temperate zone.

Some laxity has been tolerated in application of these groupings to allow freedom of action to authors who have greater familiarity with particular sites. For example, the semiarid grasslands of India and Africa are treated as tropical grasslands, while a site in desert grassland at a similar latitude in the southern USA is grouped with other North American sites.

Part II deals with the natural temperate grasslands. The largest expanses of these occur in central North America, in eastern Europe and the middle latitudes of Asia, and in northern Argentina and adjacent areas, while smaller areas occur in South Africa and Australia. The contributions to Part II in this volume are principally from North America. Important IBP studies were made in natural temperate grasslands in the USSR, but language and communication difficulties have interfered with adequate incorporation of their results. Some contributions from the USSR are included in Part III, including results in natural temperate grassland. Also, a summary of grassland studies in the USSR is included in the companion volume. The major IBP study in natural temperate grasslands in the southern hemisphere apparently took place in Argentina, but had not progressed to the data analysis stage at a date early enough to be included here.

Part III is concerned with semi-natural meadows and pastures in the temperate zone. All IBP research in semi-natural temperate grasslands, of which we are aware, took place in the northern hemisphere in Europe and Asia. It was planned to present IBP results on semi-natural meadows and pastures in two separate parts. However, deficiencies of contributions to the part on pastures necessitated last-minute adjustments to combine the available material on pastures with the section on meadows. This has resulted in an inadequate treatment of semi-natural pastures in this volume and failure to recognise some substantial IBP research.

IBP activities in tropical grasslands (Part IV) were concentrated in India, but the discussion also incorporates results from sites in Africa and from one in the western hemisphere. No contribution appears from South America. Unfortunately, this synthesis does not include some important studies that were carried out independently of IBP, in particular the very important one in Serengeti National Park in Tanzania. Communication difficulties are cited as the major reason for these gaps.

In Part V a group of workers in Australia give a detailed account of a seeded grassland ecosystem in east-central Australia and make reference to only one other study (in Israel) that has produced comparable data.

In Part VI an ecosystem analysis in croplands in Poland is reported without comparison with studies elsewhere.

Use and management of grasslands

The natural grassland areas of the world have been used primarily either as cropland or as rangeland. Tillage is most intensive, and began in earlier times, near the forest margin; an increasingly larger proportion of the land is devoted to grazing along the climatic gradient from forest to desert. In some regions tilled agriculture has invaded too far into the semiarid zone, with resulting crop failures and severe erosion of soil by wind during dry phases of the weather cycle. These lands have often been returned to perennial grass, either by seeding or by natural succession. Overgrazing has caused depletion of the grass cover in many areas and is a hazard to the continued productivity of rangeland, particularly in semiarid areas. Overgrazing has caused some grassland landscapes to become desertic in character. The major management practice used in rangeland is control over the numbers of livestock and their distribution on range. In some areas control of invasion of woody plants is necessitated by the use of fire and herbicides. Grazing capacities in natural grassland vary from as high as 2 hectares/animal unit (450 kg bovine animal or 5 ewes) to 20 hectares or more/animal unit in some unproductive semiarid areas.

Semi-natural grasslands are used for hay or pasture. The management regimes applied determine the character of the plant cover. While grazing and mowing restrict reinvasion by trees and shrubs, additional practices that are frequently used include fire and fertilisers and control of flood water. Because the climate within which semi-natural grasslands occur is more favourable to plant growth, they are more productive of forage than are natural grasslands.

Appendix 1. *Activities of the IBP/PT Grassland Working Group*

a. Details of meetings, 1967–75

Date	Location	Length of programme (days)	No. of papers presented	Attendance: No. of individuals	Attendance: No. of countries	Nature and purpose
12–14 Sept. 1967	Saskatoon, Canada	3	—	27	5	*Ad hoc* meeting to coordinate methods (1)[a]
5–10 Sept. 1969	Saskatoon, Canada	4½	46	86	12	General meeting to review progress in research (2)
18–19 Apr. 1970	Surfers' Paradise, Australia	1	—	38	8	General meeting to plan international data synthesis
28–30 Sept. 1970	Rome, Italy	1	—	22	9	General meeting to plan international data synthesis
29 Dec.–3 Jan. 1971–2	Lamto, Ivory Coast	4½	31	41	11	General meeting to review African research and plan international synthesis (3)
14–26 Aug. 1972	Fort Collins, U.S.A.	12	15–20	62	19	Workshop with Tundra Group to initiate international synthesis (4)
5–6 Sept. 1972	Seattle, U.S.A.	1	—	20–30	12–15	Meeting of Editorial Committee
2–7 Oct. 1972	Montpellier, France	1	—	6	6	Meeting of Editorial Committee
29–30 Mar. 1973	Paris, France	2	—	6	4	Meeting of Editorial Committee
6–12 July 1973	Warsaw, Poland	6	19	53	13	General meeting to review European research and for editorial planning
17–22 Jan. 1974	Varanasi, India	6	21	31	8	General meeting to review tropical research and for editorial planning (5)
13 Mar. 1974	London, U.K.	½	—	6	4	Meeting of editors to coordinate volumes
17–18 Apr. 1975	Fort Collins, U.S.A.	2	—	5	3	Meeting of editors to coordinate volumes

[a] Published proceedings: (1) Coupland & Van Dyne (1970: pp. 200–3); (2) Coupland & Van Dyne (1970: 1–199, 204–8); (3) IBP/PT Grassland Working Group (1973); (4) IBP/PT Grassland and Tundra Working Groups (1972); (5) IBP/PT Grassland Working Group, 1974).

b. Numbers of participants in general meetings by country

Country	Saskatoon, Canada Sept. 1967	Saskatoon, Canada Sept. 1969	Surfers Paradise, Australia Apr. 1970	Rome, Italy Sept. 1970	Lamto, Ivory Coast Dec.–Jan. 1971–2	Ft. Collins, U.S.A. Aug. 1972	Warsaw, Poland July 1973	Varanasi, India Jan. 1974	Total
Afghanistan	—	—	—	—	—	—	—	1	1
Argentina	—	1	—	—	—	2	—	—	3
Australia	—	1	17	1	—	4	1	—	24
Austria	—	—	—	—	—	1	—	—	1
Bulgaria	—	—	—	—	—	—	3	—	3
Canada	17	33	1	2	1	4	1	1	60
Czechoslovakia	—	1	—	1	—	—	2	—	4
France	—	2	—	1	3	4	2	1	13
Germany (West)	—	—	—	—	—	1	2	—	3
Ghana	—	—	—	—	1	—	—	—	1
Hungary	—	—	—	—	—	—	1	—	1
Iceland	1	—	—	—	—	—	—	—	1
India	—	—	1	—	—	3	1	20	25
Iran	—	—	1	—	—	—	—	—	1
Ireland	—	—	—	—	—	1	—	—	1
Israel	—	—	—	1	—	1	—	—	2
Ivory Coast	—	—	—	—	26	—	—	—	26
Japan	—	3	—	—	—	1	—	—	4
Mexico	1	—	—	—	—	1	—	—	2

b. Numbers of participants in general meetings by country (cont.)

Country	Saskatoon, Canada Sept. 1967	Saskatoon, Canada Sept. 1969	Surfers Paradise, Australia Apr. 1970	Rome, Italy Sept. 1970	Lamto, Ivory Coast Dec.–Jan. 1971–2	Ft Collins, U.S.A. Aug. 1972	Warsaw, Poland July 1973	Varanasi, India Jan. 1974	Total
Netherlands	—	1	—	1	—	2	—	—	4
New Zealand	—	—	2	—	—	—	—	—	2
Nigeria	—	—	—	—	1	—	—	1	2
Norway	—	1	—	1	—	2	1	—	5
Phillipines	—	—	—	—	—	—	—	1	1
Poland	—	1	—	2	—	2	24	—	29
Senegal	—	—	—	—	1	—	—	—	1
South Africa	—	1	—	—	—	2	—	—	3
Sweden	—	—	—	—	—	2	—	—	2
Tanzania	—	—	—	—	—	1	—	—	1
Uganda	—	—	—	—	1	—	—	1	2
U.K.	2	2	3	2	1	2	1	—	13
U.S.A.	5	38	10	6	2	25	11	4	101
Upper Volta	—	—	—	—	1	—	—	—	1
U.S.S.R.	—	—	—	—	—	—	1	—	1
Zaire	—	—	—	—	2	—	—	—	2
S.C.I.B.P.	1	1	—	2	—	1	2	1	8
U.N. agencies	—	—	2	3	1	—	—	—	6
Total	27	86	37	23	41	62	53	31	—

c. Membership of committees

Coordinating Committee (1969–72)

A. Breymeyer (Poland)
R. T. Coupland (Canada) – chairman
M. J. Hadley (PT Section) – *ex officio*
R. Misra (India) – corresponding member
Y. S. Nasyrov (USSR) – corresponding member
M. Numata (Japan)
A. Soriano (Argentina)
B. Speidel (West Germany) – corresponding member
G. M. Van Dyne (USA) – vice chairman
J. J. Van Wyk (South Africa)
W. M. Willoughby (Australia)

Editorial Committee (1972–6)

A. Breymeyer (Poland)
R. T. Coupland (Canada)
J. B. Cragg (PT Convener) – *ex officio*
G. Douglas (IBP Publications Committee) – *ex officio*
M. Lamotte (France): 1972–3
R. Misra (India): 1972–3
G. M. Van Dyne (USA)

Appendix 2. Researchers whose studies contributed to the synthesis presented in this volume. In some instances only the coordinator's name was available

Country	Name of site	Project address	Contributors					
			Producers	Consumers	Micro-organisms	Energy flow	Nutrient cycling	Abiotic factors
Australia	Armidale	CSIRO, Pastoral Research Laboratory, Armidale, NSW, Australia	P. J. Vickery	K. J. Hutchinson, L. King	R. L. Davidson	K. J. Hutchinson	A. R. Till	P. J. Vickery K. J. Hutchinson
Canada	Matador	University of Saskatchewan, Saskatoon, Sask., S7N 0WO, Canada	R. T. Coupland, R. E. Redmann	W. J. Maher (birds), D. H. Sheppard (small mammals), P. W. Riegert (surface invertebrates), J. W. Willard (soil invertebrates)	E. A. Paul (bacteria), V. O. Biederbeck (Actinomycetes), D. Parkinson (fungi), E. A. Paul (processes)	R. T. Coupland	R. T. Coupland, J. W. B. Stewart	E. de Jong (Soil Physics), E. A. Ripley (Meteorology, Micrometeorology)
Czechoslovakia	Lanžhot	Botanical Institute, Ecologcal Dept, Czechoslovak Academy of Sciences, Stará 18, Brno, Czechoslovakia	E. Balátová-Tuláčková, J. Jakrlová, M. Rychnovská, J. Gloser	J. Rusek	B. Úlehlová, M. Tesarova	M. Rychnovská	J. Jakrlová, B. Úlehlová	E. Balátová-Tuláčková, P. Smíd
Ethiopia	Geech Plateau	Geobotanical Inst., Federal School of Technology, Zurich bergstn 38, CH-8044, Zurich, Switzerland	F. Klötzli					F. Klötzli
Federal Republic of Germany	Solling	c/o Prof. H. Ellenberg, Geobotanical Laboratory, Untere Karspüle 2, D-3400, Göttingen, West Germany	B. Speidel, O. L. Lange, E. Geyger	W. Funke, G. Weidermann, J. Schauermann, O. Graff	L. Steubing, E. Küster, E. H. Domsch	B. Ulrich, M. Runge, R. Grimm	B. Ulrich, M. Runge, R. Grimm	F. Wilmers, W. Eber, P. Benecke, R. Graul
Finland	Tvärminne	Dept of Zoology, University of Finland, Helsinki, Finland	M. Kosonen, G. Gyllenberg	G. Gyllenberg		G. Gyllenberg		M. Kosonen, G. Gyllenberg
France	Pin-au-Haras	Laboratoire de Zoologie, 16 Rue Dufay-76100 Rouen, France		G. A. E. Ricou				
Hungary	Csévharaszt	Dept of Botany, Budapest University, Budapest, 1088 Muzeum krt 4/a, Hungary	E. Kovács-Láng				E. Kovács-Láng	E. Kovács-Láng, M. Szabo

Hungary	Újszentmargita	Inst. of Botany, Hungarian Academy of Sciences, Vácrátót, Hungary	I. Précsényi			I. Précsényi		I. Précsényi, G. Bodrogközy
India	Ambikapur	Dept of Botany, Gov't College, Ambikapur, M.P., India	M. L. Naik					M. L. Naik
India	Behrampur	Dept of Zoology, Berhampur University, Berhampur, Orissa, India	M. C. Dash					
India	Delhi	Dept of Botany, University of Delhi, Delhi-7, India	C. K. Varshney					C. K. Varshney
India	Jhansi	Indian Grassland & Fodder Research Inst., Jhansi, U.P., India	K. A. Shankarnarayan					K. A. Shankarnarayan
India	Jodhpur	Central Arid Zone Research Inst., Jodhpur, Rajasthan, India	R. K. Gupta	L. D. Ahuja		S. K. Saxena		R. K. Gupta
India	Kurukshetra	Dept of Botany, Kurukshetra University, Kurukshetra-132119, India	J. S. Singh					J. S. Singh
India	Pilani	Dept of Biological Science, Birla Institute of Technology & Science, Pilani-333031, Rajasthan, India	M. C. Joshi					M. C. Joshi
India	Rajkot	Dept of Biosciences, Saurashtra University, Rajkot, India	S. C. Pandeya					S. C. Pandeya
India	Ratlam	School of Studies in Botany, Vikram University, Ujjain, M.P., India	S. K. Billore					S. K. Billore
India	Sagar	Dept of Botany, University of Saugar, Sagar, M.P., India	S. K. Jain		S. B. Saksena			S. K. Jain
India	Ujjain	School of Studies in Botany, Vikram University, Ujjain, M.P., India	C. M. Misra					C. M. Misra
India	Varanasi	c/o Prof. R. Misra, Dept of Botany, Banaras Hindu University, Varanasi-221005, U.P., India	R. S. Ambasht		R. S. Dwivedi	J. S. Singh	K. P. Singh	K. C. Misra
Israel	Migda	Volcani Inst. of Agricultural Research, Gilat Agricultural Research Station, Negev Mobil Post 2, Israel	R. W. Benjamin					R. W. Benjamin

Appendix 2 (*cont.*)

Country	Name of site	Project address	Contributors					
			Producers	Consumers	Micro-organisms	Energy flow	Nutrient cycling	Abiotic factors
Ivory Coast	Lamto	c/o Prof. M. Lamotte, Lamto Tropical Ecology Laboratory, B.P. 28, N'Douci, Ivory Coast *or* 46 rue d'Ulm, 75235 Paris, France	J. Cesar, J. C. Menaut, J. C. Roland, U. Monnier, R. Vuattoux	P. Lavell, G. Josens, J. Levieux, J. M. Leroux, D. Gillon, Y. Gillon, B. Darchen, P. Blandin, R. Barbault, D. Lachaise, M. Vaillaud, J. P. Vincent, C. Lecordier, F. Athias	R. Schaefer, A. Rambelli	M. L. Célérier, M. Lamotte	P. Villecourt, R. Schaefer, J. Balandreau, D. Bauzon	C. Lecordier, P. Bonvallot, G. Riou, J. L. Tournier, J. P. Bony, F. Athias
Japan	Aso-Kuju	c/o Prof. M. Numata, Laboratory of Ecology, Faculty of Science, Chiba University, Yayoi-cho, Chiba, Japan	N. Yano, R. Kayama				R. Kayama	R. Kayama
Japan	Kawatabi	c/o Prof. M. Numata, Laboratory of Ecology, Faculty of Science, Chiba University, Yayoi-cho, Chiba, Japan	Y. Shimada, H. Iwaki, B. Midorikawa, S. Iizumi, K. Koike, T. Suganuma	Y. Ito, M. Nakamura, T. Matsumoto, N. Nakamura, Y. Fijimaki, K. Hayashi, R. Igarashi, K. Hayashi	T. Saito, T. Ando	H. Iwaki, T. Okubo	R. Kayama	M. Mitsudera, I. Yamane
Japan	Nanashigure	c/o Prof. M. Numata, Laboratory of Ecology, Faculty of Science, Chiba University, Yayoi-cho, Chiba, Japan	T. Ikubo, Y. Inoue	Y. Inoue		T. Okubo		T. Okubo
Japan	Tonomine	c/o Prof. M. Numata, Laboratory of Ecology, Faculty of Science, Chiba University, Yayoi-cho, Chiba, Japan	T. Suganuma, N. Yano, R. Kayama				R. Kayama	R. Kayama

Netherlands	Terschelling	Research Inst. for Nature Management, Brockhuizen Castle, Leersum	P. Ketner					P. Ketner
Nigeria	Lagos	Dept of Agricultural Biology, University of Ibadan, Nigeria	J. K. Egunjobi	J. K. Egunjobi				
Panama	Panama	Inst. of Geography and Spatial Organisation, Polish Academy of Sciences, Karkowskie Przedmiescie 30, 00-927 Warsaw, Poland	A. Breymeyer					
Poland	Ispina, Ojców, and Showronno	Inst. of Botany, Juyellonian University, Nature Conservation Research Centre, Polish Academy of Sciences, Lubicz 46, 31-512 Kraków, Poland	K. Jankowska, M. Kotańska, A. Medwecka-Karnás, E. Baradziej					M. Karkanis, J. Klein
Poland	Kampinos and Kazuń	Inst. of Ecology, Polish Academy of Sciences, Dziekonów Lésny near Warsaw, P.O. Lomianki, Poland	T. Traczyk, Z. Wójcik	L. Andrzejewska, A. Breymeyer, E. Olechowicz, J. Petal, B. Diehl, G. Makulec, A. Kajak, H. Nowak	H. Jakubczyk			Z. Czerwiński
Poland	Turew	Dept of Agrobiology, 60-809 Poznan, Swiercrewskiego 19, Poland	C. Kukielska	Y. Karg	J. Gotebiowska	L. Ryszkowski	Z. Margowski	C. Radomski
Senegal	Fété Olé and Ferlo	c/o Prof. F. Bourlière, 5 Avenue de Tourville, 75007 Paris, France	J. C. Bille, H. Poupon	Y. Gillon, D. Gillon, M. Lepage, G. Morel, A. R. Poulet, P. Forge		J. C. Bille	J. C. Bille	J. C. Bille
South Africa	Welgevonden	Dept of Botany, Potchefstroom University of Christian Higher Education, Potchefstroom, Transvaal, South Africa	J. J. Van Wyk					
Uganda	Mweya Peninsula	Uganda Institute of Ecology, P.O. Box 22, Lake Catwa, Uganda	E. L. Edroma	E. L. Edroma				
UK	Moor House	ITE, Merlewood Research Station, Grange-over-Sands, Cumbria, England	G. I. Forrest, R. S. Clymo, M. Rawes, R. A. H. Smith	M. Rawes, J. C. Coulson, J. B. Whittaker	A. J. Holding, V. G. Collins, P. M. Latter	H. E. Jones, A. J. P. Gore, R. S. Clymo, O. W. Heal	H. E. Jones, A. J. P. Gore, R. S. Clymo, O. W. Heal	O. W. Heal

Appendix 2 (*cont.*)

Country	Name of site	Project address	Contributors					
			Producers	Consumers	Micro-organisms	Energy flow	Nutrient cycling	Abiotic factors
UK	Snowdonia	Inst. of Terrestrial Ecology, Bangor Research Station, Penrohos Road, Bangor, Caerns, UK LL57 2LQ	D. F. Perkins			D. F. Perkins	D. F. Perkins	
USA	ALE	Battelle Northwest Laboratories, Richland, Washington 99352, USA	W. H. Richard, Jr.	T. P. O'Farrell (vertebrates), L. E. Rogers (invertebrates)	R. Wildung			
USA	Bison	University of Montana, Missoula, Montana, USA	M. Morris	R. Hoffmann				
USA	Bridger	Montana State University, Dept of Botany, Bozeman, Montana, USA	T. Weaver	R. Hoffmann, B. Hegland	T. Weaver			T. Weaver
USA	Cottonwood	Animal Science Dept, South Dakota State University, Brookings, South Dakota 57006, USA	J. K. Lewis		R. D. Pengra			
USA	Dickinson	Dept of Botany, North Dakota State University, Fargo, USA	W. Whitman	E. Birney				W. Whitman
USA	Hays	Fort Hays Kansas State University, Fort Hays, Kansas, USA	G. W. Tomanek					G. W. Tomanek
USA	Jornada	Dept of Animal, Range and Wildlife Sciences, New Mexico State University, Las Cruces, New Mexico 88001, USA	R. D. Pieper, C. H. Herbal	R. L. Packard (mammals), S. L. Pimm and R. J. Raitt (birds), J. G. Watt (invertebrates)				
USA	Osage	Botany Dept, University of Oklahoma, Norman, USA	P. G. Risser	E. Birney, D. Blocker	J. Harris, R. May	P. Risser		P. Risser
USA	Pantex	Entomology Dept, Texas Technological University, Lubbock, Texas 79409, USA	R. D. Pettit	E. W. Huddleston (invertebrates), R. L. Packard (vertebrates)				

USA	Pawnee	c/o Prof. G. M. Van Dyne, Natural Resource Ecology Laboratory, Colorado State University, Fort Collins, Colorado 80523, USA	J. L. Dodd, P. L. Sims, W. K. Launeroth	R. W. Rice (large herbivores), J. W. Leetham (invertebrates), N. R. French (small mammals), R. A. Ryder, J. A. Wiens (birds), J. G. Nagy (antelope), R. M. Hanson (diets)	K. G. Doxtader, E. B. Sparrow	D. C. Coleman, J. E. Ellis	R. G. Woodman-see	F. M. Smith, J. R. Nunn
USA	San Joaquin	Pacific Southwest Forest Range Experiment Station, San Joaquin Experimental Range, Coarsegold, California, USA	D. A. Duncan	F. Shitowski, T. F. Newsman, D. Duncan				
USSR	Baraba, Karachi	Inst. of Soil and Agrochemistry, USSR Academy of Sciences, 18 Sovetskaja, Novosibirsk 99, USSR	T. A. Vagina, A. A. Titlyanova, N. G. Shatokhina	V. G. Mordko-vich	N. I. Ganti-murova, I. L. Klevenskaja	A. A. Titlyanova, N. I. Bazilevich	A. A. Titlyanova, N. I. Bazilevich	N. I. Bazilevich, L. V. Voronina, A. P. Slyadner

2. The nature of grassland

R. T. COUPLAND

The term 'grassland' is used in this volume to refer to ecosystems in which the dominant vegetative component is comprised of herbaceous species. The other three terrestrial sections of IBP also include herbaceous floristic elements. Deserts abound in annual grasses and forbs, but only for short periods when moisture conditions permit; the permanent vegetation is of shrubs and succulents. Herbaceous species are important occupants of the ground layer of woodlands, but trees characterise and dominate the vegetation of these landscapes. The vegetation of tundra ecosystems is characterised by small shrubs, herbs, mosses and lichens, each of which is the principal contributor of organic production in different situations. Herbaceous communities in the tundra can be considered as grasslands, and interest in such areas overlaps with that of the IBP Tundra Working Group. Some of the meadows discussed in Part III are in some respects similar to certain herbaceous communities of the tundra zone, both being influenced by high moisture content in the substratum. The absence of trees in tundra and natural temperate grassland is due to low temperatures in the former and to moisture deficiency in the latter. However, in the IBP organisation, studies of heathlands and peatlands of the temperate zone were associated with those of the arctic tundra, taking with them some studies of meadows. Similarly, the interests of the Woodland and Grassland Working Groups overlap in savanna, which is characteristic of vast areas of the tropics and subtropics. These other three biomes will be treated in other volumes in this series (for tundra see Bliss, Cragg, Heal & Moore, in preparation; for woodlands see Reichle, 1979; and for aridlands see Goodall & Perry, 1979; Perry & Goodall, in preparation).

Abiotic components

Natural grassland occurs in situations too arid for the development of closed forest, but not so adverse as to prevent the development of a closed perennial herbaceous layer that is lacking in desert (Coupland, 1974). Regionally, climate controls the biotic components directly, as well as indirectly through its influence on soil development. Within the grassland zone, however, local variations in topography and soils result in microclimatic conditions so different from the zonal climate that azonal ecosystems develop locally that are dominated by plants and animals not characteristic of the region. In the semiarid tropics and subtropics, specially adapted trees are scattered throughout the grassland to form a savanna (Shantz & Marbut, 1923; Beard, 1953; Moore, 1970).

The climate of natural grassland exhibits marked periodicity of precipitation, both intraseasonally and interseasonally (Borchert, 1950). Annual droughts of several weeks to several months are typical. The severity of drought increases with distance from the forest margin. It is also accentuated by the tendency of years of above-average and below-average precipitation to be grouped, so that a cyclical climatic pattern is evident, but of low predictability. Mean annual precipitation in temperate grassland usually ranges from 250 to 750 mm and in tropical and subtropical grasslands and savannas from 600 to 1500 mm (Coupland, 1974). Precipitation determines the nature and extent of natural grasslands by affecting the supply of soil moisture; most trees cannot compete effectively with grasses where upper soil layers are intermittently moist, but deeper layers are continuously dry. The relatively high winds increase the tendency towards water deficiency by their evaporative effect. The adverse effects of drought on organisms are accentuated by high temperature. The length of the active growing season is determined in the tropics and subtropics by the length of the rainy season which often ranges from 120 to 190 days and is sometimes more, while in cold temperate regions temperature becomes the controlling factor by interrupting many biological activities in winter. The semi-natural grasslands of deforested areas are, of course, under the influence of a climate in which moisture deficiency is of less consequence than in regions of natural grassland.

The soils of natural grasslands are quite different in character from those of forests, because of differences in the soil-forming processes (Weaver & Clements, 1938). In temperate grassland, soil leaching is restricted by the scarcity of percolating water and by the low solubility of minerals in a basic solution and the slowness of their release from humus. In temperate forest soil an acid medium results in slow reduction of organic matter, but in rapid mineralisation and translocation in a greater volume of solution to depths beyond the reach of roots. Consequently, temperate grassland soil has an abundance of organic matter at all depths; in the forest organic debris accumulates on the surface of the soil, but very little humus is stored within the profile. In many tropical and subtropical grasslands, soils are highly leached, and there is rapid decay with low levels of humus accumulation. These different soil-forming processes are reflected in the colour of the surface soil, which in temperate grassland varies from black near the forest margin, where the organic content is highest, to brown in semiarid areas. Temperate forest soils (which occupy semi-natural grasslands of deforested sites) are brownish to greyish in colour depending on the degree of leaching, which in cool climates leaves silica in the profile and removes iron. Tropical and subtropical soils are often reddish or yellowish in colour, because of the high content of iron left in the profile by leaching under high temperatures. Soil texture is an important modifying factor in relation to the proportion of precipitation that enters the soil and is available to plants. Sandy soils hold less moisture per unit volume, but permit more rapid percolation of precipitation water than do soils of finer texture, with a lower resultant loss by runoff and evaporation. Because of their greater moisture supply and availability of moisture to greater depth, sands support stands of tall grasses in regions where grasses of smaller stature (and less depth of rooting) occur in soils of finer texture.

Arable grasslands and croplands occur in all climates and in many types of soil. When cultivated, soils of natural temperate grassland have the capacity to release nutrients slowly over a long period of time, while temperate forest soils are relatively infertile and often must be 'improved' by liming and manuring before they are capable of yielding at acceptable levels. Some tropical and subtropical soils are very much more subject to degradation by tillage than are temperate soils.

The absence of trees in natural temperate grassland has been attributed to aridity of climate, periodic burning and the nature of the soil. Fire discriminates against trees and shrubs because buds (developed for the next season of growth) are located well above the soil surface, in a position more vulnerable to fire than those of herbaceous plants that are at or below the soil surface. Since woody stems live for many years, their destruction by fire will cancel out many years of growth. A soil layer impermeable to moisture has been cited as the reason for treelessness in parts of the tropics, where in its absence savanna develops (Beard, 1953). The resulting waterlogging of soil during the rainy season creates conditions not suitable for the growth of trees capable of surviving the dry season. Clay soil in temperate grasslands appears to prevent trees from forming adequate rooting systems (Keller, 1927; Moore, 1970), as does subsurface 'hard pan' that results from calcification in some tropical soils.

A detailed discussion of abiotic processes in grasslands has been prepared by Hinds & Van Dyne (Chapter 1 in Breymeyer & Van Dyne, 1979).

Biotic components

The organic component of a grassland is a complex of producers, consumers and decomposers that are organised into a food web. There are two interlocking parts to the food web: (i) the grazing food web (biophagic pathway), which is comprised of herbivores feeding on plants and providing the food for carnivores; and (ii) the detritus food web (saprophagic pathway), which contains the decomposer organisms and their predators (Evans, in Petrusewicz, 1967).

Flora

Grasslands are primarily comprised of herbaceous spermatophytes of two types: (i) grasses (Family Gramineae) and grass-like plants (particularly sedges of the Family Cyperaceae), referred to collectively as 'graminoids';

and (ii) forbs, which are non-grass-like herbs. Frequently, there are also dwarf shrubs that do not exceed the height of the grasses, while taller shrubs occur in some areas as isolated individuals or in patches (clusters); scattered trees characterise the tropical and subtropical savannas.

The complexity of the vegetative cover of natural grassland is much greater than is suggested by a view from above the grass canopy. The vegetation is multi-layered, both above ground and under ground (Weaver, 1954; Weaver & Albertson, 1956). The number of canopy layers varies. The uppermost one is comprised of taller grasses and forbs, while species of shorter stature, both dominants and associates, occupy the lower layer or layers. The latter are most demanding on the habitat early in the season before they are overtopped by species of the upper layer. Some forbs are confined in their vegetative growth to lower layers but thrust reproductive parts upwards, often above the grasses of the uppermost layer. The lowest layer above ground is the crust of lichens, mosses, club mosses and algae that often occurs on the soil surface among stems and leaves (litter) that have fallen from the canopy. There is a tendency for taller-growing species to root more deeply in the soil, so layering under the soil surface is also characteristic. Depth of rooting of tall and mid-grasses is usually no greater than 180 cm, while that of short grasses is 30 to 90 cm. Some forbs branch only below the deepest layer of dominant grasses. The horizontal distribution of some species is so non-uniform that they tend to form colonies.

The number of plant species occurring in a grassland is greatest in the most favourable environmental conditions and, typically, in situations in which the influence of man is least imposed. The effect of variation in topography is to increase the number of microhabitats, which increases the number of species. Thus, in the temperate zone semiarid natural grassland on relatively level terrain may be comprised of about 50 vascular species (Coupland, Ripley & Robbins, 1973), while in a subhumid, rolling area the numbers exceed 200 (Steiger, 1930). Composites (Family Compositae) usually rank next in abundance to the grasses (Gramineae) and sedges (Cyperaceae), with legumes (Leguminosae) being the forb groups of next importance.

Graminoids are particularly well adapted to dominate herbaceous communities, so that, although they typically make up less than 20 % of the number of species present in grassland, they often furnish 90 % or more of biomass in the canopy (Coupland, 1974). Although only about 4 % of the genera of spermatophytes on earth belong to the grass family, they are represented by 15 % of genera that are of cosmopolitan distribution (Hartley, 1950). In uncultivated areas, species of the tribes Andropogoneae and Paniceae are particularly important in tropical savannas and in the most favoured habitats elsewhere, while those of Aveneae and Festuceae are characteristic of cool temperate grasslands and Agrosteae and Eragrosteae of arid regions with warm winters (Hartley, 1950). Growth form is important in relation to adaptation. The ability of sod-forming species to reproduce asexually permits them to survive in situations where seed production is rare or absent. There is a tendency for bunch grasses to dominate in arid habitats and for sod-formers to dominate in moist situations. Physiological adaptations to drought include the ability to function in conditions of severe desiccation. The annual habit is not common in natural grasslands except where aridity or disturbance prevent the development of a closed cover of perennials. Only a few species of grasses are sufficiently well adapted to function as dominants in any particular grassland habitat. Frequently two or three species provide 60 % or more of the total shoot biomass.

The appearance of natural and semi-natural grassland varies during the year, depending on the relative proportions of green, dead and deteriorating leaves in the canopy and on the phenology of the component species.

Fauna

Populations of native animals in grasslands are much more diverse than is usually appreciated, since many surface species are inconspicuous and many others live under ground. The best known are the ungulates, grazing marsupials, predators of the cat and dog family, rodents, birds, lizards and snakes, and the larger insects, particularly grasshoppers and locusts (Pieper, in Dix & Beidleman, 1969). A large proportion of surface species exhibits either running or burrowing habits. Aggregation into colonies or herds is characteristic. Invertebrates are abundant, particularly in the soil.

The various herbivores, parasites and predators that comprise the consumer food web in natural and semi-natural grassland must be in balance with the energy supply from the producers or the nature of the ecosystem will change. There is a deficiency of information about the activities of many of the organisms and processes involved.

Populations of large grazing animals and predators have been severely reduced because of hunting pressure and the fact that they are considered as intruders in areas used for grazing livestock (Cloudsley-Thompson, 1969). In many places they now survive only under protection; their former ranges are densely populated and grazed to their maximum capacity above ground by domesticated animals. In some grasslands large herbivores can be as important as climate and soil in determining the floristic composition of vegetation, since plant species are not equally sensitive to grazing pressure (Dyksterhuis, 1958; Heady, 1975). Some ecotypes of grass species with inherent low productivity have developed under heavy grazing pressure. Large herbivores that browse have had a great influence in restricting the growth of trees and extending the boundaries of the grassland.

Small herbivorous mammals have also had an important influence in determining the nature of natural and semi-natural grassland, particularly where management inputs by man have modified the ecosystem (Gross, in Dix & Beidleman, 1969; Petrusewicz & Grodziński, in Reichle, Franklin & Goodall, 1975). Some species have increased their populations in response to reduced populations of their predators, while others have suffered by the direct efforts of man to exterminate them. Rodents are the most abundant mammals in some grasslands and cause range deterioration by over grazing and exposing the soil to erosion. Small mammals are treated in volume 5 in the present series (Golley, Petrusewicz & Ryszkowski, 1975).

Birds, amphibians and reptiles are also important. A great variety of birds inhabit grasslands (Glover, in Dix & Beldeman, 1969). Granivorous birds are the subject of a further IBP volume (No. 12) (Kendeigh & Pinowski, 1977). Omnivorous species are apparently more numerous than herbivores or carnivores. Passerines (perchers) are especially characteristic of the grassland habitat. Lizards, toads and box turtles predate on insects, while snakes are predators of rodents and other small vertebrates (Thomas *et al.*, in Dix & Beidleman, 1969).

Hundreds of species of invertebrates dwell together in natural or semi-natural grassland. Grasshoppers and other members of the Order Orthoptera are the most conspicuous and have the greatest effect as herbivores (Blocker, in Dix & Beidleman, 1969). Next, in order of their activity in feeding on herbage, are probably bugs (Hemiptera) and aphids and leafhoppers (Homoptera). Ants (Hymenoptera) and termites (Isoptera) are often abundant, especially in tropical and subtropical grasslands. They are so significant that a separate volume in this series (Brian, 1977) will be devoted to them. Spiders are important predators. In the drier grasslands insects are probably more important relative to other above-ground invertebrates, because of their ability to withstand desiccation. Populations of above-ground invertebrates are altered considerably with increased grazing pressure from large herbivores on vegetation; some groups decrease in abundance, others increase. Among the invertebrates in the soil, important groups are earthworms, nematodes, collembolans, mites and enchytraeids, but most censuses of these groups were done prior to IBP on intensively managed (arable and semi-natural) grasslands of the forest zone (Paris, in Dix & Beidleman, 1969).

Detailed discussions of the structure of consumer subsystems in grassland and of the processes involved in their activities have been prepared for the companion volume (Breymeyer & Van Dyne, 1979). These include considerations of small herbivores (Chapter 3, by Andrzejewska & Gyllenberg), large herbivores (Chapter 4, by Van Dyne *et al.*), invertebrate predators (Chapter 5, by Kajak) and vertebrate predators (Chapter 6, by Harris & Brown).

Micro-organisms

Microbial populations occur in the soil, on the surfaces of, and within the tissues of producers and consumers, and in the excreta of consumers. Microbial communities differ both qualitatively and quantitatively among these discontinuous micro-habitats (Coupland, Zacharuk & Paul, 1969).

The most abundant groups of micro-organisms in grassland soils are bacteria, actinomycetes and fungi (Clark & Paul, 1970). The bacterial biomass (and sometimes that of actinomycetes) is usually considered to be greater in grassland than in woodland. Fungi have fewer cells present, but of larger size; so biomass is greater than in other groups. Algae are less numerous, but still more abundant than Protozoa.

The detritus food web is a highly complex system comprised of large numbers of organisms (Clark & Paul, 1970). Populations of micro-organisms in soils are large, but many are in a state of dormancy. A number of different types of decomposer organisms are recognised on a functional basis. Reducers consume dead organic matter. Saprovores channel energy and nutrients from the detritus food web into the grazing food web by becoming food for carnivores. Micro-floral grazers recirculate materials within the detritus food web and also transfer material back to the grazing food web.

Micro-organisms have other important functions in ecosystems besides their role as reducers and decomposers. In many grasslands they fix significant amounts of atmospheric nitrogen through their symbiotic activities in nodules on the roots of various macrophytes, particularly legumes. In all grasslands they also effect non-symbiotic nitrogen fixation. Symbiotic and free-living nitrogen-fixing micro-organisms have been discussed in two other volumes in this series, respectively, volumes 7 (Nutman, 1976) and 6 (Stewart, 1976). The level of photosynthetic activity by algae in the surface soil of grasslands is, as yet, unknown

A very comprehensive review of the structure of decomposer populations and of decomposition processes in grassland soils has been prepared by Coleman & Sasson (Chapter 7, in Breymeyer & Van Dyne, 1979).

3. Problems in studying grassland ecosystems

R. T. COUPLAND

The purpose in this chapter is to discuss the problems, in assessing trophic structure and functional activity of various groups of organisms, that presented themselves to workers in the various IBP studies of grassland. Most of the participants in these studies found that considerable readjustment was necessary in their concepts and approaches. For example, many ecologists, who were formerly geobotanically or phytosociologically oriented, became community physiologists. Many physiologists were introduced to problems of making measurements in the real, functional outside world. The approach de-emphasised taxonomic considerations. Indeed, the number of species in some trophic levels was much too great to census, let alone to consider individually on a physiological basis. Accordingly, at all trophic levels functional groups of organisms were the basis of study, except where dominant species or 'key' species could be recognized. Most participants in IBP grassland studies had no experience in measuring rates of biological production, of energy flow and of nutrient cycling, particularly in natural habitats, nor was there much knowledge from which they could draw. One of the greatest problems encountered in estimating rates of processes in the field results from the necessity of estimating the activities of organisms at various trophic levels by interpreting temporal changes in standing crop. These interpretations are difficult because inputs and outputs into and out of each compartment of the ecosystem occur simultaneously between sampling dates. For example, the population of an animal species may be increasing by births and by immigration at the same time as deaths and emigration occur.

Producers and primary production

More is known concerning the structure of the producer stratum of grasslands than of the other trophic levels. Prior to IBP methods were well established for sampling populations and for characterising plant distributions. These were applied directly to the IBP sites. A variety of harvest methods had been used to estimate standing crop of shoots and under-ground parts, but these had not been standardised (Milner & Hughes, 1968). Harvest data had usually been used as a direct estimate of net primary production. Serious deficiencies existed in methods of estimating productivity. Usually, the discrepancy between yield and net production was overlooked.

The objective in IBP was to determine the rate of energy capture by primary producers and to ascertain how this is partitioned into production of shoots and roots. There are two general approaches that can be used. The most direct method is by measuring the rate of gas (CO_2) exchange between plants and the environment. Experience in the use of this approach was not sufficient at the beginning of IBP for it to be relied upon as the principal basis for estimating plant production. Also, the sophisticated instrumentation required is such that it can be used only where resources are considerable. Nevertheless, gas exchange measurements were made in several study sites to provide accessory data. Important advances were made in field techniques for the measurement of photosynthesis, both of small units of vegetation within the confined atmosphere of field chambers and by using the aerodynamic method of micro-meteorology. The other approach is the harvest or biomass method by which the rate of production of plant tissue is assessed. This method was used in all studies.

Net primary production can be estimated accurately by the harvest or biomass method only when sequential harvests are made at sufficiently frequent intervals to provide a basis for interpreting the rates of gain and loss of plant material in the intervals between harvests. Sampling of above-ground parts is by clipping, while under-ground parts are separated from the soil by washing known volumes of soil extracted from particular depths, usually by some coring device. Where measurements of herbage consumption by the major herbivores are not available, large herbivores are usually excluded from the area during the year of sampling. Consequently, the effect of grazing on primary productivity is usually based on comparisons of areas that have been protected from grazing by large ungulates for different periods of time. In order to remove the effect of clipping from the estimate, a different set of plots is clipped at each date of harvest. Estimates of herbage production for the entire growing season are obtained by summing the estimates of production for the intervals between harvests, while estimates of root production are usually based on the difference between maximum and minimum biomass values obtained from several dates of sampling.

The major difficulties in estimating net primary production by the harvest method are twofold: firstly, estimates of herbage intake by the major herbivores are difficult and expensive to make; and secondly, losses of plant biomass occur simultaneously with net gains by photosynthesis. These losses are a result of decomposition and consumption by small herbivores. In addition, the estimates are complicated by the fact that plant material drops from the canopy to the litter layer as leaves disintegrate by physical or chemical means or by destruction caused by grazing animals. Estimates of production of under-ground parts from biomass data are further complicated by the difficulty of distinguishing between living and dead roots.

In most IBP studies the estimates of production of herbage were based on the assumption that no consumption or decomposition took place between harvests nor was there any loss to litter. Consequently, these estimates are usually lower than true net shoot production. Nevertheless, in many sites herbage-production estimates were substantially increased over pre-IBP values by interpretations permitted by obtaining separate estimates, progressively through the growing season, for each producer component above ground – green shoots, dead shoots and litter. This technique was introduced prior to IBP by Wiegert & Evans (1964).

It was possible to separate data for periods when additions to the green portion of the canopy were rapid in relation to losses from those when losses exceeded gains; thus, the tendency was reduced for losses to cancel gains. By this approach, the larger the number of harvests, the more nearly does the estimate of net productivity obtained reflect true net production.

Division into taxonomic groups provided a basis for estimating herbage production by summing maximum standing crops of taxa that grow in different seasons of the year. The usefulness of this approach is reduced in areas with short growing seasons, and is greatest where the year is divided into more than one season of active growth.

The massive amount of harvest data available in some sites provided an incentive to find other means of interpretation that would account for losses from the canopy that took place between harvests. It seems probable that the most intensive effort in this direction was made in the Canadian grassland study (Coupland *et al.*, 1975). Total gain in biomass for each interval between harvests was calculated as the sum of the increase in weight of green leaves plus that in dead leaves and litter that could not be accounted for, respectively, by losses from green leaves and dead leaves. To this was added the estimated loss through decomposition by interpreting the increasing concentration of ash in dead shoots and litter as an indication of the rate of break-down and release of organic materials. This approach provided estimates of herbage production that were much greater than were obtained by the methods used by most workers. Nevertheless, these high values were corroborated by studies of the rates of addition and losses of leaves from shoots in the same plots. It was possible also to estimate losses to small herbivores because of information that was gathered concerning rates of food consumption by consumers.

Net primary production is reported in these studies as the sum of net above-ground production and net 'under-ground production'. The latter is that portion of plant material synthesised in the shoots that is transferred and retained under ground. The labour involved in obtaining biomass data for under-ground parts is greater than for above-ground parts, so often fewer harvests are made. Also, the variability of the data is usually greater. This reduces the feasibility of using the incremental method, so that in most studies under-ground production is estimated as the difference between minimum and maximum biomass for the year. This technique was developed prior to IBP by Dahlman & Kucera (1965). In most studies it was not possible to determine the difference between living roots and dead roots, so reliability of estimates could not be increased by considering periodic additions only to living biomass. However, Kucera, Dahlman & Koelling (1967) did make some progress in this direction. Another problem encountered in sampling of under-ground biomass is the loss of fine root branches and of root hairs during washing of soil from roots. This component of biomass is characterised by rapid rates of turnover, which tends to amplify the extent of losses. One study demonstrated the feasibility of using a radioactive tracer to estimate the proportion of photosynthate that is transmitted under ground. The estimates provided in this volume for net primary production above ground, while usually very conservative, are more reliable and probably less conservative than the estimates under ground. This implies the transfer of a larger proportion of photosynthates under ground than is suggested by the data.

Consumers and their activities

The objectives of studies of consumers of grassland in IBP were to measure: (i) the intake of plant material by herbivores (primary consumers); (ii) the proportion of this intake that is transferred to carnivores (secondary consumers) and from all consumers to the detritus food web as excreta, secreta and dead organic materials; and (iii) the losses from the ecosystem through consumer respiration. Such studies were undertaken in only a few sites (Coupland *et al.*, 1969). Three IBP handbooks were prepared for the guidance of workers in consumer studies (Golley & Buechner, 1969; Petrusewicz & Macfadyen, 1970; Phillipson, 1971). The large number of species involved, particularly of invertebrates, necessitated the selection of key species and the arbitrary application of the findings to the group as a whole.

Estimates of population density and biomass are more difficult in animals than plants. Where populations are sparse, only live trapping can be tolerated in the study area, so populations are estimated indirectly by capture–recapture methods. Many species of consumers go through more than one life cycle in a season, so that population data must be subdivided to indicate stage of development, with resultant problems of conversion of population data to estimates of biomass. In some groups (e.g. birds) high mobility necessitates categorisation of the population into residents and visitors. Some consumers spend only part of their life history in a grassland site, and their impact must be evaluated accordingly. Vertical mobility of some invertebrates is also a problem, since many spend part of their life cycle in each of the two sampling zones (canopy and soil), while environmental conditions sometimes cause canopy organisms to seek shelter temporarily on the soil surface where canopy sampling methods fail to pick them up. Sampling data for all consumers must be interpreted in relation to what the numbers really reveal about actual populations. This is particularly true with invertebrates where efficiency of capture (above ground) or extraction (under ground) is not always high and is variable between groups.

Food-habit studies of vertebrates can be made by a variety of methods. In the USA study some of the most reliable data were obtained by use of oesophageal-fistula and rumen-fistula techniques. Analysis of stomach contents of 'collected' animals and faecal analysis were more generally used, as well as direct observation of consuming habits. By these techniques much information was obtained with respect to food preferences in the field. However, information is deficient concerning food preferences of most invertebrates, in some instances being limited to judgement based on structure of mouth parts, while in other instances food preferences were examined in the laboratory.

Difficulties are also experienced in estimating the amount of food consumed by the various groups of animals. Direct measurements were often beyond the resources of the studies reported in this volume, in which case estimates were based on values related to consumer biomass or related to metabolic rates measured in the laboratory. For soil invertebrates very little information was available to permit estimates of the proportion of the disappearing root materials that are routed through the grazing food web.

Micro-organisms and their activities

Micro-organisms, because of their small size and diversity, present a particular challenge in ecological studies (Coupland *et al.*, 1969). While considerable information is available concerning micro-organisms with specialised roles (e.g. in the nitrogen cycle), the organisms that have received the most study do not comprise a significant portion of the micro-flora present in the soil–plant system at any one time. Studies of nitrogen fixation continued as part of IBP but, in addition, an important objective of microbiological research was the measurement of the degree of activity and functional role of micro-organisms in energy flow and nutrient transformations. An instruction manual for microbiologists working in IBP sites formed part of an IBP handbook on quantitative soil ecology (Parkinson *et al.*, in Phillipson, 1971). Such research is complicated because it is difficult to separate the organisms into functional groups and to estimate their biomass. Furthermore, because of changes in activity of the soil micro-flora after sampling, the samples must be utilised promptly.

Variations in concentrations of micro-organisms throughout the ecosystem cause difficulty in sampling. Sampling on a macro-habitat basis provides data that cannot be interpreted in terms of activities that take place in such micro-habitats as the litter layer, on the surface of the soil, the rhizosphere of the roots, and the dead leaves in the canopy. Most studies reported in this volume were limited to estimates of the general soil microbial populations. Bacteria and fungi were usually differentiated and, sometimes, actinomycetes. A major problem is that the numbers of cells detected by traditional plating methods are much smaller than the total numbers actually present. The latter were determined in some studies by direct-counting techniques. Care has to be taken, in considering population and biomass values, to state the method by which estimates were obtained.

The rate of activity of micro-organisms in their role as decomposers can be measured by carbon dioxide output. In the natural environment, however, this gas is also being emitted by consumers and producers and is being absorbed during photosynthesis. The validity of estimates of decomposition rates from carbon dioxide levels in soil is dependent on knowledge of rates of root respiration and respiration by soil animals, which is scanty for both. Accordingly, the rate of loss in weight of buried portions of litter or filter paper has been the major means of estimating decomposition rates. Relative rates of involvement of micro-organisms and small consumers in these disappearances have been assessed by excluding some consumers by placing litter in finely meshed bags or by the application of chemical treatments to which micro-organisms and small consumers respond differentially.

The use of the acetylene-reduction technique to measure rate of nitrogen fixation in the field was introduced just prior to IBP and was applied in some grassland studies.

A more detailed account of methods of studying microbial activities in grassland ecosystems has been contributed by Coleman & Sasson (Chapter 7, in Breymeyer & Van Dyne, 1979).

Systems synthesis

Interconnections of organisms in the various trophic levels to form the integrated whole that comprises the ecosystem can be illustrated in terms of the flow of energy and nutrients. If sufficient information is available, budgets or balances of the whole system can be calculated which are useful in validating the measurements of the component parts. In many IBP studies analysis did not proceed beyond the determination of dry weights of organisms. These are useful in comparing standing crops and organic production in the same trophic level in different sites, but they are only rough indications of the rates of flows of energy and nutrients within the ecosystem. *All biomass values in this volume are expressed as dry weight (including ash), unless otherwise specified.*

Where energy flows are discussed in this volume they have been interpreted from standing-crop biomass and biomass production values by determination of energy values of representative samples, combined with information concerning rates of respiration and disappearance.

Nutrient budgets have been determined by converting biomass data (standing crop and production) on the basis of chemical analyses. One of the major problems in interpreting data on nutrient cycles is the understanding of the intricate exchanges that take place within the soil complex; these affect the proportion of the amount of each element in the soil-nutrient pool that is available for plant growth. As a result, nutrient deficiencies sometimes occur in environments in which ample total amounts of each element are present. This problem has not been approached in most studies; the nutrient budgets provided here represent only the routing of nutrients between the times when they are absorbed into plants and when they are returned to the soil as animal wastes and products of decomposition.

Usage of the term 'systems synthesis' in this volume differs considerably from that of the companion volume (Breymeyer & Van Dyne, 1979). Here, the term is applied to the synthesis of information about standing crops and flows. In the companion volume, systems analysis is used in the sense of the development of mathematical simulation and optimisation models of ecosystem components or of the entire ecosystem. In both instances the system is compartmentalised and the connections between compartments are defined. However, the procedures differ. In this volume the data are processed to provide mean values for the compartments and for rates of flow, generally on a year-long basis. The development of simulation models requires mathematical descriptions of the flow functions, followed by calculation of the dynamics of compartmental values and of the dynamics and integrated values for the flows.

There was a great variety of both published and unpublished information available from IBP studies for synthesis into this volume and the companion volume (Breymeyer & Van Dyne, 1979). In the process of international coordination many progress reports were exchanged among workers. However, the authors were recruited after the period of activity of the Grassland Working Group and were not recipients of these unpublished reports. It should also be emphasised that during the period when these volumes were being prepared much of the information had not yet appeared in the open literature. Much of it was still in the form of technical reports, theses and dissertations of post-graduate students, and manuscripts of scientific papers or chapters for inclusion in IBP series of volumes. These materials were not freely or equally accessible to authors. While the objective in this volume is to incorporate information resulting from both IBP studies and those that were not so designated, the basis of recruitment of authors was more their familiarity with IBP research than their world-wide knowledge of the literature in their respective disciplines. Consequently, the synthesis presented in this volume is less complete in some respects than is desirable.

Part II. Natural temperate grasslands

4. Introduction

N. R. FRENCH

This discussion emphasises IBP investigations of natural temperate grassland ecosystems in North America, which were made in 11 sites in the USA and one in Canada (Table 4.1, Fig. 5, 1, Plates 4.1, 4.2). Although studies were carried out in natural temperate grasslands of other continents, information provided in the data bank was not sufficient to permit integration with the North American sites. Two studies discussed in Part III include sites in forest steppe of the eastern hemisphere that appear to be natural temperate grassland. These are the meadow steppe at the Karachi station in the USSR and the 'closing grassland' of the Czévharaszt area in Hungary.

It is desirable in a general description of natural temperate grassland sites to point out the basic similarities of grassland ecosystems and to contrast their characteristics with those of different ecosystem types. The constancy of grassland characteristics should be manifested in the biota of the different sites and may be expected to be related to the climatic driving variables of the system. In this chapter we will briefly examine the general characteristics of natural temperate grasslands and the major characteristics of the present climatic conditions, as well as the soils which have developed under these conditions and the biota adapted to these situations.

The 12 natural temperate grassland sites considered here (Table 4.1) occur between 32° and 51° N latitude. In elevation they range from 335 m to over 2370 m above sea level. They are characterised by a wide variety of species and genera of grasses, which effectively utilise the characteristic growing conditions of their sites to produce characteristic vegetation dominated by tall grasses (true prairie), by mid-grasses (mountain grassland), by a mixture of mid-grasses and short grasses (mixed prairie), by short grasses (short-grass prairie and desert grassland), by bunch grasses (palouse prairie) or by annual grasses. Seed consumers appear to be important at arid sites (e.g. Jornada and ALE) and small herbage consumers at mesic sites (e.g. Osage, Hays, Bridger, Matador and San Joaquin).

The climatic diagrams for these sites (Fig. 4.1) can be divided roughly into four groups. The first group consists of the true prairie site (Osage), mountain grassland (Bridger) and the southern mixed-prairie site (Hays). These sites are characterised by adequate precipitation, but with a tendency toward a bimodal distribution of precipitation throughout the year, more pronounced at higher elevations. The last in this series shows increasing dominance of the first peak of precipitation. This trend is accentuated in the second group of climatic diagrams, which include the other three mixed-prairie sites (Cottonwood, Dickinson and Matador) and the palouse, Bison, site. These diagrams

Table 4.1. *Characteristics of the North American IBP study sites in natural temperate grassland (values in parentheses are estimated)*

Site	State or province	Country	Latitude	Longitude	Grassland type	Elevation (m)	Soil
ALE	Washington	USA	46°24′N	119°33′W	Shrub steppe	365	Glaciofluvial
Bison	Montana	USA	47°19′N	114°16′W	Palouse prairie	985	Chernozem
Bridger	Montana	USA	45°47′N	110°47′W	Mountain grassland	2370	—
Cottonwood	South Dakota	USA	43°57′N	101°52′W	Mixed prairie	735	Silty clay
Dickinson	North Dakota	USA	46°54′N	102°49′W	Mixed prairie	825	Loamy fine sand
Hays	Kansas	USA	38°52′N	99°23′W	Mixed prairie	710	Calcareous
Jornada	New Mexico	USA	32°36′N	106°51′W	Desert grassland	1350	Alluvial fill
Matador	Saskatchewan	Canada	50°42′N	107°48′W	Mixed prairie	680	Lacustrine clay
Osage	Oklahoma	USA	36°57′N	96°33′W	True prairie	380	Brunizem
Pantex	Texas	USA	35°18′N	101°32′W	Short-grass prairie	1090	Silty clay loam
Pawnee	Colorado	USA	40°49′N	104°46′W	Short-grass prairie	1650	Alluvial clay
San Joaquin	California	USA	37°06′N	119°44′W	Mediterranean annual grassland	335	Coarse sandy loam

Table 4.1. (*cont.*)

Site	Annual Precipitation (mm)	Plant species	Growing season (days)	Consumers
ALE	234	*Agropyron spicatum*, *Stipa comata*, *S. thurberiana*, *Poa secunda*	116	*Eremophila alpestris*, *Perognathus parvus*, *Peromyscus maniculatus*
Bison	330	*Festuca scabrella*, *F. idahoensis*[a]	(163)	*Bison bison*, *Odocaileus hemionus*, *Microtus montanus*
Bridger	980	*Phleum alpinum*, *Festuca idahoensis*, *Agropyron subsecundum* *Lupinus argenteus* Pursh	122	*Thomomys talpoides*, *Microtus montanus*
Cottonwood	384	*Agropyron smithii*, *Stipa viridula*	163	*Ammodramus savannarum*, *Spermophilus tridecemlineatus*, *Microtus ochrogaster*
Dickinson	398	*Stipa comata*, *Bouteloua gracilis*, *Agropyron smithii*	(161)	*Spermophilus tridecemlineatus*, *Peromyscus maniculatus*
Hays	582	*Andropogon gerardi*, *A. scoparius*, *Bouteloua curtipendula*	(215)	*Peromyscus maniculatus*, *Microtus ochrogaster*
Jornada	228	*Bouteloua eriopoda*, *Sporobolus flexuosus*	164	*Dipodomys ordii*, *D. spectabilis*
Matador	338	*Agropyron dasystachyum*, *Carex eleocharis* Bailey, *Koeleria cristata*	166	*Anthus*, *Ammodramus*, *Microtus*, *Aeropedellus clavatus*, *Melanoplus femurrubrum*
Osage	930	*Andropogon gerardi*, *Panicum virgatum*	252	*Sturnella magna*, *Microtus ochrogaster*
Pantex	533	*Bouteloua gracilis*, *Aristida longiseta*	176	*Sturnella neglecta*, *Sigmodon hispidus*, *Peromyscus maniculatus*
Pawnee	311	*Bouteloua gracilis*, *Buchloe dactyloides*	172	*Calamospiza melanocorys*, *Spermophilus tridecemlineatus*
San Joaquin	559	*Bromus mollis*, *B. diandrus* Roth, *Festuca megatura*	186	*Spermophilus beecheyi*, *Thomomys bottae*

[a] The Bison site is located in a region of palouse prairie dominated by *Agropyron spicatum*, but the area chosen for intensive study is dominated by *Festuca* spp.

Plate 4.1. Ungrazed natural temperate short-grass prairie grassland dominated by *Bouteloua gracilis* and *Buchloe dactyloides* on the Pawnee study site in the USA, as it appears in September. (Photograph by R. E. Redmann.)

Plate 4.2. Ungrazed natural temperate mixed prairie grassland dominated by *Agropyron dasystachyum* and *A. smithii* at Matador in Canada, as it appears in mid-June. (Photograph by R. T. Coupland.)

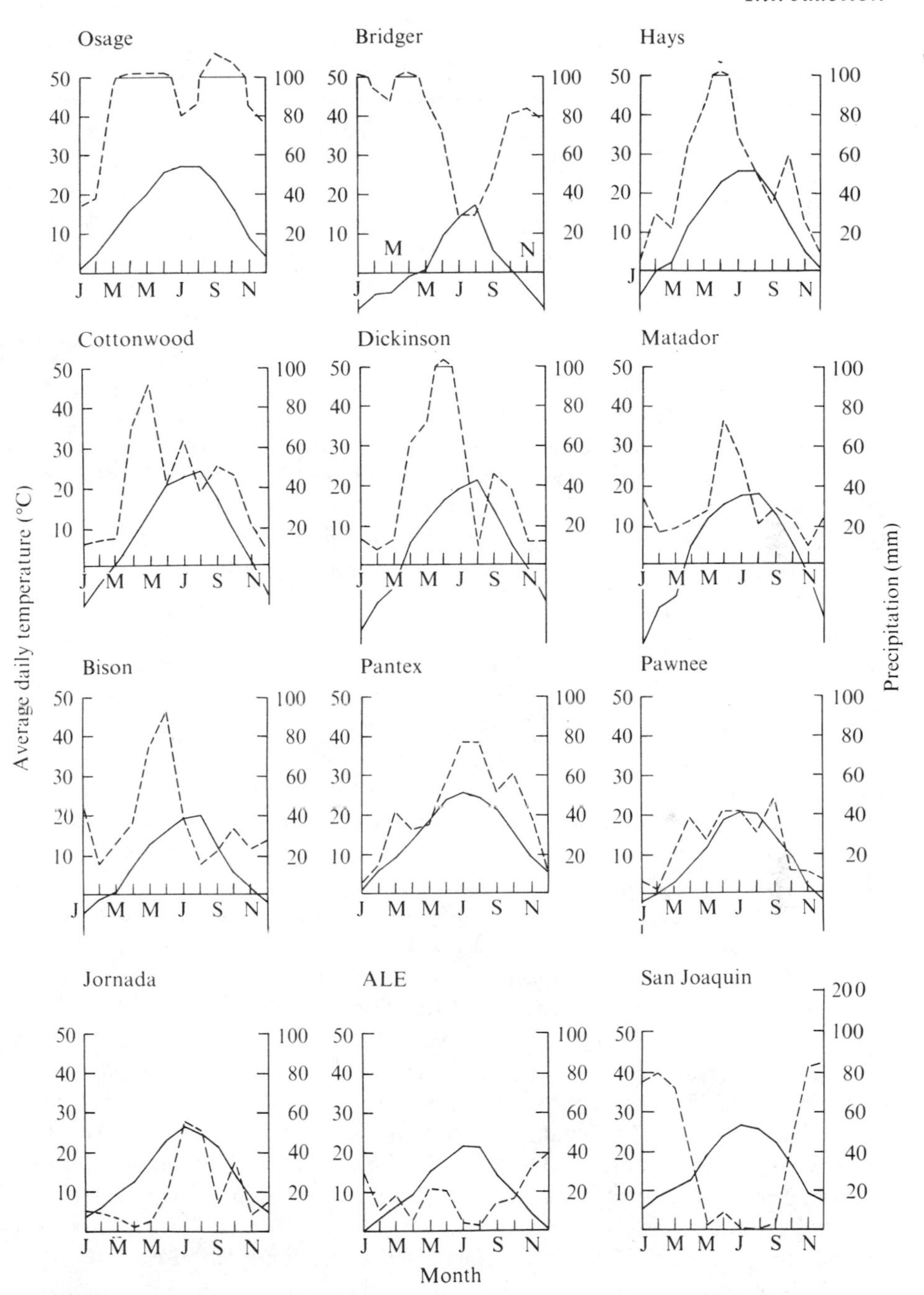

Fig. 4.1. Climatic diagrams for the natural temperate grasslands studied. ——, temperatures; – – – –, precipitation.

show an increasing tendency toward early growing-season precipitation and middle or late growing-season drought. The third group is of short-grass prairie sites (Pantex and Pawnee), which show marginal precipitation, but it is well distributed in the growing season. Finally, the last three diagrams include the two most arid sites (Jornada and ALE) and the annual grassland (San Joaquin). These show extreme late growing-season drought. Two are characterised by Mediterranean-type climatic regime (San Joaquin and ALE), in which the growing season is sustained by the precipitation that occurs in winter. The desert grassland situation (Jornada) shows the beginning of the growing season retarded until summer rainfall is adequate to combine with the amenable temperature to provide conditions that promote plant growth.

The amount of energy contained in incoming solar radiation decreases with passage through the atmosphere. As a result, it varies with the elevation of the site in question. At sea level the value will be approximately 5.4 J/cm^2/min (= 1.3 langleys/min) and on a high mountain 7.5 J at noon of a clear day (Gates, 1965). Only about 45 % of the visible radiation (between 0.4 and 0.7 μm) can be fixed in photosynthesis by green plants. The amount reaching the ground is also modified by elevation of the sun, by conditions of the sky, and by local topography.

The average daily total incoming solar radiation has been measured at a number of grassland sites and can be estimated for others on the basis of their latitudes (Fig. 4.2). Estimation, however, does not include variation due to topography and sky condition. Measured values for grassland sites indicate that the highest daily totals are attained in areas of Mediterranean-type climate (San Joaquin and ALE) and in desert grassland (Jornada), where average values attain approximately 700 langleys/day. Values greater than 600 langleys/day are attained at Dickinson, Bison, Pawnee, Cottonwood and Osage. The second group of sites generally shows greater seasonal variation in incoming solar radiation than the first group of sites. Similar maximum values for solar radiation are found at the southern Great Plains sites (Pantex and Hays), but seasonal variation is less. Lower maximum values, approximately 500 langleys/day, are characteristic of sites with more cloudy conditions or higher latitudes (Matador and Bridger).

In general, solar radiation is most intense at those sites where precipitation is most limited. Those with greatest seasonal variation and solar radiation also show the greatest seasonal variations in average temperatures. Those sites with lowest maximum solar radiation intensities are, generally, those with the greatest difference between precipitation and mean monthly temperature shown in the climatic diagram.

Soils and vegetation develop together over a very long period of time, so they are related to each other and both are related to climatic conditions. Chemical characteristics depend more on the parent material, weathered to form the mineral content of soil. In most natural temperate grasslands precipi-

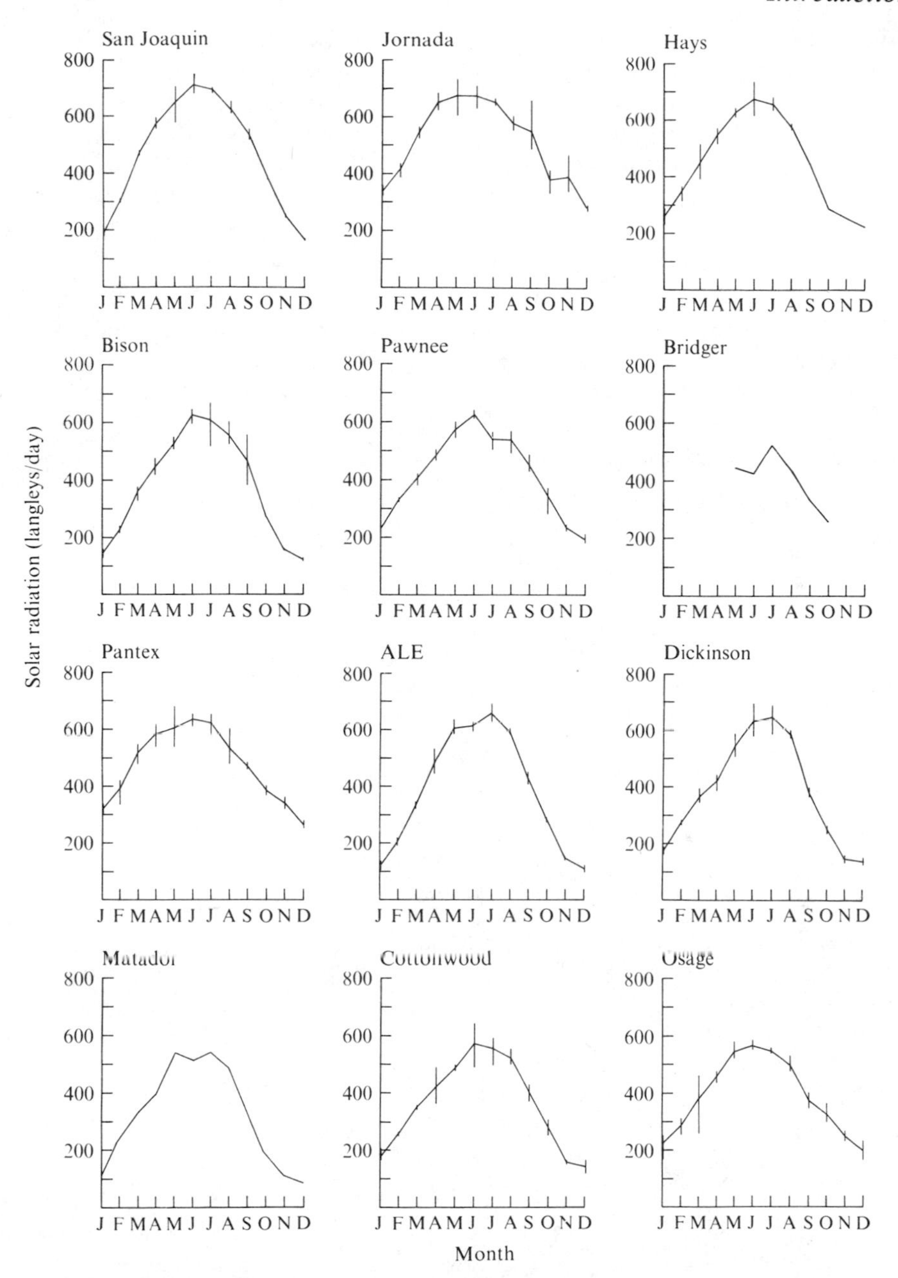

Fig. 4.2. Average daily incoming solar radiation in sites in natural temperate grasslands. Vertical lines represent ranges of monthly averages during years of IBP studies.

tation is adequate to form an organically rich surface horizon. The soil is usually near neutrality in pH, but may be basic under more arid conditions where precipitation is insufficient for leaching of the soil and there is an accumulation of salts or carbonates. The soil that has developed in most of these sites is silty clay, alluvial clay, or silty clay loam in texture. In the arid regions the sites occur in areas of alluvial fill (Jornada) or fluvial deposits (ALE). The soil of one site (Hays) is calcareous and sandy. Soils of more mesic sites are generally much higher in content of organic matter than are those of arid sites.

Abiotic processes in these same sites are discussed in detail by Hinds & Van Dyne (in Chapter 1 of Breymeyor & Van Dyne, 1979).

Hitchcock (1950) is the general authority used for scientific names of grasses in Chapters 4 to 9 of this volume. Authors are given for those plant species not included in this manual.

5. Producers

P. L. SIMS & R. T. COUPLAND

The ultimate objective of the studies discussed here was to estimate net primary production in the absence of stress and to examine the effect, on primary production, of grazing by domesticated livestock. The data upon which this chapter is based have been interpreted in much greater detail in the technical reports and preprints of the USA/IBP Grassland Biome Study (Sims & Singh, 1971, 1978*a*, *b*, *c*; Sims, Singh & Lauenroth, 1978; Duncan, 1975*a*, *b*), and from reports of the Matador Project (Coupland 1973*a*, *b*, *c*, 1974*a*, *b*; Coupland & Abouguendia, 1974; Coupland, Ripley & Robbins, 1973; Coupland *et al.*, 1975). The locations of study sites are shown in Fig. 5.1.

Plant biomass has been divided in these studies into three above-ground compartments, green (live) and dead canopy components and litter (fallen leaves and stems), and two to four under-ground compartments (roots, crowns (shoot bases), rhizomes and litter). In ungrazed areas between 55 and 86 % (mean 70 %) of total plant biomass occurs under ground, while in the canopy 15–90 % (mean 34 %) of the biomass occurs as green shoots (Table 5.1).

Dynamics of above-ground biomass

The contributions that are made to canopy biomass by species of differing physiological response are important in characterising each grassland type. In cooler regions cool-season (C_3) species contribute 75 % or more of green-shoot biomass, while in warmer regions a similar proportion is contributed by warm season (C_4) plants; the canopy of mid latitude grasslands is comprised of a mixture of these two types of plants (Table 5.2). Grazing by domesticated livestock tends to increase the proportional contribution to biomass by warm-season species where they co-exist with cool-season species; this is particularly evident in the mixed prairie of South Dakota (Cottonwood) where warm-season species are nearing their geographical limit as dominants. In warmer areas grazing results in a decrease in biomass of warm-season grasses and a corresponding increase in warm-season forbs (Osage) or in warm-season forbs and shrubs (Jornada). In the short-grass prairie (Pantex and Pawnee) warm-season succulents are conspicuous increasers as a result of grazing. Only in 1973 was there any significant growth of warm-season annuals in the annual grasslands of California.

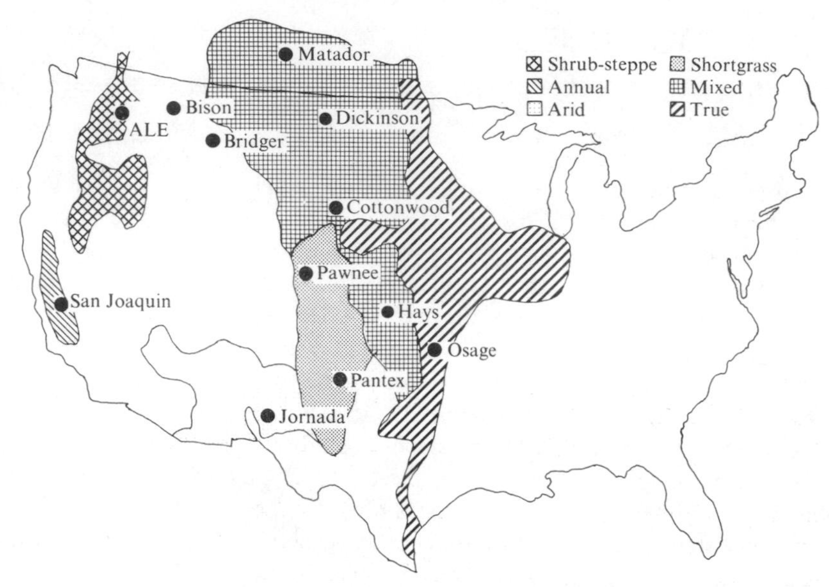

Fig. 5.1. Outline map showing locations of USA and Canadian IBP study sites in natural temperate grasslands.

Green shoots

Considerable differences occur in biomass of green shoots in various types of grassland, depending particularly on moisture regime. Time-weighted annual means of green-shoot biomass in ungrazed areas ranged between extremes of 21 and 241 g/m^2; however, except for one year in desert grassland (21 g/m^2), one year in southern short-grass prairie (25 g/m^2), one year in shrub steppe (34 g/m^2), and one year in mountain grassland (241 g/m^2), the values ranged between 53 and 178 g/m^2 (Table 5.3). Peak green-shoot biomass in ungrazed situations was more than 100 g/m^2, except in one year at Pantex, Matador and ALE and two years at Pawnee, while it exceeded 270 g/m^2 in the annual grassland, as well as in one year at Bridger and one at Osage. Considerable fluctuation occurred in shoot biomass from year to year in the same site (Table 5.3). This was especially evident in the Pantex site, where annual values for mean green-shoot biomass and peak green-shoot biomass varied by a factor of about 4 to 1 in a period of three years.

Most grasslands have a unimodal pattern of increase and decrease in green biomass (Sims *et al.*, 1978), the peak occurring earlier in the growing season in sites dominated by cool-season species and later in grasslands dominated by warm-season species. In the desert grassland the curve of green biomass is quite irregular because of irregular rainfall and a long growing season. Several small peaks in live biomass occur that closely follow major rainfall events.

Table 5.1. *Mean biomass* (g/m^2) *of producer compartments in ungrazed natural temperate grassland sites in the USA and Canada, listed in approximate order of increasing total biomass*

Site	Grassland type	Years of study	Canopy: Green shoots	Canopy: Dead shoots	Canopy: Subtotal	Litter	Under-ground parts: Crowns	Under-ground parts: Roots	Under-ground parts: Subtotal	Total
Jornada	Desert grassland	3	49	33	82	71	—	—	187	340
ALE	Shrub steppe	2	52	107	159	78	—	—	—	—
Bison	Palouse prairie	1	120	84	204	316	—	—	—	—
Pantex	Short-grass prairie	3	70	129	199	331	295	738	1033	1563
Pawnee	Short-grass prairie	3	70	65	135	251	348	1368	1716	2102
Bridger	Mountain grassland	1–2	181	21	202	114	—	—	1928	2244
Osage	True prairie	3	152	446	598	365	287	1148	1435	2398
Hays	Mixed prairie	1	79	117	196	904	—	—	1319	2419
Cottonwood	Mixed prairie	3	93	141	234	496	273	1690	1963	2693
Dickinson	Mixed prairie	1	178	369	547	457	—	—	1715	2719
Matador	Mixed prairie	3–5	75	411	486	238	141	2026[a]	2167	2891

[a] Root biomass for Matador is given only to a depth of 60 cm, so as to be comparable with USA data where sampling on the average was to a depth that included about 76 % of the roots. At Matador an estimated 23 % of under-ground parts (excluding shoot bases) occurred below a depth of 60 cm (Table 5.4).

Table 5.2. *Proportion of shoot biomass harvested throughout the sampling season that was comprised of cool-season and warm-season species in various USA natural temperate grassland sites in both ungrazed and grazed condition, listed in approximate order of declining importance of cool-season species*

			Percentage of shoot biomass contributed by					
		Mean temperature for year	Cool-season species		Warm-season species		Other species[a]	
Site	Year	(°C)	Ungrazed	Grazed	Ungrazed	Grazed	Ungrazed	Grazed
Bison	1970	7.47	100	97	0	2	0	1
San Joaquin	1973	15.20	87	71	13	29	–	–
	1974	15.43	100	100	0	0	–	–
	1975	14.46	100	100	0	0	–	–
Bridger	1970	1.21	72	70	1	0	27	30
	1972	2.90	83	83	3	0	14	17
ALE	1971	11.92	75	72	25	28	0	0
	1972	11.45	63	68	37	32	0	0
Cottonwood	1970	8.02	75	11	25	89	0	0
	1971	7.80	82	30	17	69	1	1
	1972	5.74	82	31	18	69	0	0
Dickinson	1970	4.10	61	48	39	37	0	15
Pawnee	1970	6.97	10	6	90	94	0	0
	1971	7.95	13	4	87	96	0	0
	1972	8.05	13	9	87	83	0	8
Pantex	1970	14.29	22	13	78	87	0	0
	1971	13.65	1	1	99	99	0	0
	1972	13.25	7	11	93	89	0	0
Jornada	1970	14.29	1	1	79	90	20	9
	1971	14.70	0	0	61	82	39	18
	1972	13.45	10	15	73	79	17	6
Osage	1970	15.06	5	9	95	91	0	0
	1971	15.18	0	1	100	99	0	0
	1972	14.88	0	0	100	100	0	0
Hays	1970	11.62	1	4	99	96	0	0

[a] These species were either not separated or insufficient autecological information was available to categorise them as cool- or warm-season.

Some green biomass occurs well beyond the thermic growing season. Even at the northern-most site, 1–5 g/m^2 of green shoots was found to persist under snow cover in mid-winter. During one year in the mountain grassland site snow covered much green material remaining at the end of the growing season. This over-wintered, but turned brown soon after snow-melt in the spring.

Year-to-year variations in peak green biomass depend more on amount of precipitation during the growing season than on other environmental factors that have been considered (Smoliak, 1956; Ricklefs, 1973). In the IBP sites, a linear increase in green biomass occurs between 100 and 450 mm of growing season precipitation, but there is a tendency for levelling off at higher moisture levels; similarly, a close relationship is shown between green biomass and actual evapotranspiration (Table 5.3). Because of different proportions of warm-season and cool-season species, the grasslands differ in their capacity to use water. Warm-season species have C_4 characteristics and cool-season species have C_3 characteristics, which Black (1971) has found to represent two distinct groups in relation to water usage.

Grazing by domesticated livestock has a very important impact on green biomass in these grasslands. In the 47 comparisons made in Table 5.3, reductions in green biomass of 20 % or greater occurred in 19 instances as a result of grazing in years prior to the year of sampling, while in six comparisons increases greater than 20 % occurred. The effect of grazing was apparently to increase green-shoot biomass in the Pantex site and to reduce it in the Bison, ALE, Jornada and Cottonwood sites. Elsewhere, the response was not appreciable. The reason for this differential response is not evident.

Dead shoots

The dead-shoot compartment consists of biomass produced during the current year that has not been decomposed, as well as undecomposed material produced in previous years. Not much attention has been given previously to the fate of shoot biomass in the canopy after it is no longer green. The most intensive study of this nature appears to have been undertaken at the Canadian IBP study site (Matador) (Coupland & Abouguendia, 1974). This study revealed that, for species of *Agropyron*, 10–20 % of leaf blades of the current year fall to litter by autumn (Fig. 5.2), and all over-wintered leaf blades are transferred to litter by early August. The presence of dead shoots in the canopy is a function of the rate of production and of senescence of the green shoots. This Canadian study revealed that the length of time during which leaves remain green is only from a few days to 4 weeks (Fig. 5.2). Furthermore, the transfer of green growth of the current year to the dead compartment of the canopy begins in April and continues throughout the growing season. At the same time, dead leaves are being dropped to litter and litter decomposition occurs. The trend for dead-shoot biomass in this northern site is to decrease during spring and summer and increase before snowfall. However, depending upon current weather conditions, increases in dead-shoot biomass occur as early as August or as late as November. During the winter there is usually a decrease in biomass of dead shoots; this averaged 24 % at Matador.

There is a considerable range in dead canopy biomass in ungrazed natural temperate grasslands studied (Table 5.1). Minimum values occurred in mountain grassland (time weighted mean of 21 g/m^2) and in desert grassland (33 g/m^2), while maximum amounts occurred in the true prairie site (446 g/m^2) and in the northern-most mixed prairie site (411 g/m^2). The mean relative

Table 5.3. *Green-shoot biomass (g/m²) in USA and Canadian sites of natural temperate grassland in ungrazed and grazed condition, listed in approximate order of increasing maximum standing crop of green shoots in ungrazed locations; relationships to moisture parameters are suggested*

Site	Year	Green-shoot biomass				Precipitation (mm)		Actual evapotranspiration (mm) (growing season)	Length of growing season (days)[a]
		Maximum		Mean[b]					
		Ungrazed	Grazed	Ungrazed	Grazed	Mean annual (long term)	Growing season (current year)		
Pawnee	1970	91	88	63	48	311	149	148	166
	1971	94	124	58	82		222	220	207
	1972	119	81	89	58		324	321	207
ALE	1971	118	97	71	55	234	89	89	231
	1972	98	84	34	23		109	109	283
Jornada	1970	134	74	74	54	228	166	166	335
	1971	134	34	21	11		183	183	280
	1972	108	99	53	42		324	324	333
Matador	1968	100	—	55	—	338	182	—	194
	1969	120	—	83	—		239	—	182
	1970	187	—	114	—		255	—	166
	1971	163	—	83	—		192	—	194
	1972	86	—	58	—		145	—	163

[illegible]	1970	62	134	25	62	533	184	182	271
	1971	230	214	78	103		522	517	265
	1972	236	210	107	93		359	355	272
Cottonwood	1970	164	101	95	46	384	242	232	187
	1971	205	153	87	59		485	466	213
	1972	196	147	97	75		354	340	201
Bridger	1970	144	124	121	105	980	182	96	104
	1972	201	236	241	247		529	280	141
Bison	1970	220	102	120	72	330	244	244	185
Hays	1970	225	243	79	88	582	426	409	226
Dickinson	1970	270	218	178	158	398	360	346	168
Osage	1970	270	286	142	139	930	435	335	272
	1971	336	314	173	202		782	602	270
	1972	254	311	141	156		805	620	275
San Joaquin	1973	526	413	—	—	559	611	611[c]	365
	1974	459	374	—	—		577	577[c]	365
	1975	337	340	—	—		480	480[c]	365

[a] Length of growing season applied here is the number of consecutive days with a 15-day running mean air temperature equal to or greater than 4.4°C. This is the thermal potential growing season.
[b] These are time-weighted means for the growing season.
[c] At San Joaquin an unestimated amount of water is lost to groundwater by percolation.

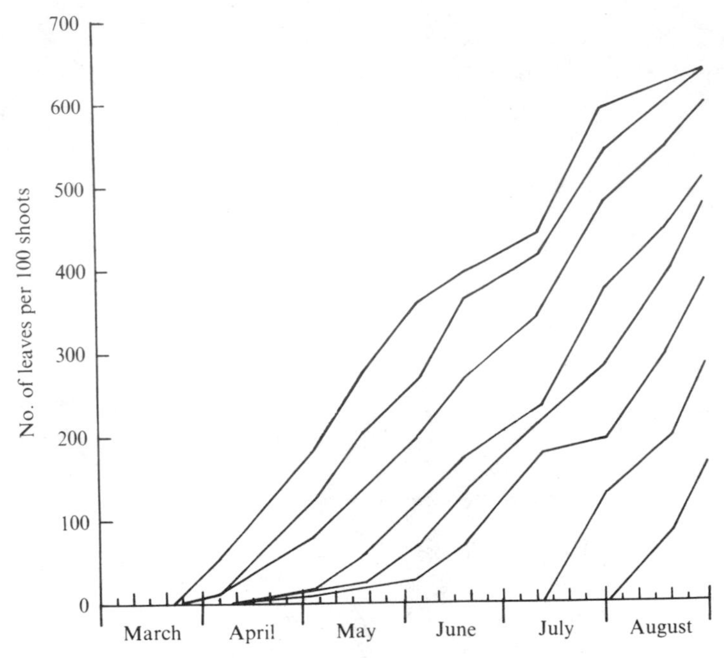

Fig. 5.2. Diagrammatic representation of the length of period during which leaf blades of *Agropyron dasystachyum*, first observed at various dates in 1972, remained in each condition class. The lengths of these periods are indicated by the horizontal distance between drawn lines that represent the cumulative numbers of leaves per 100 shoots that arrived in each leaf-blade category *after 29 March*. These lines depict, from left to right, arrival in the following categories: developing, fully developed, dying from tip, yellow, brown, grey, fractured and litter. The space to the right of each of the first seven lines indicates the time spent in this condition before passing into the next one. For example, leaves that appeared first during the first week of April were fully yellow by 10–20 May and dropped to litter between mid-July and mid-August (after Coupland & Abouguendia, 1974).

dead-shoot biomass to green-shoot biomass also varied over a wide range. Dead-shoot biomass exceeded green-shoot biomass except in the mountain grassland, desert grassland, palouse grassland and northern short-grass sites. In the shrub steppe, southern short-grass, and three mixed prairie sites mean dead-shoot biomass ranged from about one and one-half to two times mean green-shoot biomass, while this multiple was approximately three times in the true prairie site and five times in the northern-most mixed prairie site (Table 5.1). This high proportion of dead biomass occurs in the latter site, even though mean residence time of leaves of the dominant grasses in the dead-shoot compartment is as short as three months (Fig. 5.2). Dead-shoot biomass is reduced in areas that have been subjected to grazing in previous years.

In the USA sites dead-shoot biomass was separated into 'recent dead' (produced and senesced during the current growing season) and 'old dead'.

The recent-dead compartment is the interface between green biomass and old dead and, consequently, reflects the dynamics of both of these compartments. The general pattern in the recent-dead compartment is a gradual build-up as the growing season progresses, as material is transferred from the green compartment. In grassland with an abundance of both cool-season and warm-season plants, the build-up in the recent-dead compartment begins relatively early and continues at a rather steady rate throughout the growing season. In grasslands dominated by warm-season species, increases in the recent-dead compartment generally begin mid-way during the growing season, soon after the peak in green biomass. In all cases, peaks in recent dead follow peaks in the green compartment. In desert grassland (Jornada), where several peaks occurred in green biomass, there were several peaks also in the recent-dead compartment, each following a corresponding peak in green biomass. In mountain grassland the general trend described for the other sites did not apply; the peak in recent-dead biomass occurred very early in the growing season, apparently because of transfer to the recent-dead category of green shoots that over-wintered. The general pattern and dynamics of biomass in the recent-dead portion of the canopy in grazed areas is similar to, but less marked than, that in ungrazed grasslands.

In contrast to the recent-dead, the old-dead material is most abundant early in the growing season and declines as it is transferred to litter. The major decline in old-dead biomass corresponds with an increase in green biomass, suggesting that the same environmental conditions foster both growth and decomposition and fall of old-dead material.

Litter

The processes that affect the accumulation and disappearance of litter are a function of precipitation events and occur in a pulsed fashion. Litter accumulates from both recent- and old-dead canopy compartments, as well as by some transfers directly from green shoots. Litter disappears through the processes of leaching, decomposition and soil mixing. Multiple sources for litter and the pulse behaviour of litter disappearance result in an erratic pattern in litter dynamics. There is, however, a general peak in litter biomass following the peak in old-dead shoot biomass, a primary litter source.

Mean litter biomass in ungrazed locations ranges from about 70 g/m^2 in desert grassland to approximately 900 g/m^2 in southern mixed prairie (Table 5.1). It fluctuates less within a site than does canopy biomass. For example, in the study of longest duration (Matador), yearly mean litter biomass ranged from 218 to 288 g/m^2, while yearly mean canopy biomass (green plus dead) ranged from 382 to 586 g/m^2. No apparent constant relationship was found in this site between the standing crop of litter and progress through the growing season; however, the mean proportion of litter was 42 % of the total above-ground plant biomass (i.e. canopy plus litter) in 1968 and declined to 27 % in 1972. Rarely did litter comprise less than 20 % or more than 50 % of the total above-ground standing crop.

Although the standing crop of litter is lower in most grazed grasslands, this difference is less marked in short-grass prairie and desert grasslands than elsewhere. However, in the true prairie site, grazed areas have twice as much litter as ungrazed areas. This may be due to trampling by the grazing animal resulting in an increased rate of transfer, from dead shoots to litter, of decay-resistant structural material that is abundant in true prairie (Quinn & Hervey, 1970; Sims *et al.*, 1978).

Dynamics of under-ground biomass

The determination of the amount and dynamics of under-ground plant material is more difficult and subjective than for shoot biomass. One of the difficulties relates to high variability of samples because of the small surface sampling area necessitated by core sampling. This variability is associated, near the surface, with lack of horizontal uniformity in distribution of crowns and, at greater depths, with declining density of roots. In addition, considerable effort is required to remove plant parts from soil in the large number of samples required to reduce sampling error. These difficulties have resulted in very little information being available with regard to temporal fluctuations in biomass of under-ground plant parts. Fluctuations in under-ground biomass during the season are a function of losses that result from decomposition and root exudates, as well as translocation of photosynthates back and forth to shoots.

Under-ground plant material in these natural temperate grasslands is comprised mostly of roots, except in the uppermost soil layer, where shoot bases (crowns), rhizomes and litter (unclassifiable material) contribute to plant biomass. The mean proportion contributed in the various layers in one site is given as an example in Table 5.4. In the USA studies regular sorting of under-ground materials was into two components, crowns and roots. The proportion of under-ground biomass contributed by crowns averaged about 22 % (Table 5.5); grazing had no consistent effect in changing this proportion.

A relationship between temperature and mean biomass of total under-ground parts is suggested by the data in Table 5.5. For example, within mixed prairie there was a decrease southwards in mean total under-ground biomass from approximately 2100 g/m^2 at Matador to 1600 g/m^2 at Dickinson and Cottonwood to 1150 g/m^2 at Hays; similarly, within short-grass prairie there was a decline from Pawnee (1349 g/m^2) to Pantex (874 g/m^2). A relationship to moisture parameters is suggested in the three southern-most sites where mean total under-ground biomass declined with increasing aridity from true prairie (Osage, 1211 g/m^2) to short-grass prairie (Pantex, 874 g/m^2)

Table 5.4. *Mean contributions to biomass (in g/m²) of each category of under-ground plant material that was found in each layer of soil in the northern-most mixed prairie site (after Coupland* et al., *1975)*

Depth (cm)	Roots	Rhizomes	Shoot bases	Litter[a]	Total	Percentage in each layer
0–10	842	46	141	104	1133	33.2
10–20	312	5	0	14	331	13.1
20–30	216	0	0	5	221	9.4
30–40	184	0	0	4	188	7.7
40–50	167	0	0	4	171	7.1
50–60	124	0	0	3	127	5.2
60–70	151	0	0	4	155	6.3
70–80	103	0	0	2	105	4.3
80–90	100	0	0	2	102	4.2
90–100	60	0	0	1	61	2.5
100–110	40	0	0	1	41	1.7
110–120	38	0	0	1	39	1.6
120–130	38	0	0	1	39	1.6
130–140	27	0	0	1	28	1.1
140–150	21	0	0	1	22	0.9
Total	2423	51	141	148	2763	100.0

[a] Below 30 cm it was assumed that the proportion of litter was the same as in the 20–30 cm layer (i.e. 2.3 %).

to desert grassland (Jornada, 169 g/m²). The dynamics of under-ground plant parts is greatest in the annual grassland at San Joaquin, because of the absence of perennial structures.

The effect of grazing on total under-ground plant biomass appears to be greater in cooler areas than in warmer ones (Table 5.5). In mountain grass-land, the Dickinson and Cottonwood sites in mixed prairie, and in northern short-grass prairie (Pawnee), under-ground biomass increased as a result of grazing. Southwards the impact was less pronounced and most often resulted in a decline in biomass of under-ground parts as a result of grazing. This differential effect of grazing may be associated with the degree to which the composition of the plant cover is modified by grazing. We have already noted a marked trend towards warm-season species in grazed areas in cooler sites that is not present to the same extent southward (where cool-season species are less important constituents of the vegetation). It is probable that there is an interaction between temperature and grazing (Lewis, 1971). For example, in mountain grassland, dominated by cool-season plants, although there was no marked shift in composition as a result of grazing, under-ground biomass was greater in the grazed situation.

Table 5.5. *Biomass of underground plant parts (in g/m²)*[a] *in natural temperate grasslands in ungrazed and grazed condition, listed in approximate order of increasing biomass and related to mean yearly temperature*

		Roots				Crowns				Total mean under-ground parts		Mean tempera-ture throughout year (°C)
		Peak		Mean		Peak		Mean				
Site	Year	Ungrazed	Grazed	Ungrazed	Grazed	Ungrazed	Grazed	Ungrazed	Grazed	Ungrazed	Grazed	
Jornada	1970[b]	235	230	202	159	—	—	—	—	202	159	14.3
	1971[b]	210	231	162	146	—	—	—	—	162	146	14.7
	1972[b]	177	168	142	125	—	—	—	—	142	125	13.4
San Joaquin	1973[b]	757	956	645	753	—	—	—	—	645	753	15.2
	1974[b]	626	724	640	603	—	—	—	—	640	603	15.4
	1975[b]	558	491	430	464	—	—	—	—	430	464	14.5
Pantex	1970	1177	1368	936	1212	133	249	124	186	1060	1398	24.3
	1971	418	491	344	402	376	448	289	302	633	704	13.6
	1972	981	989	580	562	448	582	349	410	929	972	13.2
Hays	1970[b]	1939	1790	1150	1061	—	—	—	—	1150	1061	11.6
Osage	1970	1387	1158	1211	1045	343	273	294	195	1505	1240	15.2
	1971	901	997	1022	965	121	238	93	138	1115	1101	15.2
	1972	842	876	658	763	540	301	355	195	1013	958	14.9
Pawnee	1970[b]	1892	1828	1631	1701	—	—	—	—	1631	1701	7.0
	1971	1872	1593	1014	996	500	414	340	322	1354	1318	8.0
	1972	1031	1403	803	1290	308	463	258	308	1061	1598	8.0
Bridger	1972[b]	1639	2120	1583	1895	—	—	—	—	1583	1895	2.9
Dickinson	1970[b]	2128	2670	1605	2492	—	—	—	—	1605	2492	4.1
Cottonwood	1970	1308	2087	1131	1896	328	360	258	317	1389	2213	8.0
	1971[b]	2120	2884	1498	2348	—	—	—	—	1498	2348	7.8
	1972	1872	2570	1633	2289	263	330	218	284	1851	2573	5.7
Matador	1968	2134	—	1832	—	142	—	111	—	1943	—	4.1
	1969	2104	—	1898	—	162	—	152	—	2050	—	2.5
	1970	2556	—	2220	—	159	—	145	—	2365	—	1.9

[a] Depths below soil surface to which the above values apply are as follows:
60 cm at Cottonwood, Dickinson, Pantex, Pawnee and Matador; 50 cm at Bridger and Osage; 30 cm at Jornada (because caliche occurred at this depth) and at San Joaquin (because bed rock frequently occurs at this depth); 15 cm at Hays (because bed rock occurred at this depth). The proportion of total roots that occur to these depths was estimated to average 76 % in the USA sites and 77 % at Matador.

[b] Crowns were not separated from other under-ground parts in these instances.

Crowns

The standing crop of crowns is of importance in grassland communities because they are primary storage organs for carbohydrate reserves. The dynamics of litter biomass depends on translocation of materials to and from shoots and roots. The general nature of the seasonal rhythm of crown biomass is apparently different from that of either shoots or roots, and is more erratic.

Time-weighted means of crown biomass in ungrazed locations ranged from about 100 to 350 g/m², while yearly peaks were from about 120 to 540 g/m² (Table 5.5). In each of the four USA sites for which crown data are available, the mean of all biomass values was in the 240–300 g/m² range, while at Matador (where the major grass species is rhizomatous) it was 136 g/m². Usually, crown biomass showed a declining trend during the early part of the growing season, coincident with initiation of shoot growth. Afterwards it accumulated to reach a peak after that of green-shoot biomass. In mixed prairie more than one peak often was observed, the first either preceding, or being coincident with, the peak in green-shoot biomass.

Except in the true prairie site, crown biomass tended to be greater in grazed locations than in ungrazed locations, the average increase being about 20 %. There was, however, considerable variability between sites, as well as between years in the same site (Table 5.5). This variability was also evident in the true prairie site, where decreases of 66 and 55 % were measured, respectively, in 1970 and 1972, but in the intervening year an increase of 46 % occurred in crown biomass as a result of grazing.

Roots

The data that have been collected in these studies suggest that the seasonal rhythm of root biomass is erratic, but it is not certain to what degree the data have been affected by sampling errors. Intraseasonal trends are not consistent with respect to timing of maximum and minimum root biomass; however, yearly means increased coincidentally with increased shoot biomass. Mean root biomass in ungrazed locations ranged from less than 200 g/m² in desert grassland to more than 1500 g/m² in most mixed prairie situations (Table 5.5). Values for short-grass prairie and true prairie were intermediate.

Root biomass tends to be concentrated near the soil surface and declines steadily with depth (Table 5.4). In sites in which sampling was undertaken to a depth of 60 cm or more, the mean percentage of roots to this depth that occurred in the upper 20 cm of the profile ranged between about 60 and 80 %. Vertical distribution of roots was estimated to greatest depth in the Canadian site, where it was revealed that 35 % of root biomass occurred in the uppermost 10 cm of soil, while only 9 % was below a depth of 90 cm (Table 5.4).

In the USA sites occasional sampling suggested that 24 % of root biomass occurred below the depth of sampling indicated in Table 5.5. Sims *et al.*, (1978) have shown that the distribution of root biomass according to depth can be described adequately by the equation of a form $y = aX^b$: where y is the root biomass in g/m^2 within a depth interval; X is the mean of the depth interval (in centimetres); a is the value of y when $a = 1$; and b is the description of the curve.

There was a tendency towards greater concentration of root biomass in the uppermost 20 cm of soil in grazed as compared to ungrazed conditions. This may be a response to release from grazing, since the grazed treatment was not subjected to herbivory by domesticated animals during the sampling season; although it may be associated with the lower supply of soil moisture that occurs under grazing by livestock (Lewis, 1971; Stoddart, Smith & Box, 1975).

Relative biomass under ground and above ground

Ratios of under-ground plant biomass to above-ground biomass reflect the proportion of photosynthates that is translocated under ground. Mean ratios of under-ground biomass to green-shoot biomass were found to range from 2 to 13. They were lowest in desert grassland, ranged from 3 to 6 in mixed prairie, were about 6 in mountain grassland, and reached 13 in one short-grass prairie site. The data collected in these studies suggest that temperature is an important influence in determining the relative plant biomass under ground and above ground, with low temperature tending to favour under-ground accumulations. Increases in root:shoot ratios also have been shown to occur due to increasing aridity (Struik & Bray, 1970). There is also a strong relationship to grazing, which caused a considerable increase in ratios of under-ground biomass to shoot biomass in most of the sites studied. This influence was more marked in cooler sites than in warmer ones.

Shoot production

As far as is known to the authors, prior to the studies made in IBP, estimates of shoot production in natural temperate grassland were limited to those made on the basis of peak standing crops of shoot biomass or a summation of two or more clippings per growing season from the same plots. The estimates of shoot production that are presented in Table 5.6 have been determined by the summation of peak green biomass of individual species or groups of species (Odum, 1960). This method has a decided advantage over single harvest at the time of maximum standing crop because, by the time this harvest is taken, significant amounts of plant material may have senesced, become detached, and disappeared from the canopy (Hadley & Kieckhefer,

Table 5.6. *Net yearly production of dry matter (g/m²) in USA and Canadian sites in natural temperate grassland, listed in approximate order of increasing net production and compared with moisture parameters for current growing season*

Site	Year	Growing season: Precipitation (cm)	Growing season: Evapotranspiration (cm)	Net production: Shoots Ungrazed	Shoots Grazed	Crowns Ungrazed	Crowns Grazed	Roots Ungrazed	Roots Grazed	Total under ground Ungrazed	Total under ground Grazed	Total Ungrazed	Total Grazed	Efficiency of energy capture (%)[a] Ungrazed	Efficiency of energy capture (%)[a] Grazed
ALE	1971	89	89	82	54	—	—	—	—	—	—	—	—	—	—
	1972	109	109	114	99	—	—	—	—	—	—	—	—	—	—
Jornada	1970	166	166	134	97	—	—	—	—	91	130	225	227	0.12	0.13
	1971	183	183	125	51	—	—	—	—	254	182	379	233	0.24	0.15
	1972	324	324	186	180	—	—	—	—	96	57	282	237	0.16	0.12
Matador	1968	182	—	110	—	—	—	—	—	591	—	701	—	0.72	—
	1969	239	—	134	—	—	—	—	—	405	—	539	—	0.53	—
	1970	255	—	197	—	—	—	—	—	669	—	866	—	0.96	—
	1971	192	—	179	—	—	—	—	—	—	—	—	—	—	—
	1972	145	—	95	—	—	—	—	—	—	—	—	—	—	—
Pawnee	1970	149	148	160	123	—	—	—	—	411	471	571	594	0.64	0.67
	1971	222	220	218	108	215	200	471	522	686	722	904	830	0.80	0.73
	1972	324	321	138	77	235	88	372	341	607	429	745	506	0.98	1.03
Cottonwood	1970	242	232	212	150	167	113	349	577	516	690	728	850	0.79	0.95
	1971	485	466	255	205	—	—	—	—	821	1019	1076	1224	1.15	1.32
	1972	354	340	279	177	115	49	148	296	263	345	542	522	0.60	0.60
Bridger	1970	182	96	168	145	—	—	—	—	—	—	—	—	—	—
	1972	529	280	330	344	—	—	—	—	471	573	801	917	1.02	1.18
Bison	1970	244	244	272	170	—	—	—	—	—	—	—	—	—	—
Osage	1970	435	335	331	434	380	166	222	336	602	502	933	936	0.77	0.67
	1971	782	602	416	523	70	140	361	281	431	421	847	944	0.70	0.72
	1972	805	620	290	370	407	390	185	593	592	983	882	1353	0.70	1.09
Pantex	1970	184	182	155	155	0	0	417	410	417	410	572	565	0.41	0.41
	1971	522	517	289	218	297	633	311	354	608	987	897	1205	0.64	0.87
	1972	359	355	327	302	295	327	581	641	876	968	1203	1270	0.98	1.03
San Joaquin	1973	611	611[b]	526	580	—	—	—	—	581	568	1107	1148	—	—
	1974	577	577[b]	459	375	—	—	—	—	451	353	910	728	—	—
	1975	480	480[b]	337	361	—	—	—	—	361	317	698	678	—	—
Dickinson	1970	360	346	351	302	—	—	—	—	932	958	1283	1260	1.43	1.41
Hays	1970	426	409	363	372	—	—	—	—	1062	855	1425	1227	1.20	1.03

[a] Calculated on the basis of photosynthetically active radiation (45 % of global) for the growing season.
[b] At San Joaquin an unestimated amount of water was lost by percolation.

1963; Kucera, Dahlman & Koelling, 1967). This method assumes that, at the time of peak community standing crop, all species in the community are not at peak current-season biomass. However, even the sum of the peaks of current-season biomass for the various species yields a conservative estimate of shoot production, because some material has senesced and detached before the peak of each individual species has been attained. However, in a recent study comparing 31 computational techniques for calculating net shoot production, Singh, Lauenroth & Steinhorst (1975) have indicated that the summation of peak green biomass of individual species is among the best methods to determine net shoot production.

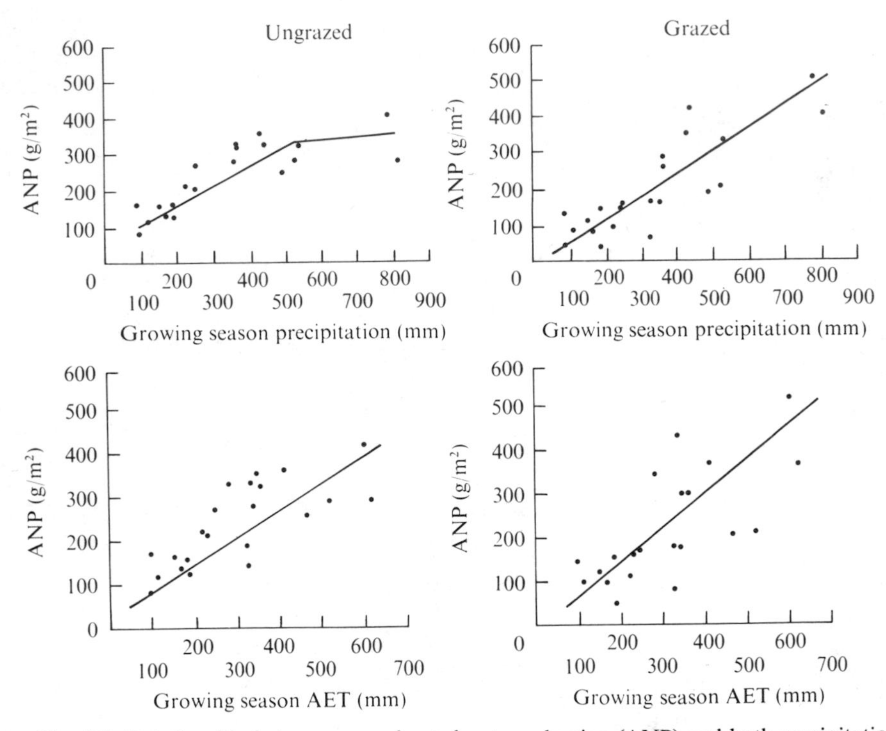

Fig. 5.3. Relationship between annual net shoot production (ANP) and both precipitation and actual evapotranspiration (AET) in ungrazed (left) and grazed (right) natural temperate grassland sites (after Sims & Singh, 1978*b*).

By this approach, estimates of annual shoot production in the absence of grazing ranged from 82 to 526 g/m² from site to site and year to year (Table 5.6). In the absence of grazing there is a linear increase in shoot production with increasing precipitation (Fig. 5.3) up to a level of about 500 mm of grazing season precipitation; it then levels off. The relationship to annual precipitation is similar. The annual grassland at San Joaquin did not fit this general trend. In a review of net above-ground productivity in temperate grassland communities around the world, Ricklefs (1973) concluded that shoot production increases rapidly as annual precipitation increases up to about 1200 mm and tends to level off. In the studies reported here, annual net shoot production also increased in a linear pattern with increased actual evapotranspiration (Fig. 5.3), not unexpectedly since actual evapotranspiration is an index to actual water use.

In the grazed condition, net shoot production was reduced in almost all sites (Table 5.6). A notable exception was the true prairie site (Osage) where net shoot production increased markedly as a result of grazing. The data suggest an increasing detrimental effect of grazing on shoot production with increasing aridity and an increase in shoot production resulting from grazing of the sites that have most favourable moisture conditions. Perhaps this is related to the release of nutrients that would otherwise accumulate in litter in these areas. Clipping results also have revealed increased shoot production in true prairie (Aldous, 1930; Biswell & Weaver, 1933) and in western Kansas (Cressler, 1942; Lacey, 1942; Tomanek, 1948).

When the data from grazed IBP sites are related to precipitation values increases in shoot production continue to a higher level of precipitation than in ungrazed sites (Fig. 5.3). Perhaps this is due to a lower biomass of dead shoots in grazed as compared to ungrazed situations in moist sites where dead canopy biomass may have a shading effect sufficient to limit shoot production.

Peak rates of increase in green biomass can be used as indices of the capacity of various grasslands to accumulate photosynthate under stressed and unstressed conditions. Such values are presented for the sites under discussion, in both grazed and ungrazed condition, in Table 5.7. The mean peak daily rate measured in ungrazed perennial grasslands was 3.8 g/m^2, compared to 3.2 g/m^2 in grazed situations. Peak values were higher on the average in annual grassland (7.8 g/m^2 daily) followed by mountain grassland, mixed prairie and true prairie, where they were usually in excess of 5 g/m^2 (except in the northern margin (Matador)). Values for short-grass prairie, desert grassland and shrub steppe ranged from an average of about 3.5 g/m^2 to near 1 g/m^2. The growing season for the annual grassland is from November to May and is very definitely delineated by precipitation and not temperature. In this regard it is similar to desert grassland.

In the Canadian study considerable effort was expended in accounting for the fate of green leaf biomass. This evolved into a study of rate of leaf development and deterioration of grass shoots and revealed that turnover of green-leaf biomass and dead-leaf biomass in the canopy was much more rapid than had been supposed (Fig. 5.2). This analysis (Coupland & Abouguendia, 1974) suggested that estimates of shoot production made by sum-

Table 5.7. *Peak mean daily accumulations in green-shoot biomass (in g/m²) in natural temperate grassland,*[a] *listed in approximate order of increasing rate*

Site	Grassland type	Year	Ungrazed		Grazed	
			Peak rate	Time interval	Peak rate	Time interval
ALE	Shrub-steppe	1971	1.05	8 Apr.–28 Apr.	0.40	8 Apr.–28 Apr.
		1972	1.21	28 Mar.–18 Apr.	0.79	28 Mar.–19 Apr.
Matador	Mixed prairie	1968	1.43	3 June–24 June		
		1969	2.11	5 May–21 May		
		1970	2.38	21 May–22 June		
		1971	2.98	15 June–5 July		
		1972	0.83	22 May–19 June		
Jornada	Desert grassland	1970	2.90	10 Aug.–20 Aug.	3.36	31 July–11 Aug.
		1971	1.50	13 July–4 Aug.	0.58	24 Aug.–17 Sept.
		1972	3.43	3 July–19 July	1.91	31 Aug.–23 Sept.
Pawnee	Short-grass prairie	1970	2.62	1 June–18 June	1.70	1 June–18 June
		1971	4.38	1 June–14 June	5.38	1 June–14 June
		1972	2.42	5 June–26 June	1.00	17 May–1 June
Pantex	Short-grass prairie	1970	1.53	27 July–10 Aug.	5.00	24 Aug.–5 Sept.
		1971	6.26	13 July–2 Aug.	1.42	13 July–2 Aug.
		1972	3.56	22 Apr.–22 May	2.70	20 June–14 July
Bison	Palouse prairie	1970	4.46	15 May–30 May	3.20	15 May–30 May
Cottonwood	Mixed prairie	1970	5.35	20 May–6 June	2.14	6 June–20 June
		1971	6.50	9 June–23 June	4.15	21 May–9 June
		1972	3.69	18 May–1 June	3.00	18 May–1 June
Bridger	Mountain grassland	1970	5.75	30 June–8 July	3.22	29 June–8 July
		1972	6.07	27 June–10 July	4.84	15 June–28 June
Osage	True prairie	1970	5.93	1 June–17 June	5.35	2 July–16 July
		1971	6.21	2 June–10 July	6.00	12 May–2 June
		1972	4.75	15 May–6 June	5.09	5 July–6 Aug.
Dickinson	Mixed prairie	1970	5.92	10 June–24 June	5.61	11 June–24 June
Hays	Mixed prairie	1970	6.46	1 June–16 June	4.18	15 May–1 June
San Joaquin	Mediterranean annual grassland	1973	5.16	13 Mar.–12 Apr.	7.58	13 Mar.–12 Apr.
		1974	10.23	3 Apr.–25 Apr.	7.09	3 Apr.–25 Apr.
		1975	8.00	22 Apr.–5 May	8.85	22 Apr.–5 May

[a] Based on the sum of increases for individual species or groups of species between the sampling dates indicated.

ming peak standing crops of individual species were too conservative to result in reasonable estimates. Consequently, an attempt was made to account for all losses between harvests. This resulted in an estimated mean annual shoot production of 495 g/m², more than three times that (143 g/m²) arrived at by the method that was applied uniformly throughout USA sites in Table 5.6.

Under-ground production

Annual crown production was estimated in the several USA sites by summing the statistically significant positive increases in crown biomass within a growing season. On the average, crown production was estimated to be 39 % of total under-ground production in ungrazed sites and 33 % in grazed sites (Table 5.6).

Several difficulties are involved in estimating root production. These not only relate to errors in sampling of root biomass, but also to problems in separation of living and dead roots and in the identification of roots in relation to species. Other factors include insufficient knowledge concerning the rates of decomposition of dead roots at different times of the year and amounts of losses from root biomass as a result of root secretions, sloughing of root hairs, root caps and corticle layers, and grazing by soil fauna. Studies by Samtevich (1965) have suggested that root secretions alone may lead to considerable underestimation of root production. He estimated that roots of wheat plants lose 350 g/m² of dry weight of mucilage to the soil during development, and he suggests that losses may be even greater in perennial plant communities. Therefore, estimates of root production should be viewed with the constraints imposed by the available procedure. The estimates of root production presented in Table 5.6 are based on a summation of the significant positive increases in root biomass throughout the year in each layer sampled. The decision to use this method was reached after testing four methods in the Osage site (Sims & Singh 1978*b*) and was also based on a process study in the annual grasslands (R. G. Woodmansee & D. A. Duncan, personal communication).

Total under-ground production (the sum of crown and root production) in the absence of grazing by livestock ranged from a mean annual value of about 150 g/m² in desert grassland to values of about 1000 g/m² in some mixed prairie sites (Table 5.6) in the absence of grazing by domesticated animals. The response to grazing was quite variable, but on the average there was 10 % greater under-ground production in grazed situations as compared to ungrazed ones.

Net primary production

Estimates of net primary production (the sum of shoot production and underground production) in the various ungrazed sites averaged in the 700 to 900 g/m² range except in desert grassland (295 g/m²) and at Dickinson (1283 g/m²) and Hays (1425 g/m²) in mixed prairie (Table 5.6). However, note should be taken that the latter two estimates were each based on studies only in one year in which moisture conditions were very favourable. No consistent effect of grazing on net primary production was observed. The greatest mean reduction as a result of grazing occurred in the desert grassland site and the greatest mean increase was in the true prairie site. However, in each of these sites there was no appreciable effect in one year of the three years studied.

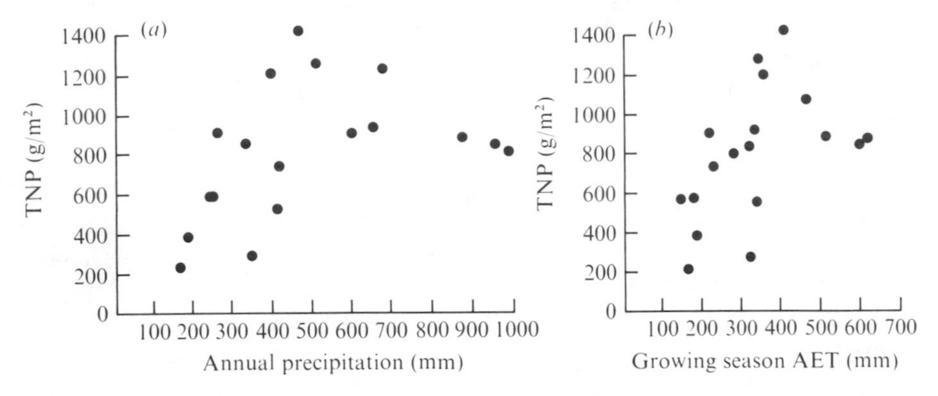

Fig. 5.4. Relationship between total net production (TNP) and current year annual precipitation (*a*) and growing season evapotranspiration (*b*) in ungrazed natural temperate grasslands.

The relationships between net primary production and two moisture parameters are shown in Figure 5.4. As for shoots, there is an apparent positive linear relationship between total net productivity and annual precipitation up to approximately 500 mm of precipitation, but no apparent relationship above this level. A similar trend of production occurs in relation to actual evapotranspiration during the growing season.

Energy flow

Efficiency of energy capture by primary producers, based on photosynthetically active radiation, in ungrazed situations ranged from an average of 0.17 % in desert grassland to values in the 0.85–1.4 % range in the USA mixed prairie and mountain sites. Short-grass prairie and true prairie were in the 0.65–0.8 % range. While the value for the Canadian mixed prairie site is low (mean of 0.74 %) on the basis of the methods that have been applied in estimating production in USA sites, when the local interpretation of production is taken into account, the efficiency level approximates 1 %. The low efficiency of energy capture in the southern sites (such as Osage), even under conditions of favourable moisture supply, seems to conflict with the finding of Black (1971) that warm-season species are generally more efficient in net production than cool-season species. These calculations of efficiency of net production must be evaluated in relation to whether or not the methods for estimating net productivity were equally applicable in all sites.

These levels of efficiency of production appear to be about one-third of the potential that could be attained under optimum management. Cooper (1970) found that canopies of temperate grasses (some cultivated) had efficiencies of energy conversion of 2.2–6.4 % during maximum growth periods and 1.3–3.0 % for the entire growing season. This suggests considerable potential for increase in production by the addition of nutrients and water. Lauenroth & Sims (1973) have observed that in short-grass prairie in Colorado production can be tripled with addition of nitrogen and the maintenance of an adequate water supply. Naturally, the practicability of application of these amendments must be considered in relation to their energy and economic costs.

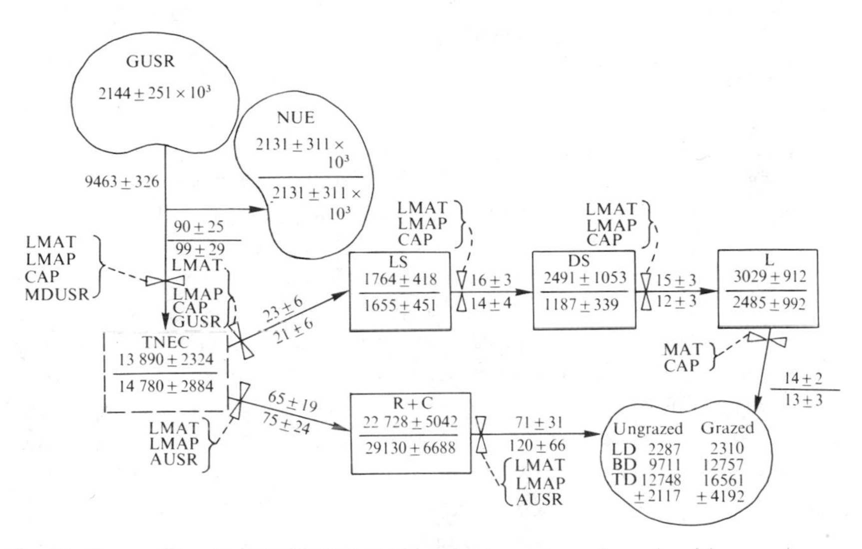

Fig. 5.5. Energy flow (in kJ/m^2) through the producer system of a natural temperate grassland in the USA. The numbers in the boxes represent the mean standing crop of the primary producer compartments; the mean standing crop of energy in live shoots (LS), standing dead (DS), litter (L), and under-ground parts including roots (R) and crowns (C). Numbers on the arrows are daily flow rates. Numbers in the GUSR block are the usable solar radiation in the growing season (GUSR) and those in the NUE block are the unused portion (NUE). Numbers in the ungrazed/grazed block are the amounts of energy dissipated from the producer system in the form of litter disappearance (LD), under-ground disappearance (BD), and total disappearance (TD). All numbers above the lines pertain to ungrazed grasslands and those below the lines to grazed grasslands. Numbers in blocks with broken lines are the total net energy captured (TNEC) by plants. Bow ties on the arrows represent the rate-controlling abiotic variables as determined by regression analysis. LMAT, long-term mean annual temperature; LMAP, long-term mean annual precipitation; CAP, current annual precipitation; GUSR, growing season usable incident solar radiation; MDUSR, mean daily usable incident solar radiation during the current growing season; AUSR, annual usable incident solar radiation; MAT, mean annual temperature (after Sims & Singh, 1976*d*).

Table 5.8. *A comparison between various energy-flow values (in kJ/m²), obtained as output from the model in Fig. 5.5, with those estimated in the Canadian site*[a]

	Average USA grassland		Matador grassland (ungrazed)
	Ungrazed	Grazed	
Standing crop			
Green shoots	1764	1655	1323
Under-ground plant parts	22728	29130	42667
Photosynthetically-active solar radiation during growing season			
Total	2144000	2144000	1796000
Daily mean	9463	9463	7357
Net energy capture during growing season			
Total	13890	14780	20556
Mean daily rate			
Above ground	23	21	36
Below ground	65	75	48
Efficiency of energy capture (seasonal mean, %)	0.65	0.69	1.14

[a] Matador values take into account losses from the canopy during growth, as well as root values to a depth of 150 cm.

A model depicting mean standing crop of energy in the various producer compartments and rates of transfer between compartments in an average USA natural temperate grassland is illustrated in Figure 5.5. In Table 5.8 energy values for ungrazed and grazed situations in this 'average' USA grassland are compared with those from the Canadian study (Matador) when all accountable losses have been estimated.

Turnover rate

The method of Dahlman & Kucera (1965) was applied in calculating turnover rates of crowns and roots separately. This consists of dividing annual increment by peak biomass value. The resulting estimates are somewhat subjective, for the same reasons as are those of production. This method, however, seems to produce turnover values that are comparable to those obtained by several other methods, including the 'litter-bag' and 'paired-plots' techniques employed during the 1971 and 1972 seasons in the Dickinson site (Abouguendia, 1973). Values calculated by the Dahlman & Kucera method suggest that crowns turn over more rapidly in warmer than in cooler areas. Turnover rate was most rapid (82 % of the standing crop each year) in true prairie, intermediate (53 %) in short-grass prairie and slowest in mixed prairie (35 %). Thus, the highest rate of turnover occurs in true prairie, where mean standing crop of crowns is least, but production of crown biomass is greatest. The rate of crown turnover increased as usable incident solar radiation and annual actual evapotranspiration increased. Although the data indicate an erratic response to grazing in the rate of turnover of crown biomass, turnover appeared to be more rapid in grazed areas. The rate of turnover of under-ground biomass seems to vary with depth. For example, at Dickinson (Abouguendia, 1973) the lower layers of under-ground biomass were found to be, generally, more dynamic than the upper layers with annual turnover rates ranging from 33 % for the top 20 cm to 62 % for the 80–100 cm layer.

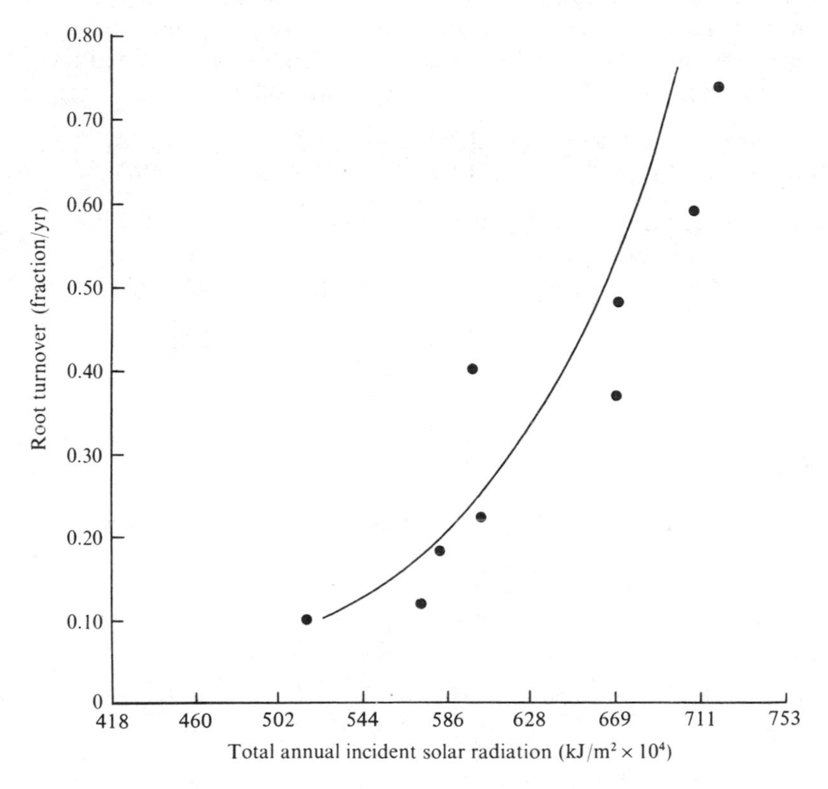

Fig. 5.6. Relationship of rate of turnover of roots to total annual solar radiation in ungrazed natural temperate grassland (after Sims & Singh, 1978*b*).

Rate of turnover of roots was slowest in mixed prairie (18 % each year), as was the case with crowns. However, the relative rates were reversed in true prairie and short-grass prairie, with faster turnover occurring in short-grass prairie (49 %), while true prairie was intermediate (30 %). Consequently, the root compartment of mixed prairie appeared to be the least dynamic and also exhibited a maximum mean standing crop of root material. In contrast, the root biomass of short-grass prairie appeared to be highly dynamic, with almost 50 % of the roots being replaced each year. The rate of turnover found in the true prairie Osage site is similar to that reported by Dahlman & Kucera (1965, 1969) for true prairie in Missouri. The latter workers found no significant differences in rate of turnover of root biomass at different depths. Data from these studies indicate no significant effects of grazing on rate of turnover of roots.

Annual usable incident solar radiation showed a positive curvi-linear relationship to rate of turnover of root biomass (Fig. 5.6). Correlation and regression analysis indicated that total annual usable incident solar radiation (X_1) and growing-season actual evapotranspiration (X_2) explained 73 % of the variability in root turnover for ungrazed grasslands ($P < 0.05$) and 60 % in the grazed grasslands ($P = 0.06$) according to the following equations:

$$\text{Root turnover (ungrazed)} = -1.36812+0.000002X_1+0.00057X_2$$
$$\text{Root turnover (grazed)} = -1.10726+0.000002X_1+0.00104X_2$$

Although root turnover is a complex phenomenon, reflecting both the processes of growth and decomposition, it has been shown that net production is usually determined by the amount of water used (Sims & Singh, 1978*b*).

6. Consumers

M. I. DYER

A great deal is known about certain consumers and their functions in native grasslands, mostly because they are so apparent owing to their prominence in an ecosystem that has such a low plant structure. However, there are some surprises to be considered that result from examining consumers in a total systems context. Through IBP we have discovered that some of those consumers that are so prominent visually do not have roles commensurate with their high visibility. On the other hand there are many consumers, for instance nematodes, that are seldom dealt with but which are extremely important in grassland ecosystems. Because many publications available to the serious reader deal with single species or isolated ecological associations, it follows that intensive examination of those inherent associations tends to over-emphasise certain species or community components in the context of the entire ecosystem. For instance, there has been a great amount of work in recent years examining predator–prey relationships of a one predator–one prey nature in parts of the grasslands. Such studies, while of interest to certain disciplines tend to over-emphasise the role of predators as, for example, when we consider the whole system. Also, such work is done with the often misleading notion that since these organisms exist on the end of the trophic food organisations of a particular system, their good health and maintenance reflect satisfactory conditions throughout the remainder of the ecosystem. Because of lag effects such emphasis cannot always be supported.

Therefore, it is important and of interest to examine simultaneously all consumer types in an ecosystem in order to obtain an integrated picture of the system associations, often different from that obtained in a series of individual, non-integrated studies. It is toward this end, then, that IBP lends additional information to the science of consumer ecology in natural temperate grasslands. Since most of the material reported here comes from the US/IBP and the Canadian IBP efforts, the details developed in the remainder of this chapter will be directed toward results of these two studies.

During the data-collection years of the IBP programmes in North America, the so-called 'state variables', i.e. those system components comprised of pertinent species or biotic groups that are measured in terms of weight per unit area, were measured in standardised ways. At the same time that information on herbage components and weather data were being obtained, various kinds and quanta of consumers were also being sampled and measured. For certain above-ground consumers, destructive sampling techniques, especially for arthropods, were devised; other consumers were censused at various intervals to obtain estimates of their total abundance in each of the various grassland sites. For under-ground consumers destructive sampling techniques were necessary. For these purposes, coring tools were used to obtain samples of soil at various depths to permit extraction of annelids, arthropods and nematodes. Upon processing and storage of these state-variable data in central computer banks, the data were scrutinized and ultimately aggregated into various 'functional groups'. Because of the enormous task in considering species-by-species description of consumers in ecosystems, the species no longer becomes the basic unit for dealing with system function (at least at this stage of development in ecosystem analysis). Instead, it is more useful to examine the larger functional aggregations, such as trophic position, food requirements (herbivorous, carnivorous, omnivorous or detritivorous) and physiological state (poikilothermic or homeothermic).

This is the reason that species lists or community compositions are not emphasised in this chapter, except where there is a noteworthy major function that is unique or where a species provides the epitome of a particular system function.

Most of the animals in the natural temperate grassland ecosystem are indigenous to that system and have co-evolved with the micro- and macroflora (Gilbert & Raven, 1975). This is not universally true, however, and the notable examples, of course, are domesticated animals, such as cattle and sheep, introduced by man. Also, the present grassland system reflects major changes in native consumer distributions, such as the elimination of past components some of which are now extinct genotypes (various canid populations), and others have been forced into other habitats more remote to man. Notable examples are some of the ungulate populations, such as the North American wapiti (*Cervus canadensis* Erxleben) and the North American bison (*Bison bison* L.) in the herbivore components, and the near total elimination of the major predator, the grizzly bear (*Ursus horribilis* Ord.). Because there are no detailed records (only historical descriptions exist), it is not possible to elaborate the meaning of such changes in time. However, notwithstanding these major changes, the grasslands still seem to constitute a stable biome and it is this present-day condition that concerns us here.

When one examines the flux of energy through these natural grasslands, it is often surprising to scientists and lay people alike to realize how little actually passes through the whole set of consumers. Much of the energy fixed by the primary producers falls as litter or is translocated into the underground components where it is processed by biophages and saprophages. In a grazed short-grass prairie at Pawnee, an estimated 64 % of the net fixed photosynthate moves along these pathways and saprophagic micro-organisms release about 90 % of the energy contained (see Table 8.4 and Andrews *et al.*, 1974). In the absence of grazing by livestock at Matador, the proportion of the energy fixed by primary producers that goes through the above-ground consumer chain approximates to only 1.5 % (Chapter 8, this volume), while the relative activities of microbial decomposers and invertebrates under ground can be judged by the estimate that the biomass of the former (to a depth of 30 cm) totals 29 times the latter (Table 8.1).

Because it is difficult, if not impossible, to pick out small groups of principal consumer components with which we can characterize grassland functions, I have concentrated on quite broad functional groupings of above- and under-ground fauna for discussions in this chapter. These broad groupings, then, deal with community characteristics of consumers as energy processors and, as such, indicate certain community strategies of arriving at the present mix of consumers rather than species or other purely taxonomic level strategies. Thus, the descriptions of consumers functioning in grasslands become abstractions which deal with how the various animal species have co-evolved with other animals, the primary food base (centred in the plant community), and perhaps the saprophagic community, although we have little information to deal with concerning the latter.

Consumer biomass distribution

Perhaps the most basic and well-known concept of community organisation is that of trophic order (Lindeman, 1942; Whittaker, 1970). For the grasslands of the USA, N. R. French & R. K. Steinhorst (personal communication) have assembled biomass data obtained for the US/IBP Grassland Biome study. Data have been extracted from these compilations in order to examine consumers in greater detail than dealt with in that effort. Also, information from the Canadian IBP Matador Project (Sheppard, 1973; Riegert, Varley & Willard, 1974; Sadler & Maher, 1974, Willard, 1974; Coupland *et al.*, 1975)

Table 6.1. *Biomass of consumers in North American temperate grasslands. Above-ground (AG) and under-ground (UG) values are given for eight sites in six types of grasslands*

Site	Grassland type	Biomass (mg/m²)		Ratio	Total biomass mg/(m²)	%	
		AG	UG	AG:UG		AG	UG
ALE	Northwest shrub-steppe	79	1516	1:19.1	1596	5.0	95.0
Bridger	Mountain grassland	527	1629	1: 3.1	2156	24.4	75.6
Cottonwood	Central mixed prairie	262	2567	1: 9.8	2830	9.3	90.7
Jornada	Desert grassland	96	1158	1:12.1	1254	7.6	92.4
Matador	Northern mixed prairie	267	4365	1:16.4	4632	5.8	94.2
Osage	True prairie	228	1863	1: 8.2	2091	10.9	89.1
Pantex	Southern short-grass prairie	458	823	1: 1.8	1281	35.7	64.3
Pawnee	Northern short-grass prairie	106	852	1: 8.1	958	11.0	89.0
Mean of six types		246	1745	1: 7.1	1991	12.4	87.6
Mean of eight sites		253	1847	1: 7.3	2099	12.0	88.0

is added so that a broader aspect of North American grasslands can be achieved. These data are presented in Table 6.1.

In line with the stated goals of examining community relationships, I have broken down the total consumer biomass into above- and under-ground quantities (Table 6.1) for six grassland types that are represented by eight study areas. The average biomass is 2.0 g/m^2 for these six types; the range is from 0.9 g/m^2 in the northern short-grass prairie to 4.6 g/m^2 in the northern mixed prairie, a spread of -38% to $+130\%$ of the mean value for the grasslands under consideration.

These standing-crop values are remarkably low in contrast to the biomass in the producer component. From the same sites an average value of 2033 g/m^2 has been reported (see Chapter 5, this volume) for plant biomass, both above ground and under ground. Thus, it easily can be seen that with this producer: consumer ratio (about 1000:1) the grasslands have a low, flat trophic pyramid, a fact noted by N. R. French & R. K. Steinhorst (personal communication), who also have noted that the base for the pyramid is broader in the southern grasslands than in the northern ones, i.e. the ratio of the consumers to producers becomes larger as one progresses northward.

One of the conventional views in ecology is that the plant community is instrumental in setting the structure for consumers (i.e. plant species composition, structural type and height, and available energy) which determines, for the most part, what types of consumers will exist (Whittaker, 1970). Certainly this is true for the grasslands, for one can easily determine that animals adapted for grazing functions (more specifically grazing on grass and low forbs and half-shrubs) best fit this vegetation type. Such an observation at the outset seems trivial, but it really is not, especially when one begins to examine in detail how consumers must function to exist in grasslands. Cursory and initial examination indicates somewhat low standing crops and ostensibly low production values; however, this observation is restricted to the above-ground portion of the ecosystem and is definitely misleading. As noted in the previous chapter, most of the producer biomass is concentrated under ground and the amount of fixed energy is really relatively large. This condition is also reflected by the consumers. The ratio of the above-ground to under-ground consumer biomass is shown in Table 6.1 for the various grassland types and sites. Clearly, the largest fraction of consumers in this biome exists under ground. Ratios range from 1:19 to 1:3. There is considerable variability expressed in the data and it is difficult to define a pattern. But, it is possible that the consumer community is responding to dry and hot conditions much as the plant community by showing a concentration of underground organisms in relation to the total biomass present in the community (see Wallwork, 1976, page 312). Certainly the arid grasslands do not support above-ground grazers as readily as grasslands in more ameliorated areas, except, of course, for the growing season.

It is instructive to look further at how the consumer community is distributed according to primary and secondary roles in each of the above-ground and under-ground strata. Unfortunately for this analysis, there are insufficient data available to complete the record for each grassland type, but an initial view of what may be occurring is suggested for several sites. Generally, the record is best for the primary consumers; the main thing that is lacking, in most instances, is a record of under-ground secondary consumer biomass, or at least an accurate allocation of functional role of the various individuals from samples where the under-ground portion is available. The data available are shown in Table 6.2.

Table 6.2. *Comparisons of above-ground (AG) to under-ground (UG) consumer ratios for primary and secondary trophic levels in natural temperate grassland*

Site	Grassland type	Primary consumer AG:UG ratio	Secondary consumer AG:UG ratio	Total
ALE	Northwest shrub-steppe	1:40.0	–	–
Bridger	Mountain grassland	1:3.8	1:26.6	1:4.5
Cottonwood	Central mixed prairie	1:8.6	1:11.4	1:7.1
Jornada	Desert grassland	1:58.8	–	–
Osage	True prairie	1:9.1	–	–
Pantex	Southern short-grass prairie	1:1.1	–	–
Pawnee	Northern short-grass prairie	1:7.3	1:12.2	1:8.1
Averages[a]		1:4.6	1:7.2	–

[a] Average values are results of weighted observations from original data.

Above-ground to under-ground ratios for primary consumers are available from six grassland types and seven sites; the North American average is 1:4.6. Again, this follows plant patterns in that most of the primary consumers live under ground and are ostensibly concentrated in the root system where they graze on live roots or detritus. The ratios change for the secondary consumers; there is even a larger difference between the above-ground and under-ground fractions, the mean ratio being 1:7.2 (Table 6.2). This suggests a difference between energy-partitioning strategies for these trophic levels in different habitats. Obviously, in the under-ground system there is a higher concentration of prey items per unit area of soil surface and predators have evolved to take advantage of this situation by showing greater biomass per unit area in contrast to the above-ground predators. It would seem that there are vastly different spatial and temporal problems to solve for above-ground predators than for their under-ground counterparts, thus the difference in the ratios. Predators above ground must deal with larger prey items that are less numerous and thus scattered over wider areas. Indeed the distribution of the prey items probably becomes more patchy, thus making the problem of locating the prey more difficult. Selective advantages, which depart from stochastic or probabilistic dictates, might well be expected to govern survival abilities of under-ground predators. However, it is interesting to speculate on the functions of these two phenomena; in the under-ground subsystem there is seemingly a greater density of predators in association with grazers. This association apparently combines to provide for a more rapid turnover of consumer biomass. This rapid turnover, while perhaps being a function of size to begin with, very likely eventually affects the producer community more directly and with a shorter time lag in any response to biotic or abiotic events than would be observed for the above-ground system. It is apparent that differing grasslands affect the ways in which these two trophic systems operate, soil and abiotic conditions being the best candidates for ultimate controllers on the type of association that develops. The mountain grassland shows a much higher ratio for the secondary consumers than noted for the mixed prairie or the short-grass prairie. However, even with the massive effort that has gone into providing these data in the IBP networks, there still is too little information to permit full evaluation of how these communities and ecosystems develop and maintain such trophic partitioning of the available energy.

Yet another way to view these biomass and trophic-oriented perspectives is by partitioning all the consumers into above-ground and under-ground categories, and then looking at the secondary to primary consumer ratios. Again, this gives a measure of the trophic relationships, but treats the two subsystems, above ground and under ground, separately. This may be a more realistic way to partition these consumer functions, since there is little direct competition for resources between the above- and under-ground components. The main interactions will come through the primary producer section, a process, however, which produces a major lag effect and is difficult to observe and measure.

The main supposition contained within this analysis is that the fauna have been faithfully sampled, identified and allocated to a trophic role. In some instances this is easily done (for the larger consumers). However, for many of the micro-arthropods and nematodes there can be some doubt about the ability to make such decisions, especially in view of the fact that the food habits of many species in the grasslands are not yet identified; nor are all the species yet described in some cases. Nonetheless, attempts have been made to segregate function and conclusions must be made on them until supplanted in the future. Above-ground allocations have been made for most natural temperate grassland types. A reasonable secondary to primary consumer mean ratio is 1:7.8, the range of values of from 1:1.6 for the desert grasslands to 1:3.5 for the mountain grasslands when large herbivores are considered

Table 6.3. *Comparisons of secondary to primary consumer ratios (S:P) in above-ground and under-ground components in natural temperate grassland*

Site	Grassland type	Ratio above ground S:P	Ratio below ground S:P	Total S:P
ALE	Northwest shrub-steppe	1:2.95	–	–
Bridger	Mountain grassland	1:8.61[b]	–	1:3.63[b]
		1:34.69[c]	1:3.44	1:4.57[c]
Cottonwood	Central mixed prairie	1:7.1	1:1.33	1:1.36
Jornada	Desert grassland	1:1.61[b]	–	–
Osage	True prairie	1:2.17	–	–
Pantex	Southern short-grass prairie	1:1.97	–	–
Pawnee	Northern short-grass prairie	1:6.72[b]	–	1:3.70[b]
		1:6.97[c]	1:3.49	1:3.72[c]
Averages[a]		1:7.8	–	–

[a] Average values are results of weighted observations from original data.
[b] Large herbivores are ignored.
[c] Large herbivores are included.

(Table 6.3). But, if the large herbivores are ignored, the ratio drops to 1:8.6, a value which is still the largest for the grasslands considered here. Under-ground secondary to primary consumer ratios are somewhat lower, although there are not many data available to make a firm conclusion. Here again, it seems that the under-ground system may be more efficient in maintaining these trophic ratios, a matter most likely made possible by the unique energy compartmentalisation and spatial conditions within the soil community. Such a presentation tells nothing about how these associations developed; rather, it simply provides information about how the association might continue to maintain itself. These closely associated biota, with a relatively large and readily available energy source and pool of nutrients, may be able to set up and hold together their trophic proportions more efficiently because there is in essence a favourable microcosm formed under ground.

Relationships among North American natural grasslands

The grasslands studied in the USA and Canadian IBP ventures (Chapter 4, this volume) are spatially quite separated and are subject to unique differences from the standpoint of macro-climate, but still contain the common thread of being grasslands *per se*. However, from the standpoint of consumer evolution, notably the large ungulates, there are major differences. The grasslands east of the Rocky Mountain chain shared quite common evolutionary development in association with large herds of ungulate herbivores, notably bison and pronghorn antelope (*Antilocapra americana* Ord) and, earlier in history, herds of wapiti. However, it is apparent that certain grasslands that have been considered, such as the northwest shrub-steppe and perhaps the annual grasslands of California, had no such parallel development. Implicitly, this means that certain grassland processes, which may have developed owing to certain plant–animal interactions (Freeland & Janzen, 1974) or to competition in a broad sense between the large and small herbivores, might not be present in the grassland without the historical ungulate association. Without such grazing processes, one might expect to find different community forms or differences in the way energy is partitioned in the trophic pyramid describing that particular community. Hence, in order to define the effects of the large ungulates, the US/IBP, in its early stages, set out to examine the perturbing effects of large-herbivore grazing as practised at present through use of cattle and sheep in the various grasslands. The results to date are somewhat inconclusive. The problem that exists at this time, since the grasslands are no longer subject to the same grazing intensity and patterns that occurred with native herds (Peden *et al.*, 1974), is what constitutes 'normal' grazing. The main problem encountered with these perturbation experiments arose when attempting to evaluate the grazing effect. Some experiments were conducted, and some state-variable data were collected, in areas being grazed in a current-year condition, whereas other information was taken from treatments where grazing had occurred until the previous year. This latter situation then constituted a 'recovery' from grazing, and the data so collected probably were not consistent with those obtained from concurrent grazing.

However, notwithstanding these differences, grasslands subjected to various intensities of grazing responded by increasing the amount of photosynthate translocated under ground, and ultimately the ratio of under-ground to above-ground plant biomass. Root and crown biomass increases with grazing up to a point (Sims & Singh, 1971). This is especially true for northern grasslands (Lewis, 1971). Thus, under such conditions, one would expect that under-ground consumers, the invertebrate groups discussed previously, would soon respond by building up their numbers and biomass to take advantage of this increased level of available energy. From process information we have available to us, it is possible to construct simulation models (Cole, 1976) which give us this result (an increase in under-ground consumer biomass), but it is exceedingly difficult to obtain satisfactory field data from which we can validate this hypothesis. Thus, this extremely interesting and important subject must go unanswered for the time being, at least until a research effort can direct major attention to the problem and collect the information needed to test this idea.

It is of interest to detect commonalities among grasslands, from a consumer standpoint. It is obvious that most grasslands share more or less the same fauna (as discussed previously), but that mixes might be different according to other driving functions in the system. In such an effort to look at commonality or difference, I have taken the biomass data stemming from the USA and Canadian programmes and have grouped them according to the ways above- and under-ground consumers are allocated in that particular system as indicated by a specific site. The method for grouping uses Horn's R_o similarity index (1966) for the computation method, a method used originally for diet overlap among similar avian species, which has been used successfully for examining broader community attributes by L. D. Harris (1971) and Wiens (1973). The materials for the six North American grassland types (eight sites are used for the analysis) are shown in Fig. 6.1.

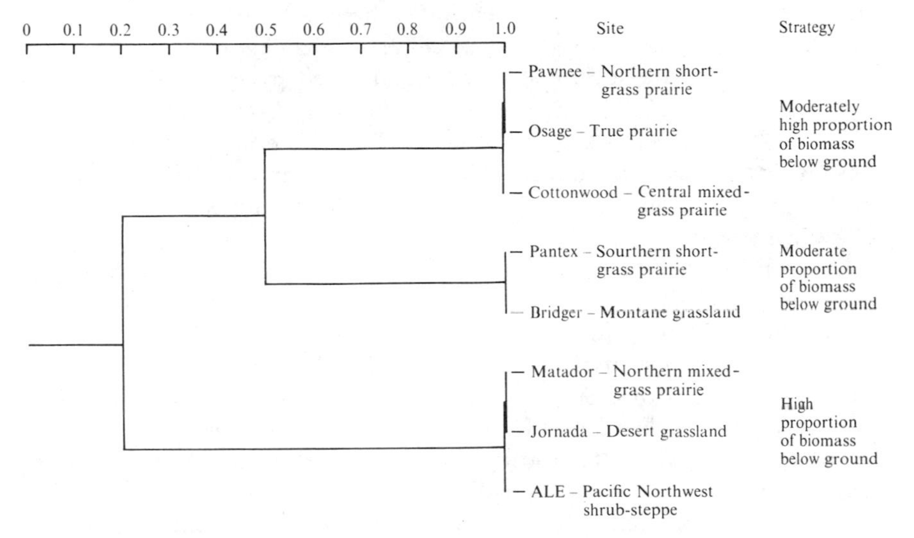

Fig. 6.1. Dendrogram of relationships among natural temperate grasslands in North America. This cluster shows relative position of the grassland types based on ratios of standing crop biomass of above-ground to under-ground consumers.

The first thing that is apparent is that several initial clusterings of estimated similar grassland types occur at very high levels of similarity (> 99 %). The northern short-grass prairie, the true prairie and the central mixed prairie are the first to combine. These are followed by the southern short-grass prairie and the mountain grassland. The third group is the northern mixed prairie and the arid grassland areas, the desert grasslands and the northwest shrub-steppe. The first two groups show a similarity at about the 50 % level and these two combine with the third association with a similarity of about 20 % (Fig. 6.1). There seem to be three distinct strategies based on this grouping of grasslands: (1) grasslands within the first association have a moderately high proportion of biomass under ground; (2) the second group of two assembles a moderate amount of biomass under ground; and (3) the last association of three have a high proportion of biomass under ground.

There seems to be no way that a simple interpretation can be given to this analysis; at least no single clear-cut parameter appears to be functioning throughout the entire grassland complex that dictates how consumer communities are sorted out in the various grasslands. For instance, it is not understood why the short-grass and mixed prairie pairs do not cluster with each other. Without substantive information, it may well be that there are, in general, important processes which create selective pressures favouring development of above-ground biota in some areas and under-ground biota in other areas. For instance, the arid sites and the northern mixed prairie site all show a high proportion of consumer biomass under ground, suggesting that harsh environments may select for this association. The Pawnee–Osage–Cottonwood association might be explained by spatial phenomena; certainly even though the sites are distant, they have been connected in time as parts of a continuous area of grassland. The clustering that makes most sense is that of the two very dry areas, ALE and Jornada. The subsequent clustering levels are even more academic and are presented mainly for heuristic interest.

Thus, an attempt to put community associations together in terms of consumer strategies of processing energy to maintain certain standing-crop relationships is disappointing at the outset. There may still be explainable relationships implicit in this presentation, but for the moment they are certainly obscure. However, even so, it is of interest to compare this picture with one obtained when a detailed examination is given to well-known fauna for which there are better-known cause-and-effect relationships for their distributions.

Wiens (1973) conducted a similar study of avifaunal conditions throughout the grasslands and considered whether bird density and biomass tended to follow the character of the plant community. He was able to characterise four plant community clusters from the avifaunal affinities: 'a set of tallgrass, mixed grass, and palouse plots...; a set of shortgrass plots, including the montane site...; the shortgrass plots at Pawnee; and the desert plots at Jornada'. He found little or no direct relationship to a similar clustering analysis conducted on floristic elements alone; thus, he concluded that for the most part floristic relationships alone are not very useful in predicting avian community patterns of distribution and density. The reason is that birds respond as much to habitat structural conditions as any feature yet found (Hilden, 1965; Wiens, 1969; Cody, 1974).

While not strictly comparable, the cluster analysis attempting to identify relationships among *all* consumer biota on the basis of above- and underground standing-crop ratios does show some degree of similarity to that obtained for a specific consumer group, viz. birds. Arid areas are identifiable, as are the Pawnee–Osage–Cottonwood elements, even though apparent artifacts emerge (as noted previously). A feature that is perhaps even more important comes to light. Wiens (1973) noted that for the avifaunal elements, the clustering levels were substantially higher than those reported for vegetational groups or small mammals studied in these grasslands. The levels for all consumer components reported here are even higher than reported by Wiens (1973); thus, it is possible that variability in community function is less for some components than it is for others and, because of this, inherently gives tighter associations. Therefore, continued detailed examination of how consumers are partitioned in the various grassland communities continues to be a worthwhile endeavour to test hypotheses of consumer function in these ecosystems.

For rodent fauna, three distinct sets of relationships among various grassland sites have been described (Harris, 1971). Principally they are a short-grass–mixed prairie–true prairie complex, desert grassland, and mountain grassland fauna. There are significant differences emerging from this representation in contrast to the one presented here in respect to the affiliations of the fauna. The principal difference is the grouping of the mountain system quite independently of the others. One can see the differences defined in the Pawnee–Osage–Cottonwood systems in contrast to the arid grassland system, but here the mountain grassland clustered with the mixed prairie and eventually the short-grass and true prairie complex. Again, the point is made that clustering for similarity using small specific faunal groups may, and does indeed in this case, give major differences in contrast to treating the entire faunal community. Thus, it is apparent that, depending upon one's viewpoint, completely different descriptive impressions can be obtained for a given biotic community by using differing sets of consumer groups.

Matrix of consumer trophic web

While the grasslands in general are relatively simple ecosystems in contrast to many other systems with greater numbers of species and more diverse structures and forms, there still are reasonably large numbers of consumer species. For instance, R. Andrews (personal communication) estimates that there are 1200 insect species for which there are biomass distribution data available from the USA grasslands and Kumar *et al.* (1967) identified approximately 1600 species of invertebrates, mainly arthropods, from the Pawnee site. Riegert & Varley (1973*a*) found 20 species of grasshoppers at Matador, even though they represented only 18 % of the above ground invertebrate biomass (Riegert, Varley & Willard, 1974). Under ground, micro-arthropods and nematodes are most abundant. There is no good assessment of the numbers of species of these groups in grasslands; many micro-arthropods apparently have not been described according to species yet, and indeed, it is uncertain whether genera and families of these organisms are well organised in the literature.

Obviously, in contrast to the micro-fauna, fewer species of larger organisms, such as mammals, birds and reptiles, inhabit these grasslands. Maher (1974*a*) reported 198 species of birds in the northern mixed prairie (Matador), of which 102 species are summer residents and 10 remain in winter; an additional 16 species are summer residents or visitants that do not breed in the grassland (as far as is known), and seven are winter residents; 73 species are spring and fall transients. Presently some 40 species of birds are regarded to be inhabitants of USA grasslands, although many more species pass through on the way to other breeding and wintering grounds (Wiens & Dyer, 1975). Thirty small mammal species have been shown to be important in grasslands (Harris, 1971), about a half-dozen snakes and lizards constitute important reptilian fauna, and four species of large ungulate herbivores including domesticated livestock, can be included in present grassland fauna. Thus, on the whole, the numbers of species of vertebrates are not large in contrast to the invertebrates.

The task of assigning functional roles in the consumer community becomes formidable. Harris & Paur (1972) selected 36 consumers from the short-grass prairie (Pawnee site) and obtained data from 112 dietary items from which to build a consumer diet matrix. Herbivore–carnivore relationships showed a clear dichotomy, but for each group a considerable degree of overlap occurred, sometimes as high as 100 %. Since consumers in the grassland system in their various trophic compartments are ultimately energy limited (Hairston, Smith & Slobodkin, 1960; Slobodkin, Smith & Hairston, 1967), a finite amount of material is available to the various consumer groups. Ostensibly each has worked out strategies for obtaining food to maintain a specific population structure. But during the process of developing a life style that will, over the long run, make these food types available, it is evident that there is a great deal of overlap developing among the species potentially competing for energy and nutrient-rich resources. Such overlap may be misleading, however. Since much of the information dealt with by Harris & Paur (1972) and N. R. French (personal communication) covers a growing season (rarely are there data for *all* consumer components within a year or a growing season), the conclusions may be reached that many species compete directly for food. However, as Pulliam & Enders (1971) noted, there may be important differences in the timing of resource utilisation or coupling between consumers and their foods. If so, we do not generally have sufficient information about competition values between various consumers to predict population fluctuations for a large number of key species.

The functional role of consumers

During IBP studies, the role of consumers in various ecosystems was examined intensively. While many of these studies emphasized utility, namely harvest potential, other questions were asked about the interactions between various consumer species and between producers and consumers. In grasslands, because of the nature of the trophic biomass pyramid, i.e. the steep reduction in biomass of the various levels of consumers in contrast to the producers (N. R. French & R. K. Steinhorst, personal communication), it has been difficult to attach a major role to consumers. Strictly speaking, biomass levels of consumers are so much lower than producers or decomposers that it is difficult to fit an important position for consumers. However, several works have pointed out the role of consumers as regulators in other ecosystems (Chew, 1974; Mattson & Addy, 1975) and the grassland consumers have similar functions. For instance, large-herbivore grazing pressure results in a change in plant root-to-shoot ratios; the amount of root biomass increases significantly in relation to shoots in grazed areas when contrasted to ungrazed areas (Lewis, 1971). The reasons for this increase are based on mechanical clipping effects provided by herbivores wherein plants, under some conditions, react by producing more root tissue (Bokhari & Singh, 1974), but there are also physiological effects prompted in the plant by introduction of animal saliva into the plant wound (Reardon, Leinweber & Merrill, 1972). Apparently one of the active components in saliva is thiamine (Reardon, Leinweber & Merrill, 1974). This is of extreme interest, and has been responsible for generating much recent literature in the field of plant–animal interactions (Harris, 1974; Chew, 1974; Mattson & Addy, 1975). From field work in the mixed prairie of South Dakota, Smolik (1973) was able to demonstrate an increase of up to a 70 % in above-ground plant biomass with removal of nematodes. Laboratory experiments conducted on grasshoppers feeding on blue grama grass (*Bouteloua gracilis*) have shown a significant increase in root activity (Dyer & Bokhari, 1976). While many plant–animal co-actions have been known for several years in terms of plant toxin development in response to animal feeding (Janzen, 1971; Levin, 1971), information is now just being accumulated about energetics of such associations. That there should be definitive reactions between plants and their hosts is quite logical in view of the fact that many of these biota have evolved together over aeons in the grasslands. Thus, it is possible to postulate many patterns of plant defence to animal feeding as a function of co-evolutionary pressures (Mattson & Addy, 1975; Dyer & Bokhari, 1976). Not only are plant toxins selected for, but it is likely that those individuals which can respond energetically to enhance subsequent survival will also be the ones natural selection will favour. Thus, animal-feeding mechanisms, both clipping and introduction of biochemicals (most likely those produced in the salivary system) have profound effects on plant selection. As we discover more about these co-actions, it becomes much easier to understand the role of animals in various ecosystems. It is probable that they have not only a direct influence by helping to regulate rates at which energy flows throughout the system (through dynamics of their population turnover), but that there are also indirect associations which, over the course of relatively long periods of time, may affect energy-flow rates and partitioning to greater extents. Also, insect populations have been associated with fluctuations in flow of nutrients throughout forest ecosystems and it is probable that grassland consumers do likewise. In a strict sense, consumers probably play a very important role in determining basic turnover rates in a system and, as such, have been labelled as cybernetic regulators (Mattson & Addy, 1975). Thus, consumers probably play their greatest role as regulators of productivity in grassland ecosystems.

7. Micro-organisms

E. A. PAUL, F. E. CLARK & V. O. BIEDERBECK

The micro-organisms in grassland represent a wide variety of morphological and functional types. All utilise energy and compete for nutrients with other components of the ecosystem. Algae and photosynthetic bacteria derive their energy from the sun, whereas chemolithotrophic bacteria, such as the nitrifiers and sulphur oxidisers, obtain energy from inorganic substrates. However, the vast majority of micro-organisms in the soil–plant system utilise organic compounds formed by primary and other secondary producers. The saprobic heterotrophs, commonly called decomposers, constitute an overwhelmingly large proportion of the total micro-biota in soil. This discussion on micro-organisms of natural temperate grassland will, therefore, be devoted almost entirely to the decomposer organisms. Parasitic heterotrophs are capable of causing catastrophic losses in cultivated grasses, particularly in the cereals, but in most native grassland the host plants exist in reasonable balance with parasites which collectively constitute only a very minor portion of the total micro-biota.

In this chapter we propose to discuss the biomass or standing crop of the grassland micro-biota, the dominant types or taxa in soil and herbage habitats and, very summarily, the role of micro-organisms in the cycling of carbon and mineral nutrients. Such data are meagre from natural temperate grassland studies other than those referred to below.

Magnitude of the microbial standing crop

The microbial biomass exceeds that of the large and small mammals, birds, reptiles, insects and all other invertebrates combined, and at times approaches that of the visible biomass of primary producers. For example, the maximum vegetative canopy (green and dead standing shoots) biomass at the Canadian study site (Matador) in 1968 was estimated to be 434 g/m^2, while the maximum biomass of soil micro-organisms was 254 g/m^2; corresponding values for 1970 were 622 and 328 g/m^2. The microbial life in grassland may be designated as 'the invisible prairie', worthy of as much attention as are the components of the visible prairie.

On the assumption that microbial cells usually do not persist for extended periods after death, microscopic techniques, although tedious and subject to sampling and detection errors, can provide reasonably good estimates of biomass. Plate and extinction dilution counts for propagules capable of growth under specific conditions are known to measure only a small percentage of the microbial population. They cannot be used for biomass estimates, but do provide some indirect estimate of microbial activity. During periods of normal microbial activity, there is a relatively consistent relationship between direct counts and plate counts approaching a ratio of 20 to 1 (Trolldenier, 1973; Paul *et al.*, 1973). Within limits, therefore, plate counts can be used for intersite and interbiome comparison.

Table 7.1. *Mean fungal biomass* (g/m^2) *in ungrazed grassland at three sites* (*compiled from data of D. Parkinson* (*personal communication*), *Doxtader* (*1969*) *and J. O. Harris* (*1971*))

Site	Year	Season	Depth (cm) 0–10	Depth (cm) 0–30
Matador	1968	Apr.–Dec.	56	153
Matador	1970	Apr.–Aug.	79	215
Pawnee	1969	July–Sept.	18	38
Osage	1970	June	88	334

Fungi constitute the major portion of the microbial biomass. Table 7.1 shows fungal biomass values for three widely separated grassland sites. There is much less fungal biomass in the dry and hot sandy loam soil at Pawnee, Colorado (311 mm annual precipitation) than in the cool, semiarid heavy clay at Matador (388 mm) or in the warm but humid clay soil at Osage, Oklahoma (930 mm). At Matador and Pawnee, fungal biomass declines with depth, with the value at 20 to 30 cm being about four-fifths that in the 0–10 cm soil layer. This decline with depth is much less than that observed for the root biomass at the same sites, indicating that mycelial measurements from greater depths include a higher proportion of inactive or dead hyphae. D. Parkinson (personal communication) found that at Matador seasonal variations at the 20–30 cm depth are similar in trend to, but less extensive than, those occurring near the soil surface. The mycelial mass on *Agropyron* leaves increases markedly as the leaves become senescent and pass through progressive stages of decomposition from standing green to lying grey litter. Nevertheless, the fungal biomass on standing shoots is very small in comparison with that of litter and soil, as the total mycelial biomass on all standing live and dead shoots at mid-season amounted to only 1.0 g/m^2.

In general, the bacterial biomass (Table 7.2) is about one-quarter to one-half that of fungi, and it, too, decreases with depth. Only a minor fraction (less than 1.0 %) of the total bacterial biomass is found on the above-ground herbage, and its relative distribution within the leaf litter is similar to that noted earlier for fungal biomass. Although biomass values of bacteria tend to show positive correlation with those of fungi in many series of soil samples, this is by no means universally true. For example, Clark & Paul (1970) noted

Table 7.2. *Seasonal changes in bacterial biomass (g/m²) to a depth of 30 cm in soil at two ungrazed grassland sites (after Paul* et al., *1973; Doxtader, 1969)*

Site ...	Matador	Matador	Pawnee
Year ...	1968	1970	1969
April	31.9	38.1	–
May	70.9	43.9	–
June	76.2	31.0	–
July	58.2	31.2	32.0
August	67.0	27.2	26.0
September	42.3	22.2	22.0
October	54.7	18.3	–
Seasonal mean	57.3	30.3	26.7

that at Matador, in 1968, the bacteria approached their highest seasonal biomass in May, at which time the fungi were at their lowest.

Plate counts show that the population of bacteria in grassland soil is not uniformly distributed within the same layer of the soil–root system. There is always a very high density of organisms associated with the roots and a much lower density in the soil. However, it is more meaningful to evaluate bacterial populations in the soil ecosystem by comparing the total number of rhizosphere and non-rhizosphere bacteria on a unit area basis. Fig. 7.1 shows that the number of bacteria on roots declines more sharply with depth than the number in soil and demonstrates that, even in grassland soils, the bacteria associated with roots represent only a very minor portion of the total population in the soil–root system. On a surface area basis at Matador, bacteria on roots represented 2.7 % of the total bacteria at 0–10 cm and declined to 0.3 % at 40–50 cm, with an overall mean of 1.1 % (Paul *et al.*, 1973).

Estimates of the biomass of Protozoa at Matador have shown rhizopods to be more prominent than ciliates and flagellates in this grassland soil and have indicated that the total protozoan biomass (less than 0.5 g/m²) represents only a minute fraction of the entire microbial biomass. Soil protozoa are of considerable ecological importance, in spite of their very small biomass, because they generally feed on bacteria and other decomposer organisms.

Nematodes are typically the most numerous members of the fauna in grassland soils worldwide. The prominence of nematodes in terms of numbers and biomass in the invertebrate fauna of three North American sites is shown in Table 7.3.

At Matador, Willard (1973) found wide variations in nematode biomass, ranging from 0.32 to 1.94 g/m² during the years 1968–71, while Smolik (1973) noted considerably less variation at Cottonwood with values of 1.22, 1.01 and 1.06 g/m² for the years 1970, 1971 and 1972, respectively. In terms of numbers, the nematode populations at Matador and Pawnee are similar,

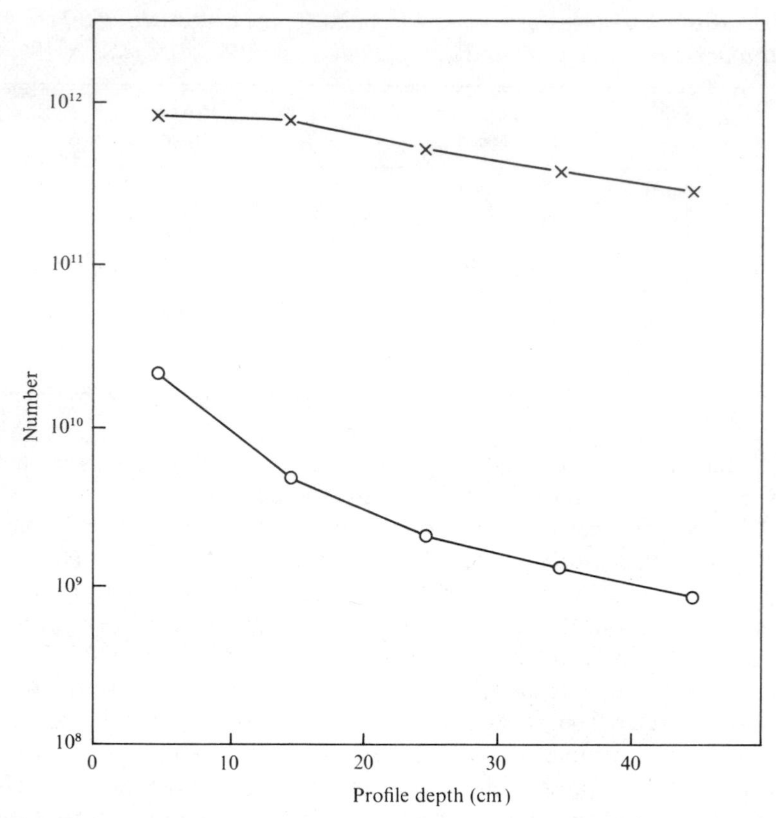

Fig. 7.1. Number of viable bacteria/m² in 10 cm segments of the soil profile. ×——×, in soil; ○——○, on roots.

Table 7.3. *Populations and biomass of soil invertebrates to a depth of 30 cm in ungrazed grassland at three sites (compiled from data of J. Leetham (personal communication), Smolik (1973) and Willard (1974)*

		Acarina		Collembola		Nematoda	
Year	Site	No./m²	mg/m²	No./m²	mg/m²	No. × 10^6/m²	mg/m²
1972	Pawnee	56766	48	4946	6	3.40	260
1970–72	Cottonwood	–	–	–	–	4.73	1090
1968–71	Matador	6153	18	14851	25	3.11	1110

but on a biomass basis, Matador and Cottonwood (both mixed prairie) are similar, while Pawnee (short-grass prairie) is much lower (Table 7.3), indicating a predominance of smaller nematodes at the Pawnee site. Acarina occur in much higher numbers at Pawnee than at Matador, but in terms of biomass the difference between the two sites is much smaller. For Collembola, both population and biomass are higher at Matador than at Pawnee. In fact, Collembola are the most abundant arthropods at Matador, whereas the arthropod population at Pawnee is dominated by Acarina. Willard (1974) found that, at Matador, nematodes represented 70.1 % of the total invertebrate biomass recovered, while enchytraeidae and micro-arthropods accounted for only 18.1 % and 11.8 %, respectively. Comparative studies by Smolik (1973) and J. Leetham (personal communication), on the effect of grazing on invertebrate populations, have indicated that neither the nematode biomass nor the biomass of major groups of micro-arthropods is markedly changed as the result of different grazing treatments.

Qualitative considerations

Discussion of the qualitative nature of the micro-biota cannot proceed independently of habitat associations. Perhaps the simplest approach is to ask what are the dominant groups or species of fungi, bacteria and invertebrates associated firstly with grassland soil and, secondly, with the herbage thereon.

Christensen & Scarborough (1969) studied 1000 dilution plate fungal isolates from the Pawnee site in August 1968 and August 1970. Over 100 species were found and taxa most commonly encountered were: *Fusarium oxysporum*, *F. moniliforme*, *F. solani*, *Aspergillus fumigatus*, *Mycelia sterilia*, *Trichoderma viride*, *Pullularia pullulans*, *Cladosporium* sp., *Penicillium janthinellum*. Species of *Fusarium* accounted for 17 % of all isolates in the two sampling years. There were also comparatively high numbers of dematiaceous and sphaeropsidaceous species; nearly one-third of the species isolated belonged to these two families. There was evidence of considerable year-to-year variation in the soil micro-fungal community. For example, the *Aspergillus fumigatus* group accounted for 7.5 % of all isolates in 1968 and 32 % in 1970. At the Tucker site in Missouri, C. L. Kucera (personal communication) found two physiologically different groups dominating the soil micro-flora. Cellulolytic Ascomycetes formed the major group, which consisted primarily of species belonging to *Trichoderma*, *Fusarium*, *Aspergillus* and *Penicillium*. A second and less prominent group consisted of non-cellulolytic Zygomycetes, especially *Mucor* spp. and *Rhizopus* spp.

Parkinson & Bhatt (1974), using the particle-washing technique at Matador, found that the fungal flora of the uppermost 30 cm of soil was dominated by six major types (Table 7.4). Species of *Fusarium* were the predominant

Table 7.4. *Dominant fungi isolated from grassland soil at Matador (Apr.–Oct. 1968, inclusive) by the soil washing technique; values represent percentage colonisation of mineral and organic particles (after Parkinson & Bhatt, 1974)*

	0–10 cm		10–20 cm		20–30 cm	
	Mineral	Organic	Mineral	Organic	Mineral	Organic
Trichoderma spp.	13.3	35.0	4.6	7.0	3.0	0.6
Fusarium spp.	44.0	47.3	41.5	45.5	46.0	40.2
Penicillium spp.	11.0	14.0	7.5	13.0	5.0	5.3
Sterile dark forms	7.3	7.0	4.0	11.3	8.3	11.3
Chrysosporium spp.	2.0	1.0	1.5	4.3	2.5	2.2
Paecilomyces spp.	3.0	3.3	3.3	5.0	0.5	2.5
Total colonisation (%)	80.6	107.6	62.4	86.1	65.3	62.1

colonisers of both organic and mineral particles at all depths and throughout the growing season. With five of the six dominant fungi, there were no major differences between the percentage colonisation of mineral and that of organic particles at three soil depths. Species of *Trichoderma*, however, were found much more frequently on organic residues, particularly in the surface soil. With increasing soil depth, the total fungal colonisation on washed particles, particularly the organic, decreased noticeably (Table 7.4). Much of this decrease was attributed to decline in frequency of occurrence of *Trichoderma* spp. with depth.

Colonisation of organic particles by sterile dark forms and by *Chrysosporium* spp. increased slightly with depth. Changes in frequency of occurrence of the six dominant fungi with season and depth were corroborated by physiological studies demonstrating differences in growth response to various nutritional, pH and temperature conditions. There was remarkably little diversity of unicellular fungi in this soil, as Spencer, Babiuk & Morrall (1971) found that four *Cryptococcus* species accounted for 92 % of all yeasts isolated.

The myco-flora of grassland herbage included species commonly found in the soil as well as others showing special affinity for the herbage materials themselves. D. E. Davidson & M. Christensen (personal communication) have shown the habitat ranges of several non-mycorrhizal fungi isolated from the roots of the dominant grass (*Bouteloua gracilis*) on the Pawnee site (Table 7.5).

D. E. Davidson & M. Christensen (personal communication) examined 10 vascular plant species at the Pawnee site for mycorrhizal fungi. All species showed the presence of *Endogone*. The proportion of root pieces (1–2 mm long) showing at least some evidence of infection ranged from 55 % for a cactus (*Opuntia polyacantha* Haw.) to roughly 90 % for *Bouteloua gracilis*, *Aster tenacetifolius* H.B.K. and *Plantago purshii* R.&S. Grazing intensity was

Table 7.5. *Fungi most frequently isolated from grass roots and their presence or absence in each root type and in soil (after D. E. Davidson & M. Christensen, personal communication)*

	Young roots	Medium aged roots	Old roots	Soil
Fusarium solani	+	+	+	+
F. oxysporum	+	+	+	+
Mycelia sterilia	+	+	+	−
Mortierella elongata	+	+	−	−
Diplodina sp.	+	+	−	−
Idriella lunata	−	+	+	+
Penicillium simplicissimum	−	+	+	+
Pyrenochaeta sp.	+	+	+	+
Phoma glomerata	+	+	−	−
Aspergillus fumigatus	+	+	+	+

not found to affect the number of root pieces showing infection. *Endogone* was not found on roots at depths greater than 50 cm.

Parkinson & Bhatt (1974) studied the fungi occurring in the litter of two species of wheat grass (*Agropyron dasystachyum* and *A. smithii*) at Matador. These grasses predominate in this ecosystem and account for three-quarters of the total green biomass. The Harley & Waid (1955) technique was used on four litter materials representing progressive stages of decomposition. The fungi dominant on standing shoots and litter (Table 7.6) were not the same as those dominant on mineral and organic particles in the underlying soil (Table 7.4). The fungal flora of the compressed litter–humus layer at the soil surface represents a transition between the *Alternaria–Cladosporium* dominated flora of the dead grass leaves and the *Fusarium–Penicillium* flora of the mineral soil. The absence of species of *Aspergillus* in Matador soil, in contrast to their common occurrence at Pawnee, is consistent with the well-established prominence of *Aspergillus* in hot, dry soils. Neither grazing at Pawnee nor cultivation at Matador caused any appreciable shift in the myco-flora from that observed in ungrazed grassland at the respective sites.

The numerical taxonomic approach applied to bacterial isolates from Matador grassland allowed the clustering of 60 % of the isolates into three genera, namely: *Arthrobacter*, accounting for 62 % of the clustered isolates; *Pseudomonas*, 23 %; and *Bacillus*, 15 % (Lowe & Paul, 1974). The high incidence of *Arthrobacter* is harmonious with the known incidence of this genus in temperate soils generally, while that of *Pseudomonas* was higher than commonly encountered in soil (Clark, 1967). Possibly pseudomonads were especially favoured by the large amount of grass-root material in Matador soil. Limited taxonomic study of the non-clustered isolates showed that they represented a large number of different genera and species.

Table 7.6. *Frequency of occurrence (%) of fungi isolated (May 1970) from* Agropyron *litter at various stages of decomposition, using the Harley & Waid technique (after Parkinson & Bhatt, 1974)*

Fungi isolated	Standing yellow	Standing grey	Lying grey	Humus + upper 1 cm soil
Absidia	–	–	–	4
Acremonium sp.	–	–	–	6
Alternaria spp.	84	58	72	4
Arthrinium phaeospermum	–	–	–	4
Cephalosporium sp.	–	4	–	4
Cladosporium malorum	10	20	8	–
Cladosporium sp.	24	14	20	2
Colletotrichum sp.	10	–	4	–
Cylindrocarpon album	–	–	–	6
Doratomyces microsporus	2	–	–	–
Epicoccum purpurascens	12	6	14	–
Fusarium oxysporum	4	8	4	15
Fusarium poae	–	10	4	8
Fusarium solani	–	–	–	4
Mortierella alpina	–	–	–	6
Mortierella nana	–	–	–	4
Penicillium lilacinum	–	–	–	4
Penicillium spp.	–	–	–	18
Sporormiella octomera	–	–	–	20
Sphaeropsidales	4	12	16	10
Sterile dark forms	4	8	20	8
Sterile hyaline forms	80	68	30	4
Trichoderma	–	–	10	16
Verticillium lateritium	–	–	–	4

Very few of the bacteria were able to decompose cellulose, suggesting that bacteria are not responsible for the initial decomposition of complex carbohydrates in grassland litter. At the Pawnee site, *Arthrobacter* was the dominant genus, while *Bacillus* ranked second. Extinction dilution counts of *Pseudomonas* showed that this genus comprised only about 1 % of the total bacterial count. At the Osage site, J. O. Harris (1971) found the bacterial flora to be dominated by a high zymogenous population of bacilli readily capable of decomposing starches and proteins. Limited study of species represented by germinating *Bacillus* spores showed about half or more to be *B. cereus*, while the remainder consisted of *B. subtilis*, *B. megaterium* and other species.

At Matador, actinomycetes account generally for one-third of the total plate count of surface soil (0–10 cm) and for more than half at depths greater than 30 cm. Using the ecological grouping procedure of Williams, Davies & Cross (1968) to sort soil actinomycetes into taxonomic units, Biederbeck (unpublished data) found that this grassland soil contains an unusually wide variety of actinomycetes, including members of the following seven genera:

Streptomyces, *Nocardia* type *Madurae*, *Nocardia* type *Asteroides*, *Micromonospora*, *Thermomonospora*, *Promicromonospora* and *Microellobosporia*. As expected, the surface population was predominantly of *Streptomyces* species, accounting for three-quarters of the isolates, with the remainder consisting primarily of *Nocardia* and *Micromonospora*. A similar order of dominance was reported by J. O. Harris (1971) for these same three genera in the grassland soil of the Osage site. Surprisingly, cultivation had effected marked shifts in the actinomycete flora of the clay soil at Matador. The grassland soil supported a considerably more heterogeneous actinomycete population than the soil cropped to cereals and none of the species occurred simultaneously as dominants in both ecosystems.

Soil organisms and the cycling of carbon and nutrients

In grassland, microbial tissues constitute a major dynamic reservoir of nutrients. Daily or seasonal changes in the microbial biomass directly affect the amounts of nitrogen, phosphorus and other nutrients available either for plant uptake or for further secondary production by soil organisms. Stewart, Halm & Cole (1973) reported that there is 0.49 g/m^2 of phosphorus in the plant canopy at Matador, 1.07 g in the under-ground plant biomass, and 1.98 g in soil micro-organisms. They postulated that two-thirds of the microbial phosphorus is released at the time of the spring thaw, thereby being made available for plant uptake during the growing season. McGill, Paul & Sorensen (1974), similarly, emphasized the role of micro-organisms in nitrogen turnover. In an inventory of the nitrogen in Matador grassland to a depth of 30 cm, they reported the nitrogen in soil micro-organisms as 13.07 g/m^2; the nitrogen content of all the above- and under-ground plant materials was reported as 22.73 g/m^2. The nitrogen in the soil organic matter (540 g/m^2) dominates the system. However, this nitrogen is known to have an extremely slow turnover rate, whereas the turnover in soil micro-organisms, in at least some cells, is accomplished in hours or days. The turnover of the total microbial tissue is variously estimated to occur from 1 to 10 times per year.

The role of micro-organisms in mediating carbon and energy flow through the ecosystem is consistent with their high biomass, enzymic versatility, and rapid growth rates. Microbial energetics are usually evaluated either by respiratory measurements or by estimating turnover time of the standing crop. Energy measurements and enzyme assays are also useful.

Morrall & Howard (1974) noted that at Matador the fungus *Pyrenophora tritica repentis*, with a biomass almost infinitesimally small, at times caused leaf spots on its host grass accounting for 0.01–5.0 % of the leaf area. Controlled-environment experiments and fungicide treatment in the field showed that primary productivity was as much as 26 % higher when the pathogen was controlled. The productivity increase was many times that which could be accounted for by direct consumption through parasites. Similarly, Smolik (1973) found that, although phytophagous nematodes contributed a biomass of less than 0.1 g/m^2, nematicide applications increased above-ground plant production by 28–59 %.

Micro-organisms are responsible for the metabolism of an overwhelmingly large portion of the net primary product. For the Pawnee site, for example, it has been estimated that microbes are responsible for 99 % of secondary production and 98 % of heterotrophic respiration.

8. Systems synthesis

R. T. COUPLAND & G. M. VAN DYNE

The data discussed in Chapters 5, 6 and 7 of this volume from the various trophic levels in the Canadian site in mixed prairie (Matador) and at the principal USA study site in short-grass prairie (Pawnee) provide a means of comparing and contrasting the rates of flow of energy and nitrogen through these ecosystems. The environments of these two sites are compared in Fig. 4.1; while both are under the influence of semiarid climate, Pawnee is warmer and drier than Matador. The estimates of activity of various organisms in these two ecosystems must be considered as preliminary in nature; it has been necessary to supplement field measurements with literature values and some guesses. Nor are direct comparisons between the two studies valid in many cases. For example, the length of the growing season at the northern site is taken as 240 days on the basis that in some years growth of some grasses was under way in late March or early April and in some years increases in green-shoot biomass were recorded after the end of October. However, on the basis of thermic and moisture limitations the growing season for Pawnee is considered to be 154 days. A more valid comparison of length of growing season between the two sites appears in Table 5.3; it appears that the thermic growing season averages 19 days longer at Pawnee (latitude 41° at an altitude of 1650 m) than at Matador (latitude 51° at an altitude of 680 m). Estimates of net primary production also are not directly comparable. At Matador the estimates included energy, apparently lost in the canopy after death of shoots that was not available to detritivores. The Pawnee estimate of net primary production, while minimal, corresponds more closely with the amount of biomass available to heterotrophs. Finally, Matador values are based on an average of three to five years of study of grassland ungrazed by domesticated animals; the estimates at Pawnee are based on one to three years of data from both ungrazed and grazed versions of the short-grass ecosystem.

Energy flow

The energy that flows through the ecosystem is derived from the short-wave radiation that falls on the surface of the vegetation and soil (global radiation). The mean daily global radiation at Pawnee is about 27000 kJ/m² during the growing season, while that at Matador is 17000. Of this, slightly less than half is considered to be photosynthetically active. However, the amount that is actually involved in photosynthesis is small and about one-third of this, in turn, is lost in plant respiration. The result is that the energy fixed in net primary production is estimated to contain only 0.3 % (Pawnee) to 0.5 %

Table 8.1. *Energy and nitrogen contents of the mean standing crop of the biotic components of the grassland ecosystem at Matador, Canada*

Component	Mean biomass (g/m²)	Energy content		Nitrogen content	
		kJ/g	kJ/m²	%	mg/m²
Plants					
Canopy					
Green shoots	75	17.58	1323	1.79	1347
Dead shoots	411	16.30	6699	0.92	3769
Litter	238	15.07	3575	1.49	3546
Under-ground parts					
0–10 cm	1134	15.42	17486	0.66	7484
0–30 cm	1688	[b]	26029	[b]	11141
0–150 cm	2767	[b]	42667	[b]	18262
Micro-organisms					
In canopy					
On green shoots					
Fungi	0.073	[b]	1.61	[b]	3.3
Bacteria[a]	0.0000015	[b]	[c]	[b]	[c]
On dead shoots					
Fungi	0.66	[b]	14.55	[b]	30
Bacteria	0.0054	[b]	0.10	[b]	0.3
In litter					
Fungi	0.63	[b]	13.89	[b]	28
Bacteria	0.0022	[b]	0.04	[b]	0.1
On under-ground parts					
Bacteria					
0–10 cm	0.14	[b]	2.61	[b]	6.7
0–30 cm	0.19	[b]	3.54	[b]	9.1
In soil					
Bacteria					
0–10 cm	16.62	18.65	310	4.8	798
0–30 cm	46.38	[b]	865	[b]	2226
Fungi					
0–10 cm	65.22	22.04	1437	4.5	2935
0–30 cm	177.06	[b]	3902	[b]	7968
Yeasts					
0–10 cm	0.0038	[b]	0.08	[b]	0.2
0–30 cm	0.0071	[b]	0.16	[b]	0.3
Protozoa					
0–10 cm	0.069	[b]	1.3	[b]	3.3
Animals					
Above ground					
Invertebrates	0.26	[b]	6.1	[b]	21
Birds	0.0024	23.79	0.06	[b]	0.2
Small mammals	0.0005	[b]	0.012	[b]	[c]
Underground					
Invertebrates					
0–10 cm	2.09	[b]	48	[b]	167
0–30 cm	4.37	[b]	100	[b]	349

[a] Values for bacteria include Actinomycetes.
[b] Values were not determined. Further calculations are based on estimated concentrations.
[c] Amount is less than 0.0005 kJ or 0.05 mg.

Table 8.2. *Comparisons of the standing crop of producers, consumers and micro-organisms in short-grass prairie at the Pawnee site (summarised from manuscripts by N. R. French & R. K. Steinhorst (for biomass and energy) and by R. G. Woodmansee, J. L. Dodd, R. A. Bowman & C. E. Dickinson (for nitrogen))*

	Means		
Biotic component	Biomass (g/m²)	Energy[a] (kJ/m²)	Nitrogen (mg/m²)
Plants			
Canopy			
Green shoots	100	1758	1790
Dead shoots	58	945	533
Litter	149	2245	2220
Under-ground parts	1973	30424	13022
Micro-organisms in soil	67	1363	3116
Animals			
Invertebrates			
Above ground			
Primary consumers	0.231	5.2	18
Secondary consumers	0.0024	0.05	0.2
Under ground			
Primary consumers	0.2291	5.2	18
Secondary consumers	0.1602	3.6	12
Birds			
Primary consumers	0.0028	0.07	0.2
Secondary consumers	0.0027	0.07	0.2
Small mammals			
Primary consumers	0.0050	0.12	0.3
Secondary consumers	0.0031	0.07	0.2
Pocket gophers	0.013	0.31	0.9
Lagomorphs	Not sampled	Not sampled	Not sampled
Reptiles	0.004	0.10	0.3
Coyotes	0.00009	0.002	[b]
Antelope	0.0082	0.20	0.6
Cattle	1.09	26	75

[a] Energy contents are calculated according to the values given for Matador in Table 8.1. Comparative values for other ecosystems are summarised by Golley (1961).
[b] Amount is less than 0.05 mg.

(Matador) of global radiation. Of the remainder about half is used in evaporation and half in heating the soil and other components of the ecosystem.

The energy contents of the biotic components of these two ecosystems are summarised in Tables 8.1 and 8.2. A comparison between the sites is presented in Table 8.3. These data suggest that the proportion of the energy reservoir of living and dead plant material that occurs under ground (to a depth of 60 cm) is 74 % at Matador and 86 % at Pawnee. They also suggest that producers contain 90 % of the organic energy at Matador and 96 % at Pawnee, while respective values for micro-organisms are 10 % and 4 %, and for animals are 0.2 % and 0.1 %.

Table 8.3. *A comparison of mean standing crop biomass and energy and nitrogen contents of grassland at Matador and Pawnee, to a depth of 60 cm in soil*

	Biomass (g/m^2)		Energy (kJ/m^2)		Nitrogen (mg/m^2)	
Component	Matador	Pawnee	Matador	Pawnee	Matador	Pawnee
Above ground						
Canopy						
Plants	486	158	8022	2703	5116	2323
Micro-organisms	0.73	?	16	?	34	?
Invertebrates	0.26	0.23	6.1	5.3	21	18
Native vertebrates	0.003	0.04	0.07	1	0.2	2.7
Cattle	—	1.09	—	26	—	75
Subtotal	487	159	8044	2735	5171	2419
Litter						
Plants	238	149	3575	2245	3546	2220
Micro-organisms	0.63	?	14	?	28	?
Subtotal	239	149	3589	2245	3574	2220
Under ground						
Plants	2174	1973	33523	30424	14348	13022
Micro-organisms	234[a]	67	4772[a]	1363	10207[a]	3116
Invertebrates	4.4[a]	0.39	100[a]	8.8	349[a]	30
Subtotal	2412	2040	38395	31796	24904	16168
Total	3138	2349	50028	36776	33649	20807
Plants	2898	2280	45120	35372	23010	17565
Micro-organisms	235[a]	67	4802[a]	1363	10269[a]	3116
Animals	4.7[a]	1.75	106[a]	41	370[a]	126
Total	3138	2349	50028	36776	33649	20807

[a] Sampled only to a depth of 30 cm.

Of the energy contained in net primary production, 57 % at Matador and 85 % at Pawnee is transferred to under-ground plant parts and is eventually released by respiration of soil organisms (Table 8.4; Figs. 8.1, 8.2). The remainder is transferred to the dead shoots as the green shoots die. About 1 % of net primary production enters the grazing food chain above ground (at both sites) and another 2 % (at Pawnee) follows this route under ground. About 1 % (Matador) of net primary production falls to the ground from the green-shoot compartment during grazing by above-ground consumers. Measurements at Matador suggest that about 15 % of net primary production is lost in the dead-shoot compartment of the canopy. This apparently, is not transferred to litter, but decomposes in the canopy. R. I. Zlotin (personal communication) has shown that 'photo-chemical' decomposition occurs in the litter layer in natural temperate grassland; perhaps this process also is involved in this apparent loss from the canopy. We do not know the extent of possible additional loss from the litter layer by non-biological processes.

Table 8.4. *Estimated flow of energy (in kJ/m²) through various components of the prairie ecosystem (Pawnee site) lightly grazed by cattle*[a] *during the 154-day growing season in 1972 (summarised from an unpublished manuscript by C. C. Coleman, R. Andrews, J. E. Ellis & J. S. Singh)*

Component	Energy input	Lost by respiration	Tissue production	Production / Consumption
Solar input				
Global radiation	4155000			
Photosynthetically active	1966000			
Primary production				
Gross	21882	7439		
Net				
Above ground			2163	
Under ground			12280	
Total net			14443	
Transfers to heterotrophs				
Above ground				
Macro-arthropods	34	23	11	0.32
Cattle	105	92	13	0.13
Carnivores	8.3	7.5	0.8	0.10
	148	123	25	
Under ground				
Herbivores				
Macro-arthropods	127	66	61	0.48
Nematodes	50	42	7.9	0.16
Carnivores	20	15	4.6	0.23
Saprophages				
Micro-organisms	12560	9632	2929	0.23
Nematodes	72	61	12	0.16
Others	9.2	6.7	2.5	0.27
Subtotal	12838	9822	3016	
Total	12985	9945	3041	
Unaccounted for	1458			

[a] One yearling steer or heifer per 10.8 ha for 180 days each year; live weight is about 200 kg at beginning of grazing season and 320 kg at the end.

About one-quarter of the energy transferred to roots is estimated to be lost as heat in root respiration. It is estimated that micro-organisms release 98 % (at Pawnee) of energy that passes through subterranean grazing and detritus food chains.

Although a surprisingly small amount of energy passes through the above-ground native consumers (42 kJ/m² at Pawnee and 285 kJ/m² at Matador, annually), the importance of these organisms in maintaining the ecological balance of the system is presumably quite high. Invertebrates play the major role in energy flow through this group, at Matador ingesting 95 % of the

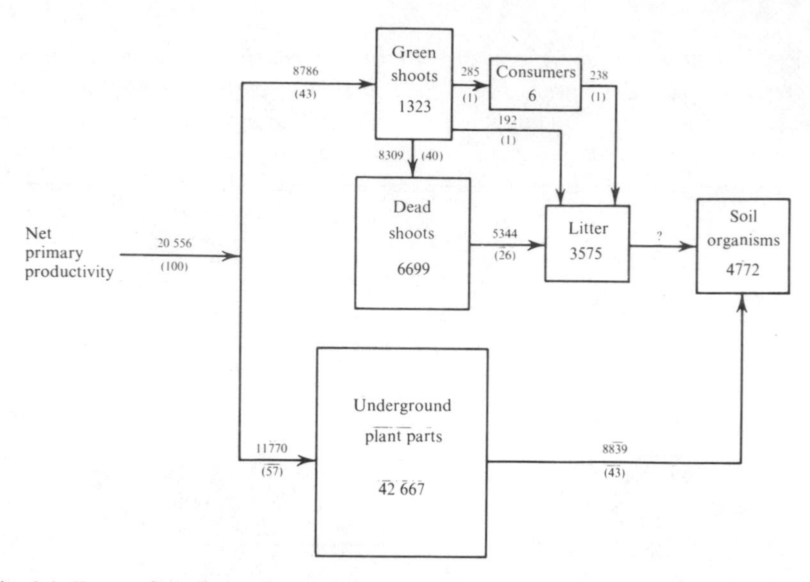

Fig. 8.1. Energy flow through the mixed prairie ecosystem at Matador. The values above the lines, on the arrows, are annual rates of energy flow (in kJ/m²), while those below the lines represent the percentages of net primary production that follow various routes. The values in the boxes represent the energy content (in kJ/m²) of the mean standing crop.

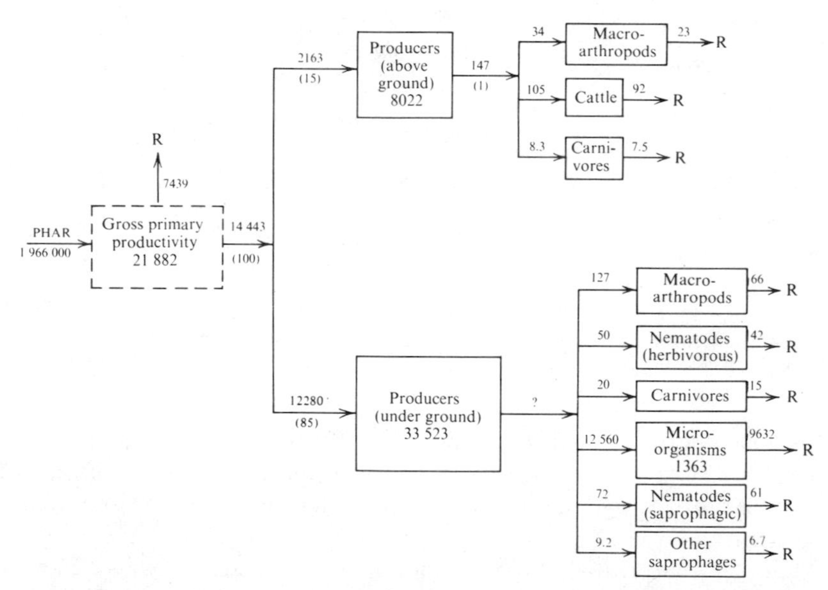

Fig. 8.2. Energy flow through the short-grass prairie ecosystem at the Pawnee site. The values above the lines on the arrows are annual rates of energy flow (in kJ/m²), while those below the lines represent the percentages of net primary production that follow various routes. The values in the boxes represent the energy content (in kJ/m²) of the mean standing crop. PHAR, photosynthetically active radiation.

Table 8.5. *Percentage of energy ingested by above-ground animals that flows between various components of the Matador grassland ecosystem. The total annual flow from producers to above-ground consumers is estimated to be 285 kJ/m²*

Component	Flow	Source
Producers to invertebrates	94.6	Riegert, Varley & Willard, 1974
Producers to birds	4.2	Sadler & Maher, 1974
Producers to small mammals	1.2	Sheppard, 1972
Invertebrates to soil organisms	81.8	Riegert, Varley & Willard, 1974
Invertebrates to small mammals	0.05	Sheppard, 1972
Invertebrates to birds	0.1	Sadler & Maher, 1974
Small mammals to soil organisms	0.3	Sheppard, 1972
Birds to soil organisms	1.5	Sadler & Maher, 1974

amount (while birds ingest 4 % and small mammals 1 %) (Table 8.5). About six-sevenths of the energy ingested by invertebrates is passed on to soil organisms and only a very small amount (a fraction of 1 %) is passed to each of the small mammals and bird groups.

The data available from the Pawnee and Matador studies permit some estimates relating to the relative amounts of energy that flow through native herbivores and livestock in grazed versions of these ecosystems. For example, at Matador the assessment of range specialists is that the proper rate of grazing is 0.84 ha per animal unit month (Lodge & Campbell, 1971). Considering the forage requirement for an animal unit of cattle to be 10 kg/day, the rate of ingestion by cattle would be 35 g/m² annually, which would contain 615 kJ. This compares with an annual ingestion rate of 105 kJ/m² by cattle in lightly-grazed range at Pawnee (Table 8.4). Based on laboratory measurements of consumption rates, the populations of grasshoppers in rangeland at Matador were estimated to ingest herbage containing 155 kJ annually (Riegert & Varley, 1973*b*), while at Pawnee potential annual consumption rates (estimated by Mitchell & Pfadt (1974) for high density populations) ranged from 268 to 368 kJ/m². In addition, the grazing action of grasshoppers is so wasteful that one to five times as much green herbage as is consumed passes directly to litter (Andrzejewska *et al.*, 1967; Mitchell & Pfadt, 1974). Thus, in these ecosystems the amount of green herbage consumed and detached by grasshoppers appears to approach or exceed the amount ingested by livestock. Annual secondary production by cattle in this Matador grassland is estimated to contain 38 kJ/m² (compared to 13 kJ/m² in the lightly-grazed situation at Pawnee), while that of grasshoppers contains 21 kJ/m².

The measurements and estimates resulting from these studies permit some interesting comparisons relating to the efficiency of secondary production.

Estimates of the efficiency of conversion of organic intake to tissue (in terms of energy) by soil nematodes and saprophages range from 16 to 27 % at Pawnee, while that of cattle is 13 % (Table 8.4). Based on laboratory studies, the grasshoppers at Matador have an efficiency of 13 % (Riegert & Varley, 1973*b*) and at Pawnee 23–32 % (Mitchell & Pfadt, 1974), while that of spiders at Matador is 27 % (Riegert, Varley & Dunn, 1974). The apparent efficiency of cattle at Matador (as described above) is only 6 %. Both studies suggest, therefore, that native herbivores are more efficient as secondary producers than are cattle. The high estimates for arthropod efficiency at Pawnee (Table 8.4) require further consideration, in view of the lower values obtained at Matador.

Another interesting feature of these data is the relationship between tissue production and standing crop of heterotrophs. At Pawnee production of soil micro-organisms (Table 8.4) is estimated to be more than double the standing crop (Table 8.2), while for soil invertebrates production is 10 times the standing crop in terms of energy. Similarly, at Matador tissue production in grasshoppers, birds and small mammals was found to be, respectively, 5, 2 and 5 times standing crop values in terms of energy. However, for cattle, tissue production during the grazing season is only between one-third and one-half the standing crop (Coupland *et al.*, 1975).

The rate of utilisation in relation to standing crop of different groups of heterotrophs is quite variable (Tables 8.2, 8.4). Energy flow into the micro-organism activity is apparently 11 times cell contents, while energy flow into soil invertebrates approximates to 30 times body contents. For cattle, energy consumption during the six-month grazing season is between 5 and 6 times body energy at both sites.

A characteristic of these ecosystems which seems surprising is the relatively small proportion of net primary production that enters the grazing food chain. In both studies native herbivores apparently consume only 1–2 % of net primary production, while a similar proportion of under-ground net primary production is consumed by herbivores (at Pawnee) (Table 8.4; Figs. 8.1, 8.2). Even cattle use a small proportion of herbage, the value under light grazing (about half the recommended stocking density) at Pawnee being about 5 % of shoot production, while at Matador the recommended intensity of use (moderate grazing) would involve an intake by cattle of 7 % of shoot production of ungrazed grassland.

Nutrient cycling

Sufficient data are available to calculate approximately the rate of flow of nitrogen through some of the organic components of the grassland ecosystem at Matador and Pawnee.

Standing crop data are presented in Tables 8.1, 8.2 and 8.3. The total standing crop of organisms (including surface plant litter) is estimated to contain about 30 g/m^2 of nitrogen at Matador (to a depth of 30 cm in the soil) and 21 g at Pawnee (to a depth of 60 cm). At Matador plants contain about 65 % of the nitrogen, micro-organisms 34 % and invertebrates 1.2 %; corresponding values for the Pawnee ecosystem are 85, 15 and 0.2 %. The proportion of nitrogen in the canopy and litter layers combined is estimated to be 22 % at Pawnee and 29 % at Matador. Soil micro-organisms contain a similar amount of nitrogen as do under-ground plant parts at Matador, while at Pawnee plants contain 4–7 times as much as microbial tissue. The amount of nitrogen in the soil that is readily available to plants (NO_3–N) at Matador varies intraseasonally from about 1.5 g/m^2 to a depth of 30 cm (2 g/m^2 to 60 cm) in spring to about 0.4 g/m^2 (0.7 g/m^2 to 60 cm) in autumn. Nitrogen

Table 8.6. *Standing crop and flows of nitrogen (in g/m^2) in short-grass prairie (Pawnee site) 1973*[a]

Ecosystem component	Maximum standing crop	Annual flow
Autotrophs		
Green shoots	2.65	
to dead shoots		2.03
to above-ground herbivores		0.02
to litter		0.30
to roots		0.40
Dead shoots	0.39	
to above-ground herbivores		0.01
to litter		2.28
Litter	6.00	
to micro-flora		2.41
Crowns	5.10	
to litter		0.51
Roots	20.72	
to shoots		4.19
to crowns		0.51
to micro-flora		0.84
Heterotrophs		
Above-ground vertebrates[b] and invertebrates		
to litter	0.01	0.02
Soil invertebrates	0.10	
Soil micro-flora	3.00	
Soil organic matter	333.00	
Soil NH_4	0.65	
Soil NO_3	0.01	

[a] Estimates are minimal, since they are based on sampling only to a depth of 36 cm below the soil surface.

[b] If grazed, the nitrogen content of the cattle would be about 0.25 g/m^2 and the flow of nitrogen would be about 0.25 g/m^2.

content of soil organic matter to a depth of 15 cm is 0.32 %; it contains 540 g/m^2 to a depth of 30 cm. In addition, a considerable amount is present at greater depths, that in the 30–60 cm layer being at least 40 % of that in the 0–30 cm layer. At Pawnee the soil organic content of nitrogen (to a depth of 36 cm) is estimated to be 333 g/m^2. Consequently, the amount of nitrogen present in the standing crop of organisms (including fallen leaves and dead, but intact, roots) probably approximates to 5–10 % of the total organic nitrogen present in the ecosystem.

It is difficult to estimate the rates of flow of nitrogen through the various components of these ecosystems because of its mobility within plants. For example, net primary biomass production (of shoots plus under-ground parts) is estimated to be 1164 g/m^2 in the Matador ecosystem. If we assume that the flow from green shoots to dead shoots, from shoots to roots, and from live roots to dead roots carries half of the nitrogen present in green leaves, then the annual amount of nitrogen flowing through producers approximates to 10.5 g/m^2.

Estimates of the rates of flow of nitrogen through various components of the short-grass ecosystem at Pawnee are presented in Table 8.6, which suggests that the annual flow through plants is about 3 g/m^2.

Another method of estimating rate of flow of nitrogen through plants is by considering the rate of turnover of plant biomass. The apparent annual rate of turnover of shoots at Matador is 68 % (annual production 495 g/m^2, mean standing crop in canopy and litter 724 g/m^2), while that of under-ground parts is 24 % (annual production 669 g/m^2, mean standing crop 2767 g/m^2). If rate of turnover of nitrogen is proportional to that of biomass then it appears that only 8.5 g/m^2 of nitrogen flows through plants annually in this balanced system. Similar reasoning for the Pawnee ecosystem suggests an annual flow rate of 7.0 g/m^2 of nitrogen.

Clark *et al.* (in Chapter 8 of Breymeyer & Van Dyne, 1979) have discussed in much greater detail their analysis of nitrogen cycling based on studies in the Pawnee site and of phosphorus cycling based on research in several natural grasslands in North America.

9. Use and management

R. T. COUPLAND

Man has used natural temperate grasslands for two principal purposes to help satisfy his needs for food and clothing. Where conditions of soil, climate and topography are suitable for the production of grain, hay and fibre crops, the native sod has been ploughed and brought under cultivation. Elsewhere livestock ranching has been practised on an extensive basis. Of major concern has been, and still is, the determination of where tillage should stop and ranching begin. Curtis (1956) and Clark (1956) have discussed man's role in modifying these grasslands.

Many land-use problems have arisen in regions of natural temperate grassland because of deficiency of precipitation. While moisture conditions near the margin with forest are favourable for farming, critical limits for production of crops are reached in semiarid regions (Muehbeier, 1958). For example, in the grassland of central North America periods of drought become more frequent, of longer duration and of greater severity with increasing distance westward from the deciduous forest region. This results in declining productivity of the grasslands and is reflected in transition in colour of soil from black in the belt near forest, to dark brown (chestnut) in the central belt, to brown along the western edge (near the Rocky Mountains). The proportion of land that is cultivated is greatest and crop yield is most dependable in the eastern-most belt of grassland (i.e. the true prairie, which includes the 'corn belt'). In the dark-brown soil zone the risks to arable agriculture are greater, so that a greater proportion of the grassland has been left untilled. In the brown soil zone less than half of the landscape has been cultivated. Nevertheless this is where man's intervention as an agriculturalist caused the greatest problem, resulting in the creation of a 'dust bowl' in the 1930s. Here precipitation is low and variable, with driest years grouped to form extended dry periods (Tannehill, 1947), which results in crop failures. Other factors also contribute to uncertainty in success of seeded crops. High temperatures in summer, highly variable temperatures in spring and fall, and high wind velocities are serious hazards to crops. The wind increases evaporation of scarce water, intensifies drought, and causes erosion of soil. The soils form a complex pattern, differing in their capacity to absorb and hold water, to resist erosion, and to produce crops.

During the settlement of the North American grasslands some areas were tilled in which crop production could not continue on a sustained basis for ecological and economic reasons. Errors of this kind did not result only from lack of information concerning the adversities of the climate. The greater vigour of native vegetation in sandy areas sometimes lured settlers away from areas of finer-textured soil with better moisture-holding characteristics. As a consequence, much of the land that was cultivated formerly in the Great Plains of North America has now been returned to perennial grass and is used for ranching purposes.

However, the argument between the range manager and the agronomist continues. Periodic shifts occur in the relative economic advantage in respect of the two land uses, due to fluctuations in weather conditions. During moist periods the optimistic tend to forget the hazards of farming in dry regions. Where land has been retired from the production of annual crops, consideration is given to growing perennial exotic forage species, often without due consideration of their performance relative to native species. The higher yields of grasses seeded in tilled land, as compared to the yield of forage in native range, has caused agronomists to be misled into concluding that the higher yield is a result of seeding more productive species. However, the fact that native grasses yield similarly to exotics when seeded in recently-tilled land (Dubbs, 1966; Whitman, Peterson & Conlon, 1961) suggests that tillage is the factor that causes the increase in forage yield. Supporting evidence for this point of view is the increase in yield of native grassland after tillage alone without reseeding (Thatcher, 1966; Houston & Adams, 1971; Rauzi, 1975). The rate of mineralisation of nitrogen in seeded (arable) grasslands is much higher than in untilled grasslands in the first few years after establishment. Two reasons that have been suggested for increases in forage yields after tillage are: (i) stimulation of ammonifying and nitrifying micro-organisms (McCalla, 1967); and (ii) improved precipitation use efficiency (Wight & Siddoway, 1972).

Dry-land farming in temperate grassland regions is a relatively new venture (Coupland, 1974*c*). Extensive cultivation of semiarid and much of the dry subhumid natural grassland began only in the nineteenth century. Because these regions are very important sources of food for humans, it is essential that they be conserved and maintained in a productive condition. It seems improbable that the present management systems, which have been arbitrarily planned on an economic basis, will prove to be sustainable on an ecological basis. The resource must be conserved in a condition that will permit its use in perpetuity.

Tillage completely destroys the natural cover of higher plants and greatly modifies the consumer and microbial components of the ecosystem. Modern cropping practices are designed to maximize the supply of moisture in the year of cropping. This is achieved by cultivating the land as a black fallow every second or third year, depending on the degree of aridity of the climate. This permits storage of moisture from precipitation, but it also accelerates the rate of mineralisation of nutrients within the soil. Not only is the recently-produced organic matter degraded rapidly, but the reserve of the organic matter that was developed under centuries of natural grassland cover is rapidly broken down. Several reports (Salter & Green, 1933; Salter, Lewis & Slipher, 1941; Newton, Wyatt & Brown, 1945; Haas & Evans, 1957; Hobbs & Brown, 1957) have indicated losses of carbon, nitrogen or organic matter from the soil during the first several decades of cultivation at the rate of 1–2 % per year of the original levels. The half-life of the organic matter in prairie soils is probably of the order of 500–1000 years, with only a small proportion of the organic content turning over in a very few years (Martel & Paul, 1974). It seems reasonable to doubt that humus from annual crops will have this high degree of resistance to degradation. Consequently, under cultivation, reduction in stability of the soil resource may occur to a point where a relatively large proportion of the annually-produced organic material turns over rapidly and the organic content of the soil is only a small fraction of that in the native grassland. An indication of how rapidly organic matter turns over in cropland was observed in a study at Matador (Coupland, 1974*a*). In the three months between seeding and harvest in a wheat field the loss of organic biomass in the uppermost 30 cm of soil was greater than that added by transfer to roots.

The biologist is concerned because land-management practices are based almost exclusively on harvestable yield. Thus, the rates of application of commercial fertilisers increase as the fertility of the soil declines through loss of organic matter. This meets the need for supplying the nutrient requirements of plants, but it does not provide for any other functions that the organic component of the soil may perform in sustaining the system. Another concern is that the long-term effect of the use of pesticides in cropland has not been sufficiently evaluated to forecast the extent to which the chemicals can be incorporated into a sustained cropping system.

In natural grasslands that are used for grazing by domesticated animals, particularly those that are untillable, it is of the utmost importance to maintain a vigorous cover of native grasses. Overgrazing results in changes in the floristic composition of grassland, because of differential responses of species to grazing pressure (Curtis, 1956; Clark, 1956). Dyksterhuis (1949) has characterized range plants, in relation to their response to grazing, as 'decreasers', 'increasers' and 'invaders'. Overgrazing results in disappearance of the most desirable species ('decreasers') and relative increases in less desirable and lower-yielding species (increasers). Eventually, the community is invaded by exotic species, also often of unpalatable nature (invaders). These varying responses of species have been used by Dyksterhuis to develop a means of comparing the condition of a rangeland with its potential climax condition. Ranges are classified as in 'excellent', 'good', 'fair' or 'poor' condition according to the extent to which the relative contributions of the various species to the canopy deviate from those expected in the absence of grazing by domesticated herbivores. This assessment is made by using information concerning differences in plant cover of grazed and ungrazed versions of the same 'range site' (i.e. in the same conditions of soil, moisture regime and salinity). Following Dyksterhuis' system the closer the plant cover is to climax the greater is the recommended rate of stocking. Stocking rates of deteriorated ('poor' or 'fair') ranges are fixed at a lower level than is warranted by the amount of forage yield, so as to provide for a measure of recovery to a condition ('good' or 'excellent') closer to the potential possible.

The control of the number of animals grazing in rangeland is the most important single tool available to the rangeland manager (Heady, 1975; Stoddart *et al.*, 1975). Few studies have been made over a long enough period to evaluate accurately the sustainability of range production under various rates of grazing. Much of the available evidence suggests that in the short term, except in deteriorated range, gains per hectare are greater under heavier rates of grazing than under moderate or light rates. However, ranchers have found moderate rates of grazing to be more economical because gains per animal are greater and the level of investment in animals and the labour input are less (Van Dyne *et al.*, in Chapter 4 of Breymeyer & Van Dyne, 1979).

It is generally agreed that natural temperate grassland is affected less by grazing if it is rested for a proportion of each year. Consequently, several types of rotational systems have been tested in comparison with continuous grazing (Heady, 1975). These have indicated that range condition may improve under rotational grazing at a level of intensity that does not permit improvement under continuous grazing. However, in most tests the higher production of forage in the rotational system has not been reflected in greater gains in weight of livestock. In a summary of 29 tests, Driscoll (1967) concluded that no advantage to livestock gains was found in nine, 12 favoured continuous grazing, and only eight favoured rotational grazing. The explanation which is commonly given for failure to achieve better animal performance under rotational grazing is that the foliage becomes over-mature before grazing takes place in the fields where grazing is deferred; consequently, its nutritive value is reduced. Many tests have not been of sufficient duration to measure adequately the cumulative long-term effect of increased forage yield under rotational grazing in increasing stocking capacity.

The general opinion of managers of natural temperate rangelands is that the degree of herbage removed by the grazing animal has more influence on range vegetation than has any other grazing factor (Heady, 1975). Commonly it is considered appropriate to utilise about half of the forage yield. This measure is related, of course, to the maximum standing crop of current year's growth. We have seen in Chapter 8 that the amount of herbage removed in moderate grazing is probably no more than 10 % of net herbage production in the absence of grazing. However, even at these low rates of herbage removal the vigour of the range is usually much less in the first few years of protection than is that of areas that have been protected for several years. Consequently, it appears that other factors involved in grazing are very important in affecting the well-being of the vegetation. Trampling compacts soil, thus decreasing the amount of precipitation that becomes soil water and increasing the extent of soil erosion. The resulting reduction in supply of soil moisture causes a reduction in vigour of the plant cover which, in turn, provides less protection of the soil surface against erosion and evaporative losses (Johnston, 1962). Grazing indirectly affects other biotic components of the system in addition to the grazed plants. For example, the populations of certain herbivores are altered as compared with those of ungrazed areas. Of particular note during IBP studies was an increase in density of grasshoppers in grazed areas (Riegert & Varley, 1973*a*) and of some species of birds (Maher, 1974*b*).

Suggestions have been made that the stocking densities in some natural rangelands will have to be increased to meet the growing demands for more meat and animal products. It is to be hoped that such proposals will be submitted to very careful appraisal in the light of what is ecologically possible. It seems reasonable to urge that in planning future management regimes all components of the ecosystem be considered. As man becomes increasingly exposed to the pressures of urban living the natural grasslands are becoming more important as recreational areas. Thus, as well as food output the well-being of wildlife in rangeland is of concern to society. It seems clear that eventually a compromise will have to be reached between maximum production and total conservation (Odum, 1969), but let us hope that this will occur before severe irreversible changes are imposed on natural temperate grassland ecosystems by use at a more intensive level than can be sustained.

Natural temperate grasslands have not preserved well under increasing pressure by man. Where climate is suitable to crop production, the conversion from grassland to cropland was more complete than the corresponding conversion from forest, because the plant cover and generally more favourable topography presented fewer obstacles to the plough (Curtis, 1956). Where agriculture is most intensively practised remnants of the original vegetation are rare. In the opinion of Allen (1966) one of the greatest needs, and one of the greatest challenges in wilderness restoration and preservation, is that of creating a major national park in the North American grassland. Even where the spread of arable agriculture has been impeded by adverse climate, unfavourable soils and rolling topography, great difficulty exists in restoring a primitive biota because of the extinction of some native species and the introduction of many exotics.

Part III. Semi-natural temperate meadows and pastures

Subeditor: M. Rychnovská

10. Introduction

E. BALÁTOVÁ-TULÁČKOVÁ

Semi-natural grasslands are considered here as those occurring in regions where the climate is suitable for forest growth. They can be classified according to use as meadows (haylands) and pastures (grazing lands). These semi-natural grasslands occur in normal upland situations that have been deforested by man, as well as in situations in which physical factors are unfavourable to the development of forests. Among the latter are: portions of river valleys and deltas subject to prolonged flooding, areas of salt accumulation (including sea shores), and above the tree line on mountains. Most exploited semi-natural grasslands in the forest belt are secondary communities that have developed in the sites of cleared forests in which mowing and grazing are the key factors that prevent reinvasion by trees and shrubs and interfere with redevelopment of forest.

The basic economic importance of these grasslands is related to their production of abundant fodder for livestock that is rich in proteins, minerals and vitamins. In the case of meadows there is a significant recreational role, because of their aesthetic value, both in montane and submontane areas. They are important also in both recreational and agricultural areas because of the many indirect beneficial functions that they perform. Meadow belts along the rivers act as biological filters for surface waters, incorporating fertilisers washed from croplands. In both meadows and pastures (unless overgrazed), a dense cover of herbaceous plants is effective in reducing erosion by wind in exposed areas and by water on steep slopes. Fertility characteristics of soils are often more favourable in those parts of the landscape occupied by herbaceous vegetation than in forest soils because of the reduced rate of leaching of nutrients from the less acidic edaphic environment associated with grass cover.

The production potential of the vegetation of these meadows and pastures is determined by vertical structure of vegetation and by species composition which, in turn, mainly depends on geographical location. Decisive factors are phytogeographical conditions, altitude, position in landscape relief, type of climate and soil properties (especially supply of nutrients), and also pasture stress. In meadows, the high capacity for nitrogen fixation by natural means is of importance, since it provides a source of cheap protein production. Rate of production of above-ground matter is closely related to the duration of growing season, temperature conditions, degree of solar radiation and precipitation. The phenological plasticity of individual species and their adaptability to grazing are other important factors. In dry regions precipitation can, to a certain degree, be compensated for by under-ground or flood

Table 10.1. *Characterisation of certain IBP sites in relation to their geobotanical situation* (*compiled by E. Balátová-Tuláčková, based partly on a summary of A. Medwecka-Kornaś*)

Locality (country)	Latitude (N)	Longitude (E, W[a])	Altitude (above sea level) (m)	Annual precipitation (mm)	Mean temperature (°C): Warmest month (July, August[b])	Coldest month (January)	Annual	Soil characteristics: Soil type	Soil acidity pH/H_2O (KCl, $CaCl_2$) A_1–C horizon	Organic matter (%)	C/N
Euro-Siberian region											
Atlantic European Province – maritime climate											
Wet and moist sites											
Terschelling	53°22′	5°13′	1	760	+17.6[b]	+2.4	+9.5 (1967–70)	Clay-deficient sand	6.9 KCl	4.9	—
(Netherlands)	53°22′	5°13′	1	760	+17.6[b]	+2.4	+9.5 (1967–70)	Clay-deficient sand with muddy layer	6.9 KCl	19.2	—
	53°22′	5°13′	1	760	+17.6[b]	+2.4	+9.5 (1967–70)	Clay-deficient sand with muddy layer	7.2 Cl	11.0	—
Moist and mesic sites											
Moor House N.N.R.,	54°40′	2°25′W[a]	556	1902	+10.0	0.0	+5.1 (1953–74)	Peat	3.0–4.6	—	—
Northern Pennines (United Kingdom)	54°40′	2°25′W[a]	556	1902	+10.0	0.0	+5.1 (1953–74)	Brown earth	4.6–5.6	—	—
Llyn Llydaw-Snowdonia (United Kingdom)	53°05′	4°05′W[a]	488	3810	+13.0	+2.5	+7.4	Brown earth	4.6–5.6	—	—
Pin au Haras (France)	48°44′	0°10′	190	730	+16.8	+2.5	+9.5 (1950–70)	Alluvial clay (pseudogley)	6.2 KCl	14.4	12.4
Saint Martin-de-Boscherville (France)	49°28′	0°53′	375	750	+16.9	+3.4	+9.7 (1950–70)	Calcareous, alluvial (pseudogley)	7.0 KCl	26.0	10.0
Fenno-Scandinavian Province – maritime climate											
Mesic sites											
Tvärminne (Finland)	59°51′	23°15′	425	701	+16.1	−5.4	+4.4	Deluvial earth	5.5	—	—
	59°51′	23°15′	425	701	+16.1	−5.4	+4.4	Deluvial earth	5.2	—	—
	59°51′	23°15′	475	701	+16.1	−5.4	+4.4	Deluvial earth	5.3	—	—
	59°51′	23°15′	475	701	+16.1	−5.4	+4.4	Deluvial earth	5.8	—	—

[a] Locations are East of Greenwich, unless otherwise indicated.

Locality (country)	Influence of underground water (UW) and/or inundation (IN)	Plant composition – dominants and/or other important species (Phytocoenological classification according to Braun–Blanquet school)	Literature (main sources)
Euro-Siberian region			
Atlantic European Province – maritime climate			
Wet and moist sites			
Terschelling (Netherlands)	UW, occasionally IN	*Juncus gerardii*, *Carex extensa* (Glauco–Puccinellietalia)	Ketner (1972)
	UW, almost every month IN	*Plantago maritima*, *Limonium vulgare* (Glauco–Puccinellietalia)	Ketner (1972)
	UW, IN more frequently	*Puccinellia maritima*, *Plantago maritima* (Glauco–Puccinellietalia)	Ketner (1972)
Moist and mesic sites			
Moor House N.N.R., Northern Pennines	UW	*Calluna vulgaris*, *Eriophorum vaginatum*, *Sphagnum* (Sphagnetalia magellanici, Erico–Sphagnion)	Heal & Perkins (1976), Bradshaw & Jones (1976)
(United Kingdom)	UW	*Agrostis tenuis*, *Festuca ovina* (Arrhenatheretalia)	Heal & Perkins (1976), Bradshaw & Jones (1976)
Llyn Llydaw-Snowdonia (United Kingdom)	UW	*Agrostis tenuis*, *Festuca ovina* (Arrhenatheretalia)	Heal & Perkins (1976), British Ecological Society Summer Meeting (1968)
Pin au Haras (France)	UW	*Poa trivialis*, *Lolium perenne*, *Holcus lanatus* (Arrhenatheretalia)	Ricou (1972, 1973, 1974), Masclet & Duval (1972), Hédin (1973)
Saint Martin-de-Boscherville (France)	UW, seasonally IN	*Agrostis stolonifera*, *Phleum pratense*, *Filipendula ulmaria* (Molinietalia)	Ricou & Douyer (1976)
Fenno-Scandinavian Province – maritime climate			
Mesic sites			
Tvärminne (Finland)	None	*Helictrotichon pubescens* (Arrhenatheretalia)	Gyllenberg (1969)
	None	*Calamagrostis epigeios* (Arrhenatheretalia)	Gyllenberg (1969)
	—	*Geum rivale*, *Helictotrichon pubescens* (Arrhenatheretalia)	Gyllenberg (1969)
	None	*Anthriscus silvestris* (Arrhenatheretalia)	Gyllenberg (1969)

Table 10.1 (*cont.*)

					Mean temperature (°C)			Soil characteristics			
Locality (country)	Latitude (N)	Longitude (E)	Altitude (above sea level) (m)	Annual precipitation (mm)	Warmest month (July, August[b])	Coldest month (January)	Annual	Soil type	Soil acidity pH/H_2O (KCl, $CaCl_2$) A_1–C horizon	Organic matter (%)	C/N
Central European Province – intermediate maritime–continental climate											
Wet and moist sites											
Lanžhot (Czechoslovakia)	48°50′	17°00′	155	550	+19.8	−1.5	+9.5 (1901–50)	Swampy alluvial semigley soil	6.0–6.8	10.7	29.0
	48°50′	17°00′	155	550	+19.8	−1.5	+9.5 (1901–50)	Swampy alluvial semigley soil	5.5–7.1	10.2	—
Ispina (Poland)	50°06′	20°22′	195	708	+18.2	−5.0	+7.8 (1966–70)	Swamp soil (peat gley)	4.8–5.9	56.1	21.1
	50°06′	20°22′	195	708	+18.2	−5.0	+7.8 (1966–70)	Swamp soil (gley)	5.5–6.7	6.1	8.4
Kampinos (Poland)	52°15′	21°00′	110	613	+18.9	−2.6	+7.8 —	Black humus soil	5.5–6.2	7.2	18.7
Moist and mesic sites											
Lanžhot (Czechoslovakia)	48°50′	17°00′	155	550	+19.8	−1.5	+9.5 (1901–50)	Alluvial semigley soil	6.4–7.2	8.4	20.0
Solling (W. Germany)	51°45′	9°40′	475	1100	—	—	+6.5 —	Oligotrophic brown soil	4.3 $CaCl_2$	—	—
Kazuń (Poland)	52°23′	20°17′	100	613	+18.9	−2.6	+7.8 —	Alluvial brown semigley soil	6.2–7.5	3.5	7.5
	52°23′	20°17′	100	613	+18.9	−2.6	+7.8 —	Alluvial gley soil	6.5–4.9	4.5	8.7
Ojców (Poland)	50°12′	19°48′	315	796	+15.6	−5.9	+6.4 (1965–7)	Brown-warp soil	7.5–7.7	3.1	9.0
				625	+18.5	—	ca. +8.0 (1949–58)				
Ispina (Poland)	50°06′	20°22′	196	708	+18.2	−5.0	+7.8 (1966–70)	Brown-warp soil	6.2–5.7	6.5	5.2
Mesic-dry and dry sites											
Lanžhot (Czechoslovakia)	48°50′	17°00′	155	550	+19.8	−1.5	+9.5 (1901–50)	Alluvial gleyic soil	6.5–7.5	5.2	16.0
Ojców (Poland)	50°12′	19°48′	370	796	+15.6	−5.9	+6.4 (1965–7)	35 cm deep rendzina soil	7.1–7.6	5.3	10.4
				625	+18.5	—	ca. +8.0 (1949–50)				
Skowronno (Poland)	50°36′	20°29′	290	730	+18.5	−3.5	+7.5 (1964–7)	Shallow rendzina soil	7.7–8.1	5.3	10.0

Locality (country)	Influence of underground water (UW) and/or inundation (IN)	Plant composition – dominants and/or other important species (Phytocoenological classification according to Braun–Blanquet school)	Literature (main sources)
Central European Province – intermediate maritime–continental climate			
Wet and moist sites			
Lanžhot (Czechoslovakia)	UW, great influence of IN	*Glyceria maxima* (Phragmitetalia)	Balátová-Tuláčková (1966), Rychnovská (1972), Úlehlová (1973*a*)
	UW, great influence of IN	*Phalaris arundinacea* (Magnocaricetalia)	Balátová-Tuláčková (1966), Rychnovská (1972), Úlehlová (1973*a*)
Ispina (Poland)	UW, great influence of IN	*Equisetum limosum*, *Iris pseudacorus*, *Acorus calamus* (Phragmitetalia)	Baradziej (1974), Kotańska (1975)
	UW, great influence of IN	*Carex vesicaria*, *Carex gracilis* (Magnocaricetalia)	Baradziej (1974)
Kampinos (Poland)	UW, seasonally IN	*Carex fusca*, *Carex panicea*, *Deschampsia caespitosa* (Molinietalia)	Traczyk & Czerwiński, in Breymeyer (1971)
Moist and mesic sites			
Lanžhot (Czechoslovakia)	UW, great influence of IN	*Alopecurus pratensis*, *Symphytum officinale* (Molinietalia)	Balátová-Tuláčková (1966), Rychnovská (1972), Úlehlová (1973*a*)
Solling (W. Germany)	None	*Festuca rubra*, *Agrostis tenuis*, *Trisetum flavescens* (Arrhenatheretalia)	Speidel & Weiss, in Ellenberg (1971)
Kazuń (Poland)	UW!	*Dactylis glomerata*, *Festuca pratensis* (Arrhenatheretalia)	Traczyk & Czerwiński, in Breymeyer (1971)
	UW!	*Festuca pratensis*, *Holcus lanatus* (Arrhenatheretalia)	Traczyk & Czerwiński, in Breymeyer (1971)
Ojców (Poland)	Occasionally UW (IN)	*Alopecurus pratensis*, *Festuca rubra*, *Dactylis glomerata*, *Trisetum flavescens*, *Arrhenatherum elatius* (Arrhenatheretalia)	Karkanis, in Medwecka-Kornaś (1967), Kotańska (1970), Medwecka-Kornaś (1967), Jankowska (1975)
Ispina (Poland)	Occasionally UW (IN)	*Arrhenatherum elatius*, *Centaurea jacea* (Arrhenatheretalia)	Kotańska (1975)
Mesic-dry and dry sites			
Lanžhot (Czechoslovakia)	UW, rare influence of IN	*Festuca sulcata*, *Sanguisorba officinalis* (Molinietalia)	Balátová-Tuláčková (1966), Rychnovská (1972), Úlehlová (1973*a*)
Ojców (Poland)	None	*Brachypodium pinnatum*, *Agrimonia eupatoria* (Festucetalia valesiacae)	Kotańska (1970), Jankowska (1975)
Skowronno (Poland)	None	*Brachypodium pinnatum*, *Salvia pratensis* (Festucetalia valesiacae)	Jankowska (1975)

Table 10.1 (*cont.*)

Locality (country)	Latitude (N)	Longitude (E)	Altitude (above sea level) (m)	Annual precipitation (mm)	Mean temperature (°C): Warmest month (July, August[b])	Mean temperature (°C): Coldest month (January)	Mean temperature (°C): Annual		Soil characteristics: Soil type	Soil acidity pH/H_2O (KCl, $CaCl_2$) A_1–C horizon	Organic matter (%)	C/N
Pontic-Pannonian Province – continental character of climate												
Dry sites												
Csévharaszt (Hungary)	47°17′	19°24′	135	515	+21.0	−2.4	+10.3	(1968–71)	Slightly humic sandy soil	7.2–7.8	0.9	—
Ujszentmargita (Hungary)	47°30′	21°00′	100	527	+21.4	−2.8	+9.8	—	Solonetz soil	—	—	—
South Siberian Province – extremely continental climate												
Wet and moist sites												
Baraba, Karachi (USSR)	55°17′	82°36′	107	438	+18.7	−20.5	+0.2	(1968–72)	Peat-boggy saline soil	5.9–8.8	Peat	—
Mesic (fresh) sites												
Baraba, Karachi (USSR)	55°17′	82°36′	107	438	+18.7	−20.5	+0.2	(1968–72)	Solonetz soil	6.2–9.3	10.3	—
	55°17′	82°36′	107	438	+18.7	−20.5	+0.2	(1968–72)	Saline-alkali chernozem	8.9–8.4	16.0	—
	55°17′	82°36′	107	438	+18.7	−20.5	+0.2	(1968–72)	Crustal solonetz	9.3–9.6	13.8	—
Dry sites												
Baraba, Karachi (USSR)	55°17′	82°36′	107	438	+18.7	−20.5	+0.2	(1968–72)	Chernozem	7.7–8.1	3.9	—
East Asiatic region												
Japanese–Korean Province – maritime, mountain climate												
Mesic (fresh) sites												
Kawatabi farm (Japan)	38°44′	140°15′	300–600	2335	+22.5	−1.9	+9.8	(1959–64)	Acidic volcanic ashes	5.1–5.4	27.7	12.6

Locality (country)	Influence of underground water (UW) and/or inundation (IN)	Plant composition – dominants and/or other important species (Phytocoenological classification according to Braun–Blanquet school)	Literature (main sources)
Pontic-Pannonian Province – continental character of climate			
Dry sites			
Csévharaszt (Hungary)	None	*Festuca vaginata*, *Koeleria glauca* (Festucetalia vaginatae)	Kovács-Láng (1970, 1974), Kovács-Láng & Szabó (1971)
Ujszentmargita (Hungary)	None	*Festuca pseudovina*, *Artemisia maritima* ssp. *monogyna* (Puccinellietalia)	Précsényi (1969), Bodrogközy (1965)
South Siberian Province – extremely continental climate			
Wet and moist sites			
Baraba, Karachi (USSR)	UW	*Calamagrostis neglecta*, *Carex gracilis*	Bazilevich (1974)
Mesic (fresh) sites			
Baraba, Karachi (USSR)	UW, extremely deep	*Agropyrum repens*, *Artemisia pontica*	Bazilevich (1974)
	UW, extremely deep	*Calamagrostis epigeios*, *Artemisia rupestris*	Bazilevich (1974)
	UW, extremely deep	*Puccinellia tenuifolia*, *Puccinellia distans*	Bazilevich (1974)
Dry sites			
Baraba, Karachi (USSR)	UW, extremely deep	*Phleum phleoides*, *Koeleria gracilis*	Bazilevich (1974)
East Asiatic region			
Japanese–Korean Province – maritime, mountain climate			
Mesic (fresh) sites			
Kawatabi farm (Japan)	None	*Miscanthus sinensis*, *Pteridium aquilinum* var. *latiusculum*	Numata (1975)

Plate 10.1. Semi-natural pasture dominated by *Poa trivialis* and *Lolium perenne* at Pin-au-Haras in France. (Photograph by G. Ricou.)

water, both of which influence species composition by their dynamic character. The supply of mineral nutrients is chiefly determined by the chemical composition of soil parent materials, and, in alluvial meadows, by the eutrophy of flood waters; it is also affected by agricultural practices in adjacent cropland.

In this part of the volume we will consider the results of research on temperate semi-natural grasslands that was undertaken as part of IBP contributions of 10 countries and concerning 34 types of vegetation. Details concerning these are presented in Table 10.1. All areas are situated in the northern hemisphere (in Europe and Asia) between 4° W and 141° E longitude and 38° and 60° N latitude. Most of them belong to the Euro-Siberian region, which consists of the following divisions: Atlantic-European Province with oceanic climate, Central European Province with climate of the transitional (oceanic–continental) type, Pontic-Pannonian Province with climate approaching the continental type, and South Siberian Province with characteristic continental climate.

The Atlantic-European Province is represented by three types of maritime halophilous meadows in wet to moist sites (the Netherlands), which do not freeze during winter. Precipitation is adequate (more than 720 mm annually) and the mean temperature is about 10 °C. Similar climatic conditions exist in northern France (Plate 10.1), where two types of grassland vegetation were

Plate 10.2. The moist meadow ecosystem in early summer at Lanžhot, Czechoslovakia, dominated by *Alopecurus pratensis*. This type is inundated regularly in spring and occasionally in summer. (Photograph by E. Balátová-Tuláčková.)

studied. In addition, attention is drawn to a blanket bog in northern England and to sheep-grazed grasslands in Wales and northern England, where the air temperatures are lower (5–7.5 °C) than mentioned above and the precipitation higher (1900–3800 mm). The soils have lower pH values.

The Fenno-Scandinavian Province is represented by one locality situated on a slope along the southern coast of Finland, with four types of mesic meadows. The low annual mean temperature (+4.4 °C) as well as the January mean (−5.4 °C) indicate the cool character of the climate. The amount of precipitation is almost 700 mm and the soil is acid to slightly acid.

In the Central European Province 14 types of meadow vegetation were studied. Four of these are situated in river alluvium in Czechoslovakia

Plate 10.3. The moist *Arrhenatherum elatius* meadow in the Ojców area of Poland; mown in the foreground, unmown in the background. (Photograph by Z. Denisiuk, reprinted from Medwecka-Karnas, 1967.)

(Plate 10.2), within an area of dry climate, where low rainfall (550 mm, mean temperature 9.5 °C) is compensated for to various degrees by flood water. Nine of the types studied are situated in the central and southern parts of Poland (mean precipitation varying between 600 and 750 mm, mean temperature between 7° and 8 °C) and range in character from swamps to very dry meadows (Plate 10.3). One study site is situated in a submontane area of West Germany (Plate 10.4), with moist slopes (mean annual precipitation is 1100 mm, mean temperature 6.5 °C). In most of these study sites the soil is slightly acidic to alkaline in reaction.

The Pontic-Pannonian Province is represented by two projects in Hungary, a sandy steppe and a halophilic pasture, both of dry type. Mean air temperature is higher than in the preceding locations (about 10 °C), while precipita-

Plate 10.4. The mesic *Festuca rubra* meadow studied in the Solling project in West Germany. (Photograph by H. Heller.)

Plate 10.5. A general view of the *Miscauthus sinensis* grassland (above the present tree line) on the Kawatabi Farm in Japan. (Photograph supplied by M. Numata.)

tion is lower (about 520 mm). The soil is slightly alkaline; there is no suggestion of any influence from under-ground water.

The vegetation studied in the South Siberian Province is situated in a forest-steppe region. The climate is characteristically continental (hot summer, extremely long and severe winter with deeply frozen soil). The soils are halic; the under-ground water table directly influenced one type of vegetation only. The low annual mean temperature (0 °C) is due to long, cold winters. Precipitation is lower than in Hungary (438 mm). The soils are rich in sodium sulphate and basic in reaction.

The Japanese Project is situated in the East Asia Region, in the Japan–Korea Province, in a maritime area (Plate 10.5). The amount of precipitation is very high (2335 mm annually); the mean air temperature is 9.8 °C. The reaction of soil is slightly acidic.

Scientific names of plants referred to in chapters 10 to 15 are according to the European flora by Rothmaler (1970) or the flora of the USSR (Flora URSS, 1964).

11. Primary producers in meadows

M. NUMATA

The structure of a plant community may be assessed, in relation to primary production, by the stratified clip technique (Monsi & Saeki, 1953). The profiles of the photosynthetic and non-photosynthetic systems that are obtained in this way reflect the productive structure, since the distribution of leaves and micro-environmental factors (such as light intensity, temperature and carbon-dioxide concentration) regulate photosynthesis (Monsi, 1968). There are two main types of productive structure in meadow vegetation: the herb-type (e.g. the *Chenopodium album* var. *centrorubrum* community) and the grass-type (e.g. the *Pennisetum japonicum* community) (Monsi & Saeki, 1953). Foliage is best developed in the upper part of the herb-type community and in the lower part in the grass-type system. The vertical structure of the four meadow stands in Czechoslovakia is shown in Fig. 11.1 at the time of first cut and just before the second cut (Rychnovská *et al.*, 1972). The profiles were different because of the difference in species composition and in the ratio between grasses and dicotyledons.

Various indices have been suggested to express the leafiness of plant communities. The most common of these is leaf area index (LAI). In the four Czechoslovakian meadow stands referred to above, LAI ranged from 2.1 to 4.3 at the time of first cut and 1.3 to 8.7 before the second cut (Fig. 11.1). However, in a Japanese meadow the maximum value reached was 3.3 (Table 11.1) (Koike & Yoshida, 1969). A second measure is leaf area ratio (LAR) (leaf area related to dry weight of shoots), which also reflects the productive structure of communities (Table 11.2). A third measure, green area index (GAI) has been proposed for grasslands as the sum of the upper surface of the flatly spread out green leaves and the projection area (shadow area) of the three-dimensional green plant parts (Geyger, in Ellenberg, 1971). GAI of unfertilised unmown meadow plots in Germany was 4.8 to 5.7. When a weight/surface ratio was applied to this index, values of 5 to 7 mg/cm^2 were obtained. This weight/surface ratio reflects the degree of hygromorphism or xeromorphism of a meadow stand and is modified by the state of maturity of individual plants.

The horizontal structure, as well as the vertical structure, influences primary production. To elucidate the relationship between contagious distribution and production, pattern analysis was applied to a reed stand (Ondok, in Dykyjová, 1970). Clustering indicated departure from the Poisson distribution in density and seemed to be correlated with the amount of biomass

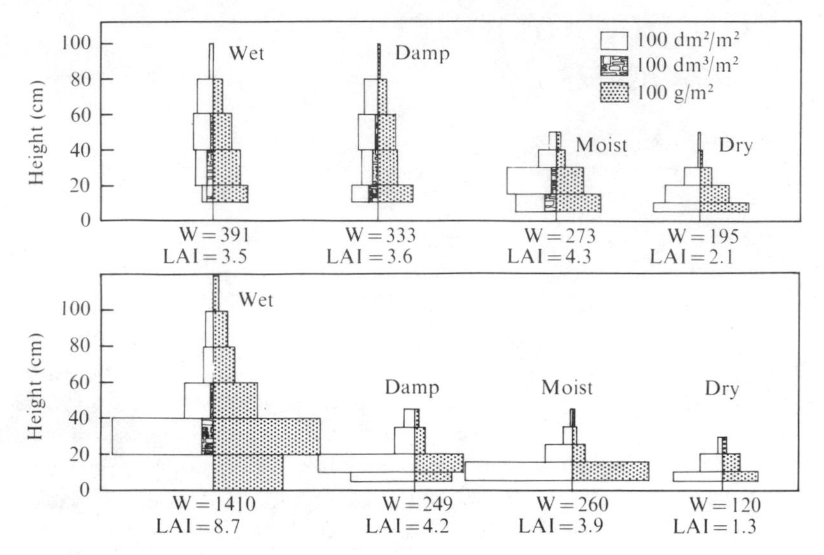

Fig. 11.1. Vertical structure of four meadow stands in Czechoslovakia. W, weight in g/m²; LAI, leaf area index. Top frame, 3 June 1970; lower frame, 13 Aug. 1970. □, leaf area; ▦, stem area; ▩, dry wt (leaves+stems). (After Rychnovská, 1972.)

Table 11.1. *Leaf area index of a high grass meadow in Japan*

	June	July	Aug.	Sept.	Oct.
Meadow as a whole	0.6	2.9	3.3	2.9	2.2
Dominant (*Miscanthus sinensis*)	0.4	2.4	2.8	2.2	2.1

Table 11.2. *Leaf area ratio (dm²/g) of four inundated meadow stands in Czechoslovakia*

Stand	First harvest	Second harvest
Dry	1.10	1.13
Moist	1.57	1.53
Damp	1.10	1.71
Wet	0.90	0.61

and litter. Some indices of dispersion (Greig-Smith, 1964) measure the homogeneity of meadow vegetation which is related, in general, to the degree of dominance of the major species and the amount of biomass (Numata, 1969). However, grasses of clumped growth habit with short rhizomes produce less homogeneous stands (Shimada *et al.*, 1973). The lower the degree of homogeneity of a meadow stand, the less precise will be the estimate of standing crop obtained per unit of sampling input.

As a structural characteristic of meadows, the index of diversity based on Shannon's formula has been applied. There exists a negative correlation between the index of diversity and the relative weight (%) of the dominant (Précsényi, 1969). Meanwhile, relative dominanace of species like the summed dominance ratio (SDR) (Numata, 1966) is used effectively as a non-destructive (non-harvesting) technique for estimating above-ground biomass (Shimada & Numata, 1971).

As thc basis of productivity studies, phenological observations are important in determining the effective growing season (Lieth, 1970) and the appropriate date for important biological events. In the Japanese IBP site, the phenology of about 100 grassland species was observed for seven years (Sugawara, Iizumi & Shimada, 1964). The growth of plants on the southern slope is better than that on the northern slope until the last 10 days of June, after which the reverse is true. However, the flowering time of the same species is similar on southern and northern slopes.

Comparisons of productivity of different meadow sites should be made in relation to their relative successional status. For this purpose Numata (1969) has devised a formula for determining the degree of succession. He has found that the biomass dynamics of meadow communities and of dominant grasses in these communities is closely related to degree of succession. Future comparisons of productivity, in such subclimax situations as are afforded by the meadow environment, should involve measurements of degree of succession.

In the studies reviewed here of primary production of semi-natural grasslands under temperate climate, the data were partly obtained under unmown conditions; that is, the vegetation was protected from mowing for the duration of study, even though such grasslands are usually maintained under the influences of mowing and, sometimes, grazing. In some instances the effect of clipping was studied, but the data presented were collected under unmown conditions.

Biomass

Above-ground parts

Above-ground biomass was measured frequently during the growing season. Values are summarised in Table 11.3. The absolute values of biomass depend on the latitude and altitude, soil fertility, weather and amount of soil moisture. An example of the Japanese site is shown in Table 11.4. In the example, the peak is located at the end of August or the beginning of September. However, two peaks of biomass have been recorded under certain weather conditions in the USSR sites (Titlyanova, 1971). As an example of salt-marsh communities, the ungrazed Netherlands stand dominated by *Puccinellia maritima* had the

Table 11.3. *Standing crop biomass in g/m² in the sites studied (compiled by M. Rychnovská and E. Balátová-Tuláčková)*

Country	Site	Dominant taxon/taxa	Above ground					Under ground (maximum or mean)	Years observed
			Live (green)		Dead				
			Spring minimum	Summer maximum	At green minimum	At green maximum	Maxi-mum		
Czechoslovakia	Lanžhot	*Festuca sulcata*	52	252	131	112	164	1556	1966–71
		Alopecurus pratensis	81	413	65	93	124	1221	1966–71
		Phalaris arundinacea	90	469	62	90	186	1664	1969–71
		Glyceria maxima	97	948	83	183	639	2640	1966–71
Finland	Tvärminne	*Helictotrichon pubescens*	—	468	—	—	—	558	1966–9
		Calamagrostis epigeios		(incl.					
		Geum rivale		dead)					
German Federal Republic	Solling	*Festuca rubra*							
		unfertilised	—	316	148	8	148	946	1968–70
		fertilised P, K	—	583	162	6	162	872	1968–70
		fertilised N, P, K	—	808	168	5	168	738	1968–70
Hungary	Csévharaszt	*Festuca vaginata*	139	172	177	185	234	863	1968–71
		Koeleria glauca						(1971)	
		Lichens							
	Újszentmargita	*Artemisia maritima* ssp. *monogyna*	430	787	3	14	121	1598	1966–71
		Festuca pseudovina						(1967)	
Japan	Kawatabi	*Miscanthus sinensis*	—	663	—	—	—	1312	1967
			347	695	—	—	—	1079	1973
	Tonomine	*Miscanthus sinensis*	370	699	—	—	—	1510	1973
	Aso-Kuju	*Miscanthus–Pleioblastus*	733	1306	—	—	—	2195	1971
	Nanashigure	*Zoysia japonica*							
		ungrazed	108	284	51	84	110	759	1973
		[illegible]	[illegible]	178	45	90	90	1003	1973

		Juncus gerardii	33	302	159	123	190	1813	1967–70
		Plantago maritima	122	438	188	167	314	2522	1968–70
		Puccinellia maritima	63	407	35	88	206	—	1967–70
Poland	Kazuń	*Dactylis glomerata*	133	435	175	91	586	—	1968
		Festuca pratensis	—	426	—	62	—	—	1968
	Kampinos	*Carex fusca*	27	196	313	241	394	823	1968
	Ojców	*Alopecurus pratensis*							
		mown	40	291	99	96	291	1664	1966–7
				(1966)		(1966)	(1966)	(1966)	
		unmown	32	332	372	208	398	—	1965–8
				(1965–7)		(1965–7)	(1965–7)		
		Brachypodium pinnatum	54	420	700	512	700	3212	1966
	Skowronno	*Brachypodium pinnatum*	13	282	344	244	378	—	1965–7
				(1965–6)		(1965–6)	(1965–6)		
	Ispina	*Arrhenatherum elatius*	—	383	—	—	—	1655	1969–71
		Carex vesicaria	32	389	164	155	203	—	1969–70
		Equisetum limosum	29	562	209	244	416	4707	1969–70
								(1969–71)	
USSR	Karachi	*Phleum phleoides*	41	220	373	254	426	2350	1968–72
		Festuca pseudovina	40	180	286	216	290	2740	1968–72
		Artemisia pontica							
		Calamagrostis epigeios	12	240	472	349	490	2610	1968–72
		Galatella biflora							
		Puccinellia distans	16	170	203	184	329	1580	1968–72
		Calamagrostis neglecta	8	290	615	534	650	14360	1968–72

Table 11.4. *Seasonal fluctuations in above-ground biomass (g/m²) in a* Miscanthus sinensis *meadow in 1967 (after Koike & Yoshida, 1969)*

	Green			Standing dead			
Date	Dominant	Others	Total	Dominant	Others	Total	Litter
1 June	73	9	82	1	—	1	1407
15 July	282	55	337	10	—	10	1218
15 Aug.	425	72	497	35	1	36	1164
2 Sept.	464	63	527	42	7	49	1504
16 Sept.	453	64	517	57	8	65	1381
16 Oct.	425	28	453	95	5	100	1373

maximum biomass (one peak) in August or September which ranged from 335 to 414 g/m² of live biomass (Ketner, 1972). An *Artemisia–Festuca* meadow in Hungary had a rather high peak of 787 g/m² at the end of June (Précsényi, 1969). The peak biomass was recorded at the end of July or the beginning of August in two meadows in Poland (Table 11.5).

In some studies fluctuations in shoot biomass were measured for three years or more. For example, in the Japanese site the maximum biomass recorded in the *Miscanthus sinensis* site ranged from 568 to 768 g/m². Annual fluctuation in total above-ground biomass in this community was relatively small, because of the stable production of the dominant, which provided 75 % of the total biomass (Shimada *et al.*, 1973). In four inundated meadow communities in Czechoslovakia there was a gradient in biomass from the

Table 11.5. *Maximum standing crop in three meadow stands in Poland (after Traczyk, in Breymeyer, 1971)*

	MS[a]		KI[b]		KII[c]	
	g/m²	%	g/m²	%	g/m²	%
Mosses	280	39	—	—	—	—
Green herbs	196	27	435	83	426	87
Dead herbs	241	34	91	17	62	13
Total	717	100	526	100	488	100

[a] MS – Strzeleckie meadows in the Kampinos Forest, unmown; dominants are *Carex fusca* and *Deschampsia caespitosa*.

[b] KI – Kazuń meadows, mown two to three times a year; dominants are *Dactylis glomerata* and *Festuca pratensis*.

[c] KII – Kazuń meadows, mown two to three times a year; dominants are *Festuca pratensis* and *Holcus lanatus*.

least productive (dry) meadow to the most productive (wet) meadow. The maximum biomass of the unmown wet meadow varied between 850 and 1250 g/m² in different years, while that of the dry, mown meadow ranged from 180 to 320 g/m².

Under-ground parts

Considerable fluctuation occurred in the standing crop of under-ground plant parts in all sites, the peaks being reached at various times in different years and in different sites. Three years of measurement in the German site (Speidel, 1973) showed peaks in October, April and June. In two Polish sites, for the soil profile as a whole maximum values were obtained in summer and the minimum in November in two sites, while in the third the maximum value was in November and the minimum in June. However, these seasonal changes were different in different layers below the soil surface (Kotańska, 1968). In the USSR sites two or three maxima (spring, summer and autumn) were observed. In salt-marsh communities in the Netherlands there was a gradual increase from May to October, so that peaks in biomass were reached in September to November.

Values given for biomass of under-ground plant parts usually do not dis-

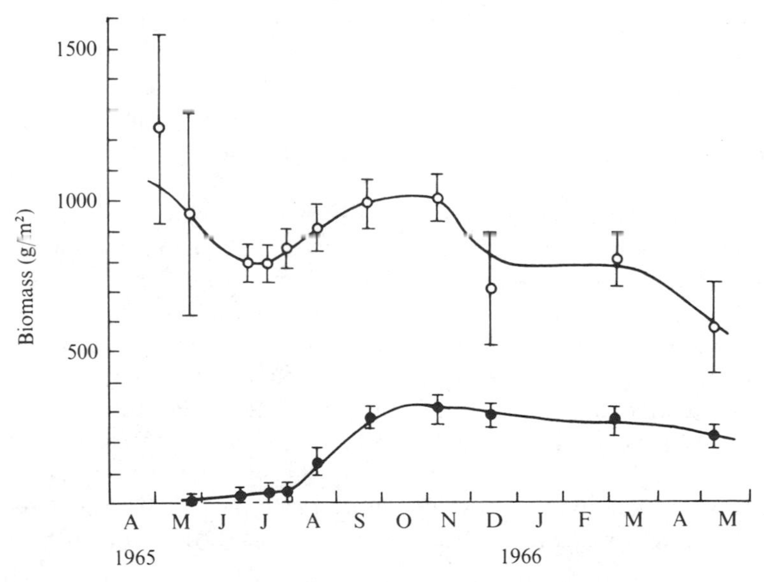

Fig. 11.2. Changes with time of standing crop (in g/m²) of rhizomes in a stand of *Miscanthus sacchariflorus*. Each value is the mean of 10 to 12 measurements. The fiducial range of each measurement is also shown with a bar. The upper curve is for old rhizomes and the lower curve for new ones. (After Mutoh *et al.*, 1968).

tinguish between tissues developed in the current year and those (both living and dead) that have accumulated from growth in previous years. However, in the Japanese study (Mutoh *et al.*, 1968) it was found possible to distinguish new rhizomes from old rhizomes on the basis of their volume in a unit of soil surface (cm^3/m^2) calculated from the average dry weight of rhizomes (g/m^2) and the bulk density (g/cm^3). The standing crop in volume of new rhizomes (100 cm^3/m^2) was found to be about one-third that of old rhizomes (2000 cm^3/m^2). Biomass of old rhizomes ranged from 770 g/m^2 in early July to 1000 g/m^2 in early November; new rhizomes appeared in late May, ceased growth in late September, and reached their maximum biomass (310 g/m^2) in early November (Fig. 11.2). The high value for the *Calamagrostis neglecta* community at the Baraba, Karachi site (Table 11.3) is explained on the basis that this community is a drying fen.

Shoot/root ratio

The relative degree of development of above-ground parts and under-ground parts is affected by interspecific and intraspecific factors, as well as by environment. In the USSR sites (Titlyanova, 1971) the shoot/root ratio is highest (0.12–0.25 on a biomass basis) in meadows of medium moisture supply and lower in drier solonetz steppe (0.06) or very moist fen (0.05–0.07).

In the *Miscanthus* meadows in Japan, which are medium in moisture content, one group of researchers estimated the shoot/root ratio to be 0.57 at the time of maximum above-ground biomass (Shimada *et al.*, in Numata, 1975), but another group made an estimate of 0.16. This great difference may be caused by sampling methods and techniques in separating live and dead roots, rhizomes and shoot bases, as well as by the density of *Miscanthus* clumps and the difference of subassociations.

In the four inundated meadow communities in Czechoslovakia (Petřík, in Rychnovská, 1972), the share of root biomass in the total maximum biomass was 60–80 % (shoot/root ratio: 0.67–0.25). Among them, the highest root biomass (3828 g/m^2) was found in the community of *Glyceria maxima* (wet meadow) and the lowest was in the community dominated by *Alopecurus pratensis* (moist meadow). In the salt-marsh communities in the Netherlands (Ketner, 1972) shoot/root ratios range from 0.15 to 0.28. In the Hungarian site (Préscényi, 1969) the shoot/root ratio was 0.17–0.38, and the percentage of 'roots' in the total biomass was referred to as the 'root importance index'.

In artificial (hydroponic) cultures of *Phragmites communis* the shoot/root ratio was 1.05 after one year and 0.38 after three years (Dykyjová, 1970).

Primary production

Maximum standing crops of green shoots ranged from 170 to 1306 g/m² of biomass in the various sites studied. The intensity of effort made to determine annual above-ground production ranged from estimating maximum standing crop to the summation of detailed estimates of increments for several periods during the season. For example, in some of the Polish stands (Jankowska, in Medwecka-Kornás, 1967), where the maximum standing crop was only 385 g/m² of dry matter, the summation of monthly increments of growth was 874 g/m². The greatest increment in one month (May) was estimated to be 205 g/m², while the second greatest monthly increment was 191 (in October). Two of the Polish meadows (at Kazuń) were mown successively on 20 June, 28 August and 5 November, 1968, with harvests of 435, 111 and 18 g/m² for one meadow and 426, 130 and 29 g/m², respectively, for the second. The total production for the season, as estimated in this way, was, thus, 30 and 37 % greater than the best single harvest (Traczyk, in Breymeyer, 1971).

The estimation of incremental changes in biomass of under-ground plant parts is more difficult than for shoots. However, if an estimate of net primary production is required, these under-ground changes must be considered. In three of the Polish sites (Kotańska, 1968) the annual increment of under-ground biomass was estimated to range from 507 to 1367 g/m². Iwaki & Midorikawa (1968) estimated increments by distinguishing rhizomes and tubers (including roots) that were formed in the current year from those formed earlier; they measured changes in weight of each group and were able to rationalise these in estimating annual increments.

In some instances the biomass produced was partitioned according to its destination. For example, in the Hungarian site 64 % of the photosynthates was estimated to have been translocated to under-ground parts; of the material retained in shoots 40 % was transferred to the litter layer before the end of the growing season (Précsényi, 1969).

Consideration was given in some of these meadow studies to the difference between gross production and net production. For example, in Japan, Iwaki, Monsi & Midorikawa (1966) estimated community respiration rates from seasonal fluctuations in weight of various plant organs. The value for respiration, thus obtained, was equivalent to 2207 g/m² of biomass in a situation where net production was estimated by the harvest method to be 1778 g/m². Thus, gross production was 24 % greater than net production. They partitioned plant respiration as follows: leaves, 39 %; stems and flowers, 44 %; and under-ground parts, 17 %. The ratio of respiration rate to gross photosynthesis was found to fluctuate seasonally, so that net production was greatest in July and gross production in August. In the Hungarian site gross photosynthesis was estimated to be 25 % greater than net photosynthesis, which was 702 g/m² of dry matter (Précsényi, 1969).

In some studies production data have been reported in terms of rates of energy flow or of flow of nutrients. Thus, in the dry meadow in Czechoslovakia the annual above-ground production contained 5243 kJ/m^2, while that of the mesophytic meadow was 3171 kJ/m^2 since it suffered from prolonged flooding (Jakrlová, in Rychnovská, 1972). In the Czechoslovakian stands the concentration of hydrolysable glycids was highest before the first harvest. The highest average amount of glycids was present in the wet meadow. The highest concentration of nitrogen was found in young shoots of the wet and moist meadows at the beginning of the growing season. The highest average amount of nitrogen was present in the communities dominated by *Alopecurus* and *Phalaris*. In all four communities, both biomass and the contents of valuable nitrogen substances and glycid reserves increased with increasing supply of water (Rychnovská, 1972). In the German site, crude ash, crude fibre and crude protein in above-ground parts, and soluble carbohydrate, crude protein and crude fat in under-ground parts were measured to determine the agriculturally utilisable production (Speidel, 1973).

Rates of turnover have been determined for some sites. For example, in the Hungarian site the turnover rate of above-ground parts was estimated to be 50 % per annum and of under-ground parts about 40 %, whereas for litter (and spring ephemerals) it was nearly 90 % (Précsényi, 1969). Thus, the above-ground biomass is replaced every two years, while about 2.5 years is required for turnover of under-ground parts. In the Netherlands salt-marsh communities turnover rate of plant parts in the surface layer of soil ranged from 17 to 39 % annually, while in subsoil values of 24–59 % were obtained (Ketner, 1972). In the USSR sites, live-root biomass was found to be 1.5–3 times that of dead roots, with the exception of the fen grassland where dead roots were several times as abundant as live roots. These data show rather rapid decomposition of dead roots in steppe and meadow, but slow decomposition in fen.

There is a positive correlation between primary productivity, energy flow and efficiency of energy capture, and these three processes have an inverse correlation with turnover time (Précsényi, 1973). Efficiency of energy capture was found to be from 0.08 to 0.41 % in two Hungarian stands. In the Netherlands salt-marsh communities annual values of from 0.44 to 0.73 % have been reported based on total radiation reaching the vegetation surface and net primary production.

Saeki (1960) has developed a mathematical model of gross community photosynthesis that has yielded estimates of production in a *Solidago altissima* community that agree well with those obtained by the harvest method. This model is based on the relationship between amount of leaves, distribution of light and rate of photosynthesis.

Effect of stresses

Fire

Fire is used for management of meadows throughout the world. They are usually burned in winter or early spring to control shrubs, weeds and insect pests. In the Japanese site it was found possible to consume 90 % of the available fuel (250 to 1000 g/m^2) in the *Miscanthus* meadow, but only 27–65 % in the *Zoysia* pasture. The effect of burning on soil temperature was found to be governed by: (i) the amount of fuel burned; (ii) the amount of duff layer left unburned; and (iii) the speed of spread of fire. In the meadow, temperature rose 25 °C to 170 °C at the soil surface and 3 °C to 7 °C at a depth of 2 cm; in the pasture the respective increases were 10 °C to 100 °C and 0 °C to 2 °C (Iwanami, 1973).

Fertilisation

Additions of nutrients in the form of fertiliser usually increase herbage production in meadows. However, Rychnovská (1972) reported that, in the Czechoslovakian site, both shoot biomass and nitrogen content of shoot biomass have not been affected by fertilising or other agrotechnical treatments; they are primarily dependent on water supply. The rate of nitrogen uptake in the untreated, flooded meadow was approximately the same as that of well-fertilised upland meadows. These results emphasise the high natural fertility of the natural alluvial meadows, where animal fodder con-

Table 11.6. *Biomass of* Miscanthus sinensis *and number of species present under various mowing regimes after three years of treatment in a* Miscanthus sinensis *meadow at Kawatabi, Japan*

Treatment[a]	I	II	III	IV
Above-ground biomass[b] of *M. sinensis* g.d.w./m^2	8	86	230	662
Under-ground biomass of *M. sinensis* g.d.w./m^2	301	507	520	941
Relative dominance of *M. sinensis*	58	94	100	88
Shoot:root ratio (per cent)	2.7	17.0	44.1	70.5
Number of species/m^2	23	44	36	28

[a] Treatments
I: Mown three times per annum, in May, July and August.
II: Mown once per annum, in July.
III: Mown in July of first and third year.
IV: Control, not mown.
[b] Four replicates of each treatment were mown in September of the third year.

taining large amounts of glycids and proteins may be produced without the addition of fertilisers.

In the German Solling study, applications of NPK fertiliser increased photosynthetic rates of *Festuca rubra* per unit of dry leaf weight and surface area of leaves, but had no effect on chlorophyll content (Ruetz, 1973).

Mowing

A three-year mowing experiment at the Japanese Kawatabi site (Numata, 1970, 1975) revealed that the following changes occurred in the characteristics of the *Miscanthus* community (Table 11.6): (i) above-ground biomass, underground biomass, and ratio of shoot biomass to root biomass decreased with increasing intensity of mowing; and (ii) relative dominance of *Miscanthus* and number of species/m^2 were depressed both by very frequent mowing and by no mowing. The species present were classified into three groups, according to their response to mowing stress, as 'decreasers', 'increasers' and 'indifferent'.

12. Consumers in meadows and pastures

As in other terrestrial ecosystems, the functional importance of animals in semi-natural temperate grasslands is, in many cases, still unknown. We are aware of some of the negative influences of herbivores on primary production, and we know the structure of some important food chains and regulation mechanisms. However, we have very few quantitative data by which the magnitude of the role of a population of one or several groups of consumers in an ecosystem might be estimated and compared with that of another ecosystem. A major reason for these deficiencies is that, because of their large numbers, even the preparation of a detailed taxonomic inventory is beyond the resources of most studies, when a quantitative inventory is also one of the basic requirements in the study of consumers in an ecosystem The nature of investigations of consumers as functional components in ecosystems is limited also by deficiencies in respect of suitable methods of sampling; disagreements exist as to the most reliable methods of sampling of various faunal groups in various situations, so there is a lack of uniformity that makes inter-site comparisons difficult when they are based on the data of different investigators (Heikinheimo & Raatikainen, 1962; Andrzejewska & Kajak, 1966; Gromadzka & Trojan, 1967). Also the relative abundance of species changes from year to year, so that observations for several years are necessary for a reliable appraisal of the relative importance of species. Unfortunately, many of the studies considered in this chapter involved monitoring fauna for only one year, and in none was this done for more than a few years. Differences in composition between sites, or in different years in the same site, may be the result of differences in abiotic factors (soil, climate, weather) and biotic factors (vegetation, predators, parasites), as well as in management practices.

In discussing the fauna of semi-natural temperate grasslands, we will consider separately those areas (meadows) in which herbage is primarily removed as hay from those in which forage is removed directly by grazing of domesticated livestock.

Meadows

H. HAAS

The principal IBP investigations of above-ground invertebrates in meadows were conducted in Germany, Japan, Poland and Romania. The groups involved are mainly insects, myriapods and spiders.

Species composition

Data from three studies are summarised in Table 12.1 to indicate relative abundance of individuals of various taxa. Consideration must be given, in studying these data, to the fact that small arthropods on the soil surface, especially spring-tails and mites, are not caught by most methods and are, therefore, not listed by most authors. In general these two groups represent the main part of the faunal populations on the soil surface of meadows.

Table 12.1. *Proportion (%) of the arthropod population that was contributed by various orders in three meadow study sites. Only values greater than 2 % are given*

Site ...	Kampinos & Kazuń	Solling	Copsa Mica
Country ...	Poland	Germany	Romania
Year observed ...	1968	1971	1970
Method of collection ...	Biocenometer	Eclector	Sweep-net
Source ...	Olechowicz, in Breymeyer, 1971	Haas, 1972	Vasiliu, 1971
Orders			
Collembola	—	—	33.9
Diptera	48.0	64.9	24.1
Homoptera	35.2[a]	8.9	12.3
Hymenoptera	10.2	14.9	7.9
Heteroptera	2.3	—	2.4
Coleoptera	—	3.6	6.3
Lepidoptera	—	—	3.9
Orthoptera	—	—	4.4
Thysanoptera	—	—	3.6
Araneae	—	5.2	—

[a] Only Homoptera–Auchenorrhyncha.

The orders Homoptera, Orthoptera and Thysanoptera are represented in all meadows by groups of insects that are in all stages herbivorous. Host-specificity is common and allows the characterisation of different types of meadows according to the species of Orthoptera, Hemiptera and Homoptera that are present (Marchand, 1953). Other abundant herbivorous groups are two families of beetles, Curculionidae and Chrysomelidae.

The principal predators on the soil surface and within the canopy are beetles (Carabidae), spiders and ants. Predatoric bugs (Nabidae) and flies (Empididae, Asilidae) play a role in some meadows in certain seasons. The abundance of ants and spiders depends to a large extent on management regimes (Table 12.2).

Parasites are represented mainly by species of Hymenoptera, which are

Table 12.2. *Biomass (in mg/m²) of invertebrate predators in Polish and Japanese meadows*

	Spiders		
	On soil surface	In canopy	Ants
Poland (Kajak *et al.*, in Breymeyer, 1971)			
Strzeleckie meadows (unmown)	175	78	71
Kazuń meadows (mown)	25	5	23
Japan[a] (Nakamura, 1972)			
15 May 1970	—	—	90
7 August 1970	—	—	19
20 May 1971	—	—	49
17 August 1971	—	—	31

[a] The Japanese values are for fresh weight.

very numerous in the German IBP area (Haas, 1972). Parasitic Diptera and Strepsiptera also occur, but do not play an important role.

Of the arthropods with saprophagous feeding habits the different beetles, spring-tails and mites are abundant. For example, in the German study area, 81 different species of Staphylinidae were encountered in sampling (Hartmann, 1974). Furthermore, snails are important consumers in some meadows. These animals are adapted to ecosystems with soils of high lime content; accordingly, they are absent from the German study area, but are very abundant in the Czechoslovakian site.

The invertebrates of meadow soils include both herbivores (root-eaters) and carnivores, but those of saprophagous feeding habits are particularly abundant. Nematodes are most numerous, several millions/m² frequently occurring (Franz, 1942). Each layer of soil has its characteristic species (Nielsen, 1949), of which most live in the uppermost layer.

A marked difference in the composition and dynamics of the soil meso-fauna according to the water regime has been found in three types of alluvial meadows in Czechoslovakia. The wet meadow is characterised by a large number of larvae of Diptera, Coleoptera and highly specialised species of Collembola. The moist meadow is typified by a large number of Collembola endowed with a great capacity of reproduction, development of which takes three to four weeks. In the dry meadow the ground fauna is characterised both by species wtih longer developmental cycles (Protura, Diplura, Pauropoda, Symphylla, larvae of Elateridae) and specialised species with shorter developmental cycles. Abundance is affected by flooding. Spring floods have little effect on numbers of soil meso-fauna; summer floods, on the other hand, often cause almost 90 % reduction (Rusek, in Dykyjová, 1970).

Earthworms (Lumbricidae) and enchytraeids have an important role in

Table 12.3. *Proportion (%) of the population of soil macro-fauna contributed by various taxa in two meadow studies. Only values greater than 2 % are given*

	Japan (Nakamura, 1972)	USSR (Mordkovich & Volkovintzer)[b]
Nematoda	—	3.7
Oligochaeta	74.0	—
Formicidae	7.0	55.7
Scarabaeidae	3.4	—
Elateridae	2.3	3.4
Carabidae	—	2.6
Curculionidae	—	24.6
Hoopea	—	4.0
Aranea	2.5	—
Myriapoda	2.2[a]	7.7

[a] Only Diplopoda.
[b] In Bazilevich (1974).

mixing the soil. In those IBP areas where earthworms were studied (Germany (Graff, 1971), France (Bouché, 1972) and Japan (Nakamura, 1972)), species of the genera *Lumbricus*, *Allolobophora*, *Dendrobaena* and *Octolasium* dominate at various depths. Among the arthropods, developmental stages of beetles and flies dominate. Ants are abundant in some meadows that are not intensively managed, such as those in Poland and the USSR (Table 12.3).

Generally, the soil animals of meadows live in distinct layers. Most of them are concentrated to the uppermost 5 cm of soil; only ants and some earthworms are found deeper than 10 cm (Mordkovich & Volkovintzer, in Bazilevich, 1974).

Vertebrates apparently have a less important role than invertebrates in all meadows studied, if judged from their relatively small biomass. In some Czechoslovakian highland meadows, for example, biomass of mammals is only 6 % and birds 1.2 % of the total standing crop biomass (2.5 g/m^2) of animals (Turček, 1972). Large mammals spend only part of their time in meadow ecosystems and are an influence only through factors related to grazing. On the other hand, mice and voles are permanent residents of all meadows in which the water table is not too high to interfere with their activities. The effect on the ecosystem of these small rodents is out of proportion to their level of consumption; where they occur in high densities their effect on primary production is apparently considerable.

The birds of meadows are made up of two groups: those that reproduce here and those that only feed in this habitat. The first group must rear their young before the first cut of hay or between cuts, or heavy losses of nestlings are suffered. After mowing, the short-grass environment makes foraging easy, so that numerous residents of other ecosystems, both insect eaters and seed eaters, are attracted to meadows.

Community dynamics

Both univoltine and polyvoltine arthropods live in meadows. The production of more than one generation each year is especially common among herbivores and parasites. Wide-ranging insects often have more generations in areas of warm climate than in cool climate (Table 12.4). Thus, *Oscinella frit*, a fly with mining larvae, has one generation in northern USSR, two in the UK, three in central Europe and four in southern USSR (Tischler, 1965).

Table 12.4. *Number of generations produced by various species of Homoptera Auchenorrhyncha in various climates (after Waloff & Solomon, 1973)*[a]

Species	Finland	Germany	UK	Italy
Graphocraerus ventralis	1	1	1	—
Elymana sulphurella	1	1	1	—
Philaenus spumarius	1	1	1	—
Deltocophalus pulicaris	1	2	—	—
Javesella pellucida	1	2	2	—
Arthaldus pascuellus	1	2	2	—
Macrosteles laevis	1	2	2	—
M. sexnotatus	1	2	2	2–3
Cicadella viridis	—	—	1	3–4

[a] Sources: Finland, Germany; Kontkanen, 1954: UK; Waloff & Solomon, 1973: Italy; Olmi, 1968.

In some populations a diapause is influenced by photoperiod. In these cases high temperatures will not effect an increase in number of generations. So *Stenocranus minutus* (Homoptera: Delphacidae) when living in the laboratory under higher temperatures and normal daylight, does not lay its eggs earlier than when it lives outdoors (Müller, 1957). A mild winter, with warm days in early spring or late autumn will not influence the speed of development and the number of generations, if day length is shorter than a critical length.

While for many species the developmental stages are found in various positions within the ecosystem, holometabolic insects often change their feeding habits during their life history. For instance, the adults of parasitic Hymenoptera nearly without exception feed on pollen and nectar; the same is true for some flies with parasitic larvae. During the development of other insects the stratum of residence is changed. For example, the larval stages of the beetle, *Sitona lineata*, feed on root nodes of legumes, while the adults live on the leaves of the same plants. Some insects live as larvae near the ground, while the adults populate the canopy.

Hemimetabolic insects seldom or never change their feeding habits or stratum of residence. All stages of the most abundant orders found in meadows (Homoptera, Heteroptera, Thysanoptera, Orthoptera) are predators or live as herbivores on the same plants throughout their life cycle.

Because of their short life span, the amount of energy flow through adults is rather small, especially in those instances where the adults consume little or no food. For example, in the German meadow study, flies (of the family Sciaridae) that were caught soon after hatching already had their abdomens full of eggs; they probably die soon after laying these eggs. On the other hand there are some invertebrates that consume significant quantities of food as adults (Table 12.5).

Table 12.5. *A comparison of annual consumption (in terms of energy) of larvae and adults in three species of insects*

Species	Stage	J/m^2	Percent of total	Source
Chorthippus montanus	Larvae	49	29	Makulec (in Breymeyer, 1971)
	Adults	122	71	Makulec (in Breymeyer, 1971)
Paprides nitidus	Larvae	—	35–40	White & Watson, 1972
	Adults	—	60–65	White & Watson, 1972
Neophilaenus lineatus	Larvae	7125	75	Hinton, 1971
	Adults	2368	25	Hinton, 1971

Some animals have a role in the nutrient cycling and energy flow processes of more than one ecosystem. Sometimes these activities in a particular ecosystem are hardly measurable. A distinction should be made between those animals that are active in more than one ecosystem because of their wide range and those that move from one ecosystem to another at different stages in their development. Some animals are transported from one ecosystem to another by passive drift.

Mammals and birds are the main animals with large ranges embracing more than one ecosystem. The large herbivores among them usually remain in forests or hedges during daylight, but obtain part of their food in meadows at night. Birds which nest elsewhere are often present in large numbers in meadows. Predators of mammals and birds also often relate to more than one ecosystem.

As compared to vertebrates, the ranges of arthropods are much smaller. Accordingly, only those that dwell in the margins of meadows wander into neighbouring ecosystems. Examples of groups that live as larvae in one ecosystem and as adults in another are the water insects (Odonata, Culicidae), the adults of which constitute an import of material to the meadow ecosystem. On the other hand, some insects of the orders Hemiptera, Diptera and Thysanoptera are active as larvae in meadows but export adults to adjacent ecosystems.

The need to hibernate also causes organisms to migrate from one ecosystem to another. The greater degree of protection offered by the forest ecosystem in the German site causes beetles (*Sitona lineata* and *Apion* sp.) to move from the meadow to hibernate (Schauermann, 1973).

The hatching time of arthropods depends to a large extent on environmental conditions. While temperature and precipitation are important influences, solar radiation, wind and atmospheric pressure also influence phenology. Several seasonal aspects can be identified by means of the presence or absence of characteristic organisms (Boness, 1953; Doskočil & Hůrka, 1962; Haas, 1972). Haas (1972) found, for several taxa (Hymenoptera, Nematocera and Cecidomyidae), a positive correlation between temperature and the number of insects captured in the German site; for others (Araneae, Sciaridae and Coleoptera), no such correlation was found.

Nutrient chain

The food-chain web of the meadow ecosystem, with so many species to consider, will probably never be entirely decoded. The influence of herbivores is not only related to the amount of herbage consumed. Studies of the leafhopper (*Cicadella viridis*) have revealed that 30 % of the herbage grazed is not consumed (Andrzejewska, in Petrusewicz, 1967). Similarly, according to one study (Andrzejewska & Wojcik, 1970) Orthoptera drop to the ground 6–10 times the amount of grass herbage that they consume (the amount depending on the height of the grass), while in other studies (Breymeyer, 1971; Gyllenberg, 1972) the total amount of primary production removed by Homoptera and Orthoptera lies between 5 and 9 %. The selective feeding of herbivores may tend to stabilise the plant association, but may also promote soil erosion (Turček, 1972).

Important groups of predators that feed upon herbivores include ants, spiders, some beetles and some flies. Intensive management diminishes the activities of ants and web spiders, because their nests and nets are destroyed (Boness, 1953; Petal, 1971).

The reduction of phytophages by predatoric arthropods is considerable. At times, 90 % of emerging flies are captured, while for the total growing season the reduction is between 25 and 40 % (Kajak & Olechowicz, 1970). The degree of reduction of Homoptera-Auchenorrhyncha is of the same order (Andrzejewska, in Breymeyer, 1971).

Predators often are eaten by other insects (Breymeyer, in Petrusewicz, 1967). This means that the food web is made even more complicated. In

Polish meadows the energy flow normally goes through three or four consumer–predator connections (Petal & Breymeyer, 1969). A full appreciation of the influence of this process on the nutrient flow web is dependent on an understanding of host–parasite relationships.

Energy flow

Most of the data available concerning energy flow through consumers in meadows relates to populations of individual species of herbivores. An example of this type of data is presented in Table 12.6. Some studies have

Table 12.6. *Values for annual flow of energy (in kJ/m²) for one species of grasshopper* (Chorthippus parallelus) *in a Finnish meadow (after Gyllenberg, 1969)*

Process	1966	1967	1968	1969
Production (*P*)	25	21	17	26
Respiration (*R*)	30	27	21	36
Consumption (*C*)	—	—	133	154
Defaecation (*FU*)	—	—	96	92
Assimilation (*A*)				
$A = P+R$	55	48	38	62
$A = C-FU$	—	—	37	62
Assimilation: Consumption (%)			28	40
Production: Consumption (%)			13	17

Table 12.7. *Efficiency (%) of assimilation and of production by several species of meadow herbivores*

Species	Assimilation	Production	Source
Orchelinum fidicinum	27	10	Smalley, 1960
Different Orthoptera	37	10–13	Wiegert (in Petrusewicz, 1967)
Chorthippus parallelus			
1968	29	13	Gyllenberg, 1970
1969	40	17	Gyllenberg, 1970
Chorthippus montana			
larvae	60	22	Makulec (in Breymeyer, 1971)
adults	60	17	Makulec (in Breymeyer, 1971)
Schistocerca gregaria	35–78	—	Davey, 1954

been concerned with the efficiency of assimilation and production in relation to the amount of food consumed. The results of several such studies are summarised in Table 12.7. The data concerning a species of grasshopper in Tables 12.6 and 12.7 suggest that in more favourable years (1969 had very favourable weather conditions compared with 1968) both the rate of energy flow through populations and the efficiency of utilisation of the consumed material is higher than in unfavourable years. Both assimilation efficiency and production efficiency differ greatly between species. For cold-blooded organisms, production is an important part of energy flow, while for warm-blooded ones production efficiency is only 1–2 % (Turner, 1970). According to Welch (1968), carnivores may be expected to have higher efficiencies than herbivores, but data are insufficient to support this view. Petal (in Breymeyer, 1971) reports production efficiencies of only 2 % for ants, but Kajak (in Petrusewicz, 1967 and in Breymeyer, 1971) found values between 5 and 30 % for spiders.

Pastures

G. A. E. RICOU

Semi-natural temperate grasslands that are used for pasturing domesticated livestock can be divided into two groups: (i) those located in plains and valleys at low elevation, that are most commonly grazed by cattle; and (ii) those of mountainous areas, that are usually grazed by sheep. IBP studies were carried out in the first type of pasture in two sites in France, while attention to the second type was principally in Snowdonia in northern Wales, in the Pieniny Mountains in Poland, and in the Kawatabi site in Japan. Some relevant data from Moor House (UK), which are being synthesised as part of the Tundra biome, also are of interest.

The capacity of these grasslands to support livestock varies considerably, depending on both natural factors and management inputs. In good pastures of the first type usually one to two mature cattle are grazed per hectare for 200 days or more per year, the standing crop during the grazing season being as high as 75 g/m^2 in unfertilised areas and double where fertilisers are used (Hedin, Kerguelen & de Montard, 1972). For example, in the Pin-au-Haras study the mean standing crop of cattle was 95.1 g/m^2 and consumption of forage was 800 g/m^2 during the growing season. In good mountain pastures up to eight sheep are grazed per hectare over a shorter grazing period. For example, the mean standing crop in the Snowdonia study was 10.6 g/m^2 and annual forage consumption was 200 g/m^2 (Heal & Perkins, 1976). The proportion of herbage production that is consumed by livestock in these pastures has been calculated to be as high as 85 %, but in some is much less, particularly in bogs where apparently it is as low as 15–25 %.

Populations and biomass

The high degree of use of pastures in areas of semi-natural temperate grassland disrupts severely natural consumer populations. This is particularly the case with vertebrates. For example, in the Pin-au-Haras site in France no small mammals were captured, despite an intensive trapping programme. Birds are always present, 47 species being found at Pin-au-Haras, but many of these spend most of their life cycles in adjoining woodland. In the Japanese study site only three species nested, with a density of one individual per hectare (Numata, 1975). Invertebrates are the principal natural consumers in pastures. The composition of invertebrate communities has been influenced by past management practices. Both climatic factors and biological control mechanisms regulate the variation in speciation and density of individuals in these communities.

Typically, species of Diptera and Homoptera are the dominant herbivores above ground, while spiders are the principal predators; under ground earthworms predominate. Companion groups vary between sites, but species of Orthoptera and Lepidoptera are not common above ground, the former because of competition by cattle and the latter because of the moist climate. During dry periods the proportion of Homoptera increases in relation to Diptera, because of the increased abundance of Jassidae in late summer. In old bogs (such as the site at Saint-Martin-de-Boscherville), which are very dry in late summer, populations of Homoptera are particularly high.

The invertebrate fauna is diverse, with about 400 epigeic (above-ground) species and 300 hypogeic (under-ground) and surfacial species being found in a typical site in northwestern France. The populations of carnivores are

Table 12.8. *Mean populations (number of individuals/m²) and standing crops (g/m²) of soil macro-fauna in three pastures in northwestern France*

Site and Year	Herbivores		Carnivores		Decomposers[a] (excluding earthworms)		Total (excluding earthworms)		Earthworms	
	No.	Biomass	No.	Biomass	No.	Biomass	No.	Biomass	No.	Biomass
Pin-au-Haras										
1969	98	0.73	111	0.33	29	0.03	238	1.09		7.83
1970	81	1.21	86	0.26	13	0.03	180	1.50	188	6.65
1971	89	2.11	88	0.43	39	0.04	216	2.58	208	7.00
Saint Martin-de-Boscherville[b]										
1972	17	0.13	75	2.05[c]	10	0.02	102	2.20		
1973	30	0.63	70	6.93[c]	72	0.27	172	7.83		
La Vielle										
1955	109	3.38	83	0.15	238	0.07	430	3.60		
1956	73	3.50	68	0.15	220	0.02	361	3.67		
1957	55	2.20	128	0.16	357	0.05	540	2.41		

[a] Only macro-faunal decomposers are included for Pin-au-Haras and Saint Martin; for La Vielle Collembola on the surface are added.

[b] Sampling only in summer and early autumn.

[c] The high biomass values for carnivores at Saint Martin are associated with summer and early autumn sampling.

Table 12.9. *Intraseasonal ranges in populations (number of individuals per m²) and standing crops (g/m²) of soil macro-fauna in the French Pin-au-Haras pasture site*

Trophic level	No.		Biomass	
	1970	1971	1970	1971
Herbivores	36–138	54–127	0.24–2.32	1.02–3.86
Carnivores	30–206	33–158	0.05–0.97	0.15–0.63
Decomposers (excluding earthworms)	3–47	10–67	0.001–0.11	0.006–0.07
Total (excluding earthworms)	70–344	134–282	0.29–3.32	1.43–4.35
Earthworms	2–564	57–343	0.08–19.48	2.21–14.34
Number of sampling dates[a]	14	11		

[a] On each date of sampling 12 × 2 samples were taken.

usually balanced with the herbivores, so that populations are regulated (Tables 12.8, 12.9). However, temporary increases do occur of one functional group in relation to the other. The presence of hedges between fields tends to increase the proportion of meso-hymenopterous parasites in relation to spiders, populations of which fluctuate more and are often great even in areas where fencing replaces hedges.

The number of species of invertebrates, especially those that occur above ground, are markedly affected by weather conditions. For example, during a series of dry years (1969–71) at Pin-au-Haras the index of diversity ($\propto = S/\log N$; where S is the number of species and N is the density of individuals) of epigeic species of meso-fauna averaged only 85, while in more normal weather (1955–7) under similar climate at La Vielle it was 145. In contrast, diversity of hypogeal species was not much different, being 76 and 67, respectively.

The soil fauna of the mountain grasslands show considerable differences from those of low elevations. Slugs play a significant role in the Snowdonia site, where their mean standing crop is 0.45 g/m²; they consume 5 % as much herbage as sheep (Lutman, 1976). Six species occur, the two most abundant having a combined mean density of 26/m². At Moor House in England the soil meso-fauna function primarily as decomposers, with 75 % of the biomass of soil fauna being made up of three species (one decomposing enchytraeid and two herbivorous tipulids); Collembola is a very important group, with 38 numerous species (all decomposers); and many species (especially decomposers) of Nematoda and Acarina occur (Cragg, 1961).

The mean number of individuals of soil macro-fauna (Table 12.8) ranges from near 100 to over 500/m² (exclusive of earthworms) in the French sites. The intraseasonal changes (Table 12.9) show no readily identifiable trends, but maximum values are often many times minimum values.

The numbers of soil meso-fauna are very great. For example, at the Pin-au-Haras site, about 70000/m^2 (not including nematodes) were found to a depth of 30 cm. Of these, two-thirds were in the litter layer and one-fifth were in the uppermost 7 cm of soil. Acarina made up 58 % of this population and Collembola 32 %. The population of soil nematodes varied from a low of 15000000/m^2 to a high (during the winter) of 55000000/m^2, with corresponding biomass values of 0.4–1.5 g/m^2. In addition, there were 28000 epigeic nematodes/m^2.

Mean standing crop of invertebrates in the French sites ranges from 1 to 8 g/m^2 exclusive of earthworms, which comprise another 7–8 g/m^2 where they have been studied (Table 12.8) (Bouche, 1975). The values vary considerably throughout the year, with maximum values 10 times or more as great as minimum values (Table 12.9), so that during active periods the biomass of earthworms reaches 25 g/m^2.

In the Pieniny site in Poland biomass values for macro-fauna and meso-fauna in a pasture were determined to be: herbivores (Auchenorrhyncha), 0.0006 g/m^2; carnivores, ants (0.053 g/m^2) (Petal, 1974), spiders (0.004 g/m^2) and Diptera (0.001 g/m^2) totalling 0.058 g/m^2; decomposers, Nematoda (2.2 g/m^2), Diptera (0.006 g/m^2) and earthworms (2.611 g/m^2) totalling 4.817 g/m^2 (Kajak, 1974).

Energetics

Sufficient data are available to estimate energy flow through various parts of the pasture ecosystem at various sites.

At Pin-au-Haras the cattle (mean standing crop of 95.1 g, 1233 kJ/m^2) consume 14 226 kJ/m^2 annually, which represents 85 % of the energy contained in herbage production. Of this, 15 % is directed into tissue and milk production, 27 % to faeces and 58 % to urine and respiration. Similarly, in the Polish site 86 % of above-ground net primary production is consumed by sheep (Kajak, 1974), that is 12534 kJ/m^2 annually (Fig. 12.1).

The studies in France and Poland (Kajak, 1974) show that a considerable portion of the energy that is consumed by domesticated animals is passed onwards through the system by excretion of faeces (27 and 60 %, respectively). Macro-fauna share in further releases of this energy. In the Polish site 32 % of the energy is lost from faeces during the first month, with Scarabaeid beetles and Diptera larvae being active in the process (Fig. 12.1). In release of energy from under-ground detritus, the proportional activity of various soil fauna throughout the year has been estimated to be: Lumbricidae, 5–7 %; Acarina, 2–8 %; and Nematoda, 2–8 %. The combined soil fauna were one-fourth to one-eighth as active as micro-organisms in this function.

Secondary production by invertebrates is considerable. For example, at Pin-au-Haras an estimated 7.75 % of shoot production is consumed by in-

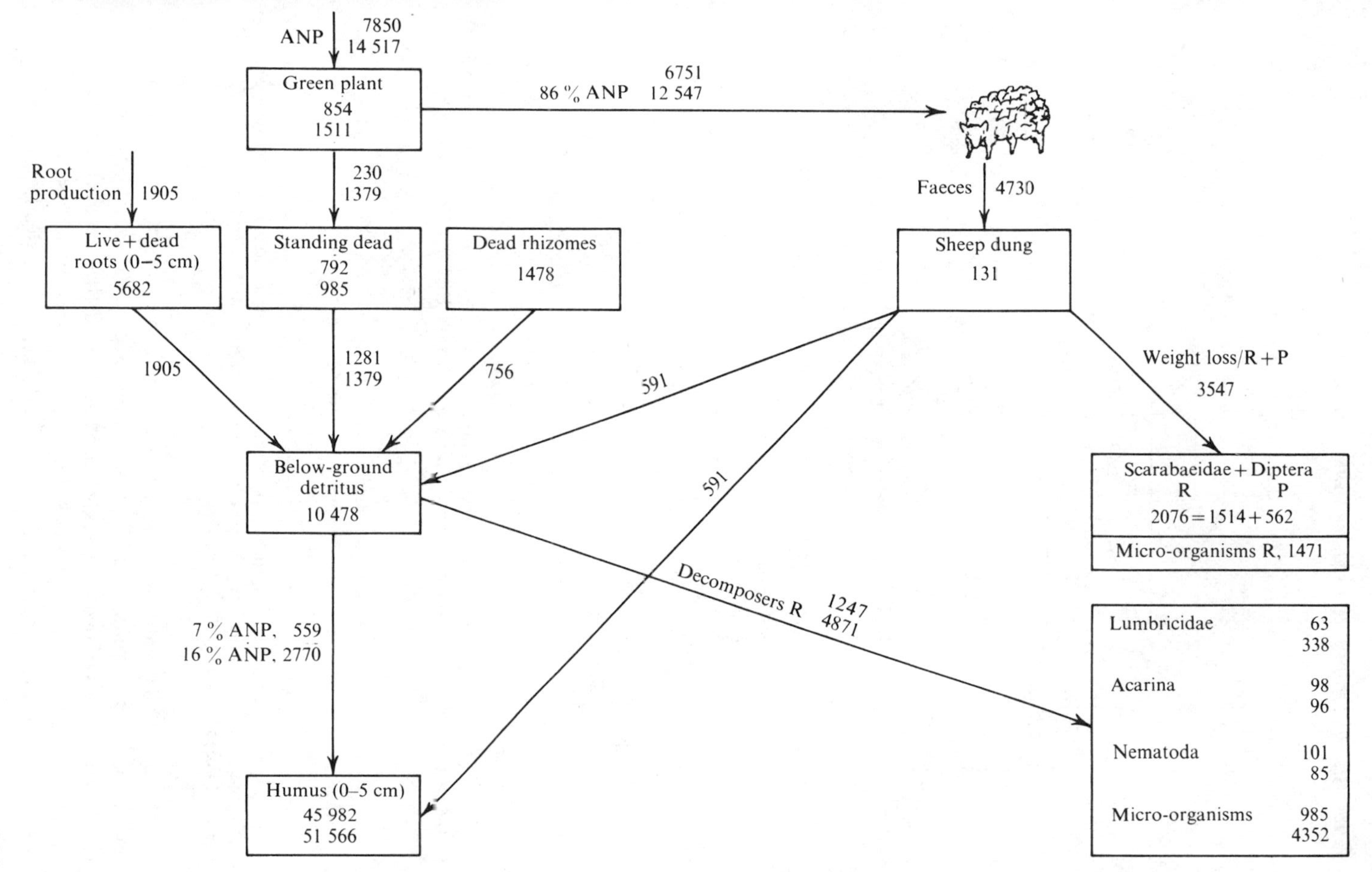

Fig. 12.1. Seasonal energy flow (kJ/m²) in Pieniny Pasture (after Kajak, 1974). Where single figures are presented these are for unfertilised pastures; where two values are given, the lower one is the value one year after manuring. ANP, green plant net production; R + P, respiration plus production.

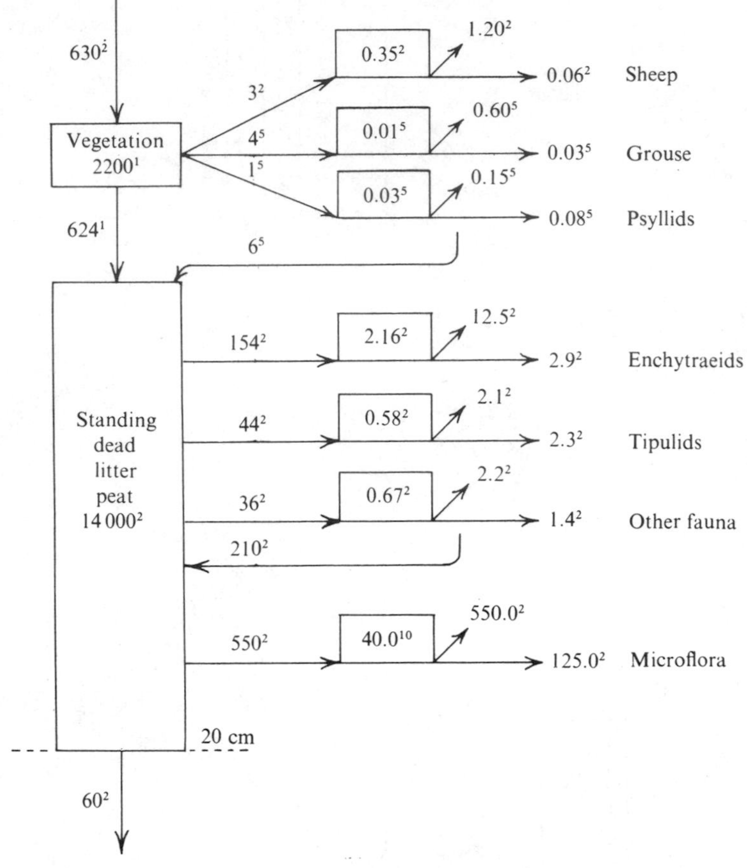

Fig. 12.2. Estimated standing crops (g/m²) and annual transfers (g/m²) of organic matter in the surface 20 cm of blanket bog at Moor House (from Heal & Perkins, 1976). For fauna and micro-flora, transfers are represented as:

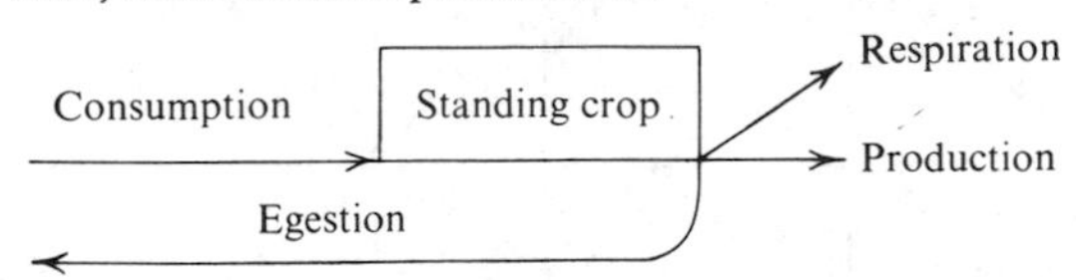

Variability in estimated values, through sampling and systematic errors and between year and between site variation, are indicated by superscripts as: ^{1}, 0.9–1.1 × mean; ^{2}, 0.5–2.0 × mean; ^{5}, 0.2–5.0 × mean; ^{10}, 0.1–10.0 × mean.

vertebrates, while annual secondary production (consumers of first and second level) is 6.8 g/m², containing 77 kJ of energy. Moreover, secondary production of consumers of decaying material, excluding earthworms, is estimated to be about 9 g/m², containing 102 kJ of energy; of this predators account for 0.38 g/m². Decaying plant and animal material (litter, dead roots, dung, excreta), containing an estimated 20188 kJ/m², are converted into 314 g of humus annually, after the action of meso-fauna, bacteria and earthworms. At Snowdonia the estimated consumption (12.8 g/m²) of slugs alone is 5 % of net primary production and secondary production of this group is approximately 40 kJ/m².

In *Agrostis–Festuca* grassland at Moor House (UK) annual secondary production of the total native fauna (vertebrates plus invertebrates) was estimated to be in the 35 to 40 g/m² range. Energy flow through tipulids and enchytraeids under ground was 310 kJ/m². Consumers were much more active in this grassland than in blanket bog nearby, where annual secondary production was estimated to be only 3 g/m², as seen in the flow diagram of this system (Fig. 12.2).

It is interesting to compare the rate of energy flow through various groups of native fauna with that of domesticated animals on the same pasture. For example, at the site in northern Wales (Snowdonia) slugs were found to consume 5 % as much herbage (12.8 g/m²) as do sheep and were 15 % efficient in converting this to body tissue (Lutman, 1976). In the French Pin-au-Haras site energy flow through earthworms was found to be 13 % of that through cattle (Ricou, 1978).

Assimilation efficiency (assimilation: ingestion) varies considerably among various faunal groups. For example, at Moor House that of sheep was found to be 50 %, while for grouse (which are probably not completely herbivorous) it was 30 %; at Pin-au-Haras the value was 65 % for cattle. The assimilation efficiency of herbivorous invertebrates is in the 20–50 % range. Typical values are: for grasshoppers, 20 % in the semi-natural *Miscanthus* area and 30 % in an area sown to *Bromus unioloides* at Kawatabi (Japan); for slugs, 35 % at Snowdonia (UK); and for tipulids, 50 % at Moor House (UK).

13. Micro-organisms in meadows

B. ÚLEHLOVÁ

The regularly cut semi-natural meadows of the temperate zone possess a number of characteristics closely related to the nature and activities of micro-organisms in these ecosystems. These meadows represent man-made and man-managed secondary labile systems. They revert with relative ease into forest when uncared for, but under conditions of intensive agriculture they are often transformed into croplands. The habitat and climatic conditions under which they occur vary considerably, as does the structure of the vegetation. This diversity is dealt with and classified from different points of view in a number of publications and reviewed by Oberdorfer (1970) and Klapp (1965). A large portion of the above-ground production is removed as hay from the ecosystem. Thus, the decomposition chain starts with the whole of under-ground litter but only a part of the above-ground production. Meadows exhibit a steady increase in amount of soil organic matter. This undergoes a rapid mineralisation as soon as the sward is ploughed (Theron, 1951; Theron & Haylett, 1953; Harmsen & Schreven, 1955; Harmsen, 1964). The soils usually contain low amounts of nitrate, while the plants are able to tolerate large amounts of nitrogen applied as mineral fertiliser; although a large part of added nitrogen is lost by denitrification (Woldendorp, 1963; Kopčanová, Řehořková & Števlíková, 1973).

In meadows humification processes mostly prevail over mineralisation and the nitrogen metabolism is rather specific. The role of micro-organisms in various soil processes and their significance to fertility of the habitat are relatively well understood (Clark & Paul, 1970). Several IBP studies had as one of their objectives the qualitative and quantitative characterisation of populations of micro-organisms and their role in energy flow and mineral cycling processes within meadow ecosystems. Most of this new information has been published recently (Jankowska, 1971; Jakubczyk, in Breymeyer, 1971; Aristovskaya, 1972; Řychnovská, 1972; Loquet, 1973; Úlehlová, 1973*a*, *b*).

Availability of substrate

The green leaves and stems of unmown or uncut plants deteriorate after death and fall to the litter layer. This layer also includes humus and the biomass of secondary producers (detritovores and micro-organisms). The thickness of this densely-packed material on the soil surface of meadows varies considerably according to environment, amount of growth and type of management regime. It is, of course, thinner in mown and grazed areas than in protected sites. It tends to accumulate in the relatively dry plant communities, such as those dominated by *Brachypodium pinnatum* and *Salvia pratensis* or by *Arrhenatherum elatius*. Minimum values for standing crop of litter often coincide with maximum biomass of leaves and stems in the canopy. Based on an estimate that about 20 % of net primary above-ground production of mown meadows and 60–80 % of unmown meadows enters the decomposition food chain directly, the annual amount of plant material available as substrate for decomposers ranges from 100 to 1500 g/m^2. Since the standing crop of under-ground plant parts has been estimated to range from 400 to 4000 g/m^2 in temperate grasslands and turnover rates have been found to be once every three or four years (Bazilevich & Kobyakova, 1971; Fiala, 1973), we can conclude that from about 100 to 950 g/m^2 of dead plant material enters the decomposition chain annually below the soil surface.

Populations

The micro-organisms in grassland ecosystems inhabit living and dead plants and animals, litter and soil. The faecal food chain may be considered as insignificant in ungrazed and mown meadows. The nature of populations of micro-organisms that inhabit plants and animals is relatively unknown. More information is available on those of litter and still more on soil micro-organisms. Micro-organisms have been shown to be much more abundant in the rhizosphere of plants than in the soil matrix generally. The effect of the rhizosphere in increasing populations has been found to be related to the stage of development of the plants, being least pronounced during the dying off process. Naplekova *et al.* (in Aristovskaya, 1972) compared the accumulation of cellulolytic micro-organisms in the rhizospheres of 28 plant families and found that grass roots produce the greatest rhizosphere effect. Their findings are in accord with those of Ten-Khak-Mun & Fedorova (in Aristovskaya, 1972) pertaining to the dominant species of the tall-grass meadows of Sachalin in the eastern part of the USSR.

Microbial populations of litter are usually higher than those of soils. Values of these are indicated in Table 13.1 for several meadow ecosystems. The diversity of environmental conditions and litter composition is reflected in the wide range in numbers of micro-organisms. Seasonal changes are quite marked and they appear to be specific for both ecosystems dominated by particular plant species and for physiological groups of micro-organisms, in spite of a common minimum in late summer when the moisture supply is low.

A summary of data on microbial counts in soils is presented in Table 13.2, pertaining to various meadow ecosystems of the Eurasian temperate zone and extending from the eastern part of the USSR (Sachalin) to France. Direct methods result in the range of $1–18 \times 10^9$, the plate methods in 10^6–10^8 micro-organisms for a gram of dry soil. The counts may be considerably higher on

Table 13.1. *Micro-organism counts per gram of litter of various meadow ecosystems (based on plate counts)*

Author	Country	Plant community dominant species	Bacteria	Actinomycetes	Fungi
Ten-Khak-Mun & Fedorova[a]	USSR (Sachalin)	High grasses	$4.7–20.4 \times 10^9$	$3.7–11.0 \times 10^7$	$86–235 \times 10^6$
Miroshnichenko *et al.* (1972)	USSR (Karelia)	*Alopecurus pratensis*			
		living plants	$1.0–2.5 \times 10^6$	—	9.6×10^3
		dead plants	$2.1–4.5 \times 10^6$	—	96×10^2
Jakubczyk[b]	Poland	*Deschampsia caespitosa*	$4–10 \times 10^7$	—	—
Úlehlová (1973*b*)	Czecho-slovakia	Alluvial meadows			
		dry	$0.5–47.3 \times 10^7$	—	—
		mesophytic	$0.7–136 \times 10^7$	—	—
		wet	$0.2–69.8 \times 10^8$	—	—
Tesařová (1975)	Czecho-slovakia	The same alluvial meadows (laboratory experiments)			
		dry	$5–22.0 \times 10^6$	$0.3–2.0 \times 10^6$	$0.3–20 \times 10^6$
		mesophytic	$5–28.0 \times 10^6$	$0.5–3.0 \times 10^6$	$1.0–5 \times 10^6$
		wet	$15–42.0 \times 10^6$	$0.5–4.0 \times 10^6$	$0.1–0.9 \times 10^6$

[a] In Aristovskaya (1972). [b] In Breymeyer (1971).

meagre agar media (with minerals and simple sugars only), on agar media lacking or low in salts, or on soil agars, than on media rich in nutrients (Mikhailova & Nikitina, in Aristovskaya, 1972). Numbers decrease with depth (Loquet, 1973; Yershov, in Aristovskaya, 1972). The heavy, alluvial soils appear to contain larger numbers of micro-organisms than other soil types, probably due to their greater inner surface. Populations fluctuate according to the prevailing conditions of the season, being at a maximum in spring, when favourable temperatures and soil moisture coincide with an ample supply of organic substrate. A decline in numbers accompanies the onset of the dry summer period, but is followed by an occasional second (but usually lower) peak in autumn when the moisture conditions improve and the fresh litter is available for colonisation. The lowest microbial counts have been observed in winter and in the summer months. The seasonal amplitude in counts extends even over three orders of magnitude (Jakubczyk, 1969).

According to a study in France (Loquet, 1973), ammonifiers and amylolytic bacteria are most abundant in soil of one meadow ecosystem and contribute equally to approximately 95 % of the population of microbes. Denitrifiers comprise most of the remaining 5 %. Anaerobic hemicellulose decomposers and aerobic cellulose decomposers are about equally abundant and each

Table 13.2. *Numbers of micro-organisms per gram of dry soil in various meadow soils. These were determined by plate counts except where otherwise indicated*

Author	Country (district)	Plant cover, soil type	Soil layer (cm)	Bacteria	Actinomycetes	Fungi
Ten Khak Mun & Fedorova[d]	USSR Sachalin	Tall grasses	0–10	$0.5–10 \times 10^7$	$0.9–5 \times 10^6$	—
Schapova[d]	USSR Primorskiy	Meadow–virgin soil podzolic, gleyic soil		$4.1–18.3 \times 10^{9a}$	—	—
	Chanka Lake	Podzolic, gleyic soil		$1.3–188 \times 10^6$	—	—
Gantimurova *et al.*[e]	USSR	Chernozem	0–20	$0.2–2.4 \times 10^6$	$0.19–1.8 \times 10^6$	—
	Karachi station		20–40	$0.05–1.2 \times 10^6$	$0.04–1.1 \times 10^6$	—
	Ob–Irtysh	Medium columnar	0–10	$0.3–3.2 \times 10^6$	$0.01–3.7 \times 10^6$	—
	interfluve	Meadow steppe	10–15	$0.4–2.9 \times 10^6$	$0.11–6.0 \times 10^6$	—
		Solonetz	15–25	$0.1–6.2 \times 10^6$	$0.01–2.2 \times 10^6$	—
			25–40	$0.06–3.4 \times 10^6$	$0.2–0.9 \times 10^6$	—
		Chernozem meadow	0–10	$0.2–5.6 \times 10^6$	$0.7–5.2 \times 10^6$	—
		Solonchakous-solonetzic soil	10–20	$0.2–4.2 \times 10^6$	$0.01–2.5 \times 10^6$	—
			20–38	$0.9–2.0 \times 10^6$	$0.08–0.6 \times 10^6$	—
		Crust meadow	0–3	$1.6–10.3 \times 10^6$	$0.3–19.7 \times 10^6$	—
		Solonetz	3–20	$0.2–6.1 \times 10^6$	$0.05–15.7 \times 10^6$	—
			20–40	$0.2–2.5 \times 10^6$	$0.02–0.4 \times 10^6$	—
		Peat-boggy soil	0–20	$1.2–16.9 \times 10^6$	$0.3–2.6 \times 10^6$	—
			20–40	$0.1–0.9 \times 10^6$	$0.07–0.3 \times 10^6$	—
Mikhailova & Nikitina[d]	USSR Zabaykal'e	Steppe meadow				
		Solonchak	0–10	$1.0 \pm 0.10 \times 10^{9a}$	—	—
	—	Chestnut		$1.8 \pm 0.34 \times 10^{9a}$	—	—
		[illegible]		$1.5 \pm 0.28 \times 10^{9a}$	—	—

		Solonchak		$365 \pm 10 \times 10^6$[c]	—	—
		Chestnut		$439 \pm 10 \times 10^6$[c]	—	—
		Chernozem		$926 \pm 16 \times 10^6$[c]	—	—
Geltzer *et al.*[e]	USSR	Meadow	0–10	$0.6–1.5 \times 10^6$	$2.4–5.6 \times 10^6$	2×10^3
	Tambov station	Chernozem	50–60	$2.8–4.1 \times 10^4$	$5.4–7.2 \times 10^4$	—
	Don lowland		0–7	$0.12–0.40 \times 10^6$	$1.0–6.7 \times 10^6$	$4–23 \times 10^3$
	European part		50–60	$0.6–2.2 \times 10^4$	1.2×10^4	—
Zaguralskaya *et al.*[d]	USSR	Drained low bog				
	Southern	much	3–23	$0.9–5.7 \times 10^6$	—	—
		little	0–22	$0.7–11.7 \times 10^6$	—	—
	Karelia	Long-ago-reclaimed transition moor	0–7	$0.4–2.5 \times 10^6$	—	—
			7–22	$0.2–0.4 \times 10^6$	—	—
		Non-meliorated peat soil	7–23	$0.8–3.3 \times 10^6$	—	—
Yershow[d]	USSR	‘Lugoovsyanichnik’ sod-podzolic loamy	0–5	$303 \pm 2.27 \times 10^6$[c]	—	—
	Karelia	soil	5–17	$370 \pm 2.40 \times 10^6$[c]	—	—
		Sod-podzolic sandy soil	0–30	$177 \pm 1.73 \times 10^6$[c]	—	—
			2–20	$111 \pm 0.81 \times 10^6$[c]	—	—
		‘Vlazhnoraznotravnik’ sod-podzolic	0–5	$16 \pm 0.68 \times 10^6$[c]	—	—
		heavy loamy soil	5–14	$82 \pm 1.10 \times 10^6$[c]	—	—
		‘Belousnik’ podzol-gleic peat soil	0–5	$57 \pm 2.39 \times 10^6$[c]	—	—
			5–17	$477 \pm 5.16 \times 10^6$[c]	—	—
		‘Schchuchnik’	0–5	$67 \pm 1.99 \times 10^6$[c]	—	—
			5–16	$43 \pm 0.93 \times 10^6$[c]	—	—
		‘Ostroosochnik’ lowland peat soil	0–6	$177 \pm 1.41 \times 10^6$[c]	—	—
			6–20	$599 \pm 3.44 \times 10^6$[c]	—	—
		Lowland gleic peat soil	0–5	$59 \pm 0.58 \times 10^6$[c]	—	—
			5–17	$16 \pm 0.22 \times 10^6$[c]	—	—
Zykina[d]	USSR	Meadow, various grasses, sandy soil		$1.5–11.6 \times 10^9$[b]	—	—
	Karelia	Meadow, various grasses, sod-podzolic gleic heavy loamy soil		$2.2–15.0 \times 10^9$[b]	—	—

Table 13.2 (*cont.*)

Author	Country (district)	Plant cover; soil type	Soil layer (cm)	Bacteria	Actinomycetes	Fungi
Jakubczyk[f]	Poland	*Deschampsia caespitosa*		$0.5–10 \times 10^7$/cm^3 soil	—	$1.0–17 \times 10^4$/cm^3 soil
Karkanis[g]	Poland Ojców	Arrhenatheretum elatioris	0–10	16.3×10^6	0.6×10^3	2×10^3
Rehořková (1974)	Czechoslovakia	*Festuca sulcata* *Alopecurus pratensis* *Deschampsia caespitosa* Alluvial gleic soils		$0.8–51.7 \times 10^7$ $1.0–41.1 \times 10^7$ $0.5–58.8 \times 10^7$	$0.1–2.6 \times 10^6$ $0.09–2.4 \times 10^6$ $0.11–1.8 \times 10^6$	$0.6–4.8 \times 10^5$ $0.5–9.7 \times 10^5$ $1–4.0 \times 10^5$
Úlehlová (1973*b*)	Czechoslovakia	Alluvial dry meadow with *Festuca sulcata* Alluvial meadow with *Alopecurus pratensis* Wet meadow with *Glyceria maxima* Alluvial gleic soils		$3.0–16.0 \times 10^7$ $3.0–25.0 \times 10^7$ $3.0–33.0 \times 10^7$	$0.3–8.7 \times 10^6$ $0.3–2.7 \times 10^6$ $0.2–2.0 \times 10^6$	$0.3–2.0 \times 10^5$ $0.4–3.4 \times 10^5$ $0.3–0.8 \times 10^5$
Jagnov (1958)	Germany	Various soils from dry to wet		$0.5–5.5 \times 10^9$[a] —	$3.1–26.1 \times 10^7$ $0.6–2.7 \times 10^7$	— —
Loquet (1973)	France le Haras-du-Pin	Lasting meadow		$0.8–7 \times 10^8$	—	—

[a] Direct count. [b] Direct method, daily determined. [c] Plate count on meat-pepton agar. [d] In Aristovskaya (1972). [e] In Bazilevich (1974). [f] In Breymeyer (1971). [g] In Medwecka-Kornas (1967).

Table 13.3. *Estimates of bacterial biomass in meadow soils*

Author	Area	Vegetation soil type	Biomass	Method of estimation
	USSR			
Schapova (in Aristovskaya, 1972)	Chanka Lake	Meadow	Min. 0.007 g/g soil, max. 0.077 g/g soil	Direct count
Mikhailova & Nikitina (in Aristovskaya, 1972)	Baykal	Unfertilised		
		Steppe meadow	100 $g/m^2/10$ cm	Generation time
		Meadow brown soil	140 $g/m^2/10$ cm	
		Chernozem	160 $g/m^2/10$ cm	
		Fertilised with P		
		Steppe meadow	310 $g/m^2/10$ cm	
		Meadow brown soil	940 $g/m^2/10$ cm	
		Chernozem	480 $g/m^2/10$ cm	
Zykina (in Aristovskaya, 1972)	Karelia	Meadow		
		Loamy fine soil	Min. 0.001 g/g soil, max. 0.0075 g/g soil	Direct count
		Loamy soil	Min. 0.0001 g/g soil, max. 0.0083 g/g soil	Direct count
Geltzer *et al.* (in Bazilevich, 1974)	Tambov	Meadow: solonetz	0.0001 g/g soil	Plate count
	Czechoslovakia	Alluvial meadow dominated by		
Úlehlová (1973*b*)	Lanžhot	*Festuca sulcata*	min. 3 g dry wt./$m^2/10$ cm, max. 15 g dry wt./$m^2/10$ cm	Plate counts in soil and litter
		Alopecurus pratensis	min. 3 g dry wt./$m^2/10$ cm, max. 26 g dry wt./$m^2/10$ cm	
		Glyceria maxima	min. 3 g dry wt./$m^2/10$ cm, max. 30 g dry wt./$m^2/10$ cm	

numbers approximately 10000/g of soil. Nitrite producers and anaerobic cellulose decomposers are one order of magnitude less abundant (each 1000/g), while nitrogen-fixing bacteria, nitrate producers and sulphate-reducing bacteria are least common (each 100/g). These estimates were based on the study of only the uppermost 5 cm of soil. This study shows the importance and the danger of characterising populations on the basis of the physiological roles of their constituents.

Biomass

The accuracy of estimates of microbial biomass is still uncertain because of frequent lack of agreement with the amount of substrate available. Microbial biomass may undergo fast changes because of potentially short generation times and easy decomposibility. Problems of calculating the microbial biomass from direct or plate counts with all factors considered have been dealt with thoroughly by Parkinson, Gray & Williams (1971), special attention being given to the presumed dimensions of microbial cells. Micro-biologists in the USSR often derive biomass of bacteria from the expression for the rate of growth developed by Ierusalimskiy (1949). Daily estimates of both direct and plate counts serve as the basis for estimating biomass and generation times. A detailed discussion of these methods has been given by Nikitina & Sharabarin, in Aristovskaya (1972).

Table 13.4. *Generation period* (*h*) *of bacteria in soil* (*from USSR studies, in Aristovskaya* (*1972*))

Author	Area	Soil	July	September
Nikitina & Sharabarin	Siberia	Forest soil	69–250	200–360
Zykina	Karelia	Sandy soil	13–75	
		Loamy soil		13–72
Schapova	Chanka Lake	Gleypodzolic soil	67 (direct count)	34 (plate count)

Estimates of microbial biomass in several meadow ecosystems are summarised in Table 13.3. The range encountered is considerable and can be related to differences in methods applied. The generation times are given in Table 13.4.

Theoretical yields of bacterial cells in a given system can be computed. A total of about 0.028 g of bacterial dry matter can be formed under aerobic or anaerobic conditions per kilojoule of energy removed from the nutrient medium, according to an analysis by Payne (1970) of the data published by a number of authors. The amounts of bacterial cells that potentially could be formed during decay of the annual production of plants are estimated in Table 13.5 for several ecosystems. For example, in the meadow dominated

Table 13.5. *Energy values (in kJ m²) of plant material and bacterial biomass in meadows*

Site	Dominant plant species	Plant energy: Above-ground parts	Plant energy: Under-ground parts	Plant energy: Total	Bacteria: Potential biomass (g/m²)	Bacteria: Potential energy
Poland						
Ojców	*Arrhenatherum elatius*	5230–5690	5523–7364	13054	368	7699
Ispina	*Arrhenatherum elatius*	6360	5272–7029	13355	375	7845
Czechoslovakia						
Lanžhot	*Festuca sulcata*	4443–8000	6527–8703	16703	168	9791
Lanžhot	*Alopecurus pratensis*	11812	4820–6427	18338	518	10837
Lanžhot	*Glyceria maxima*	16891	10845–14401	31292	873	18263

by *Glyceria maxima* the estimated potential production of dry bacterial cells is 873 g/m² annually, with an energy content of 18265 kJ. It is interesting to note, however, that the maximum bacterial biomass at any one time in the uppermost 25 cm of the soil profile in this meadow has been estimated to be only 30 g/m². Decomposition of dead plant matter may be restricted to rather short periods of favourable conditions of temperature and humidity. It follows that the yearly average for the turnover rate of microbial biomass is much lower than that attained during the short periods of microbial outburst.

14. Ecosystem synthesis of meadows

Energy flow

M. RYCHNOVSKÁ

The characteristic feature of mown meadows is a large export of biomass as fodder. The turnover of most of the remaining plant material is through the decomposition chain, in contrast to the situation in pastures where a considerable amount passes into the consumers chain. Consumption by animals in meadows has been estimated to be only 5.4–9.0 % of above-ground biomass (Breymeyer, 1971). The amount of consumption by herbivorous insects alone was found to be 2–3 % of above-ground standing crop (Ohga, in Numata, 1973). Thus, the biomass loss to consumers in meadows can hardly be ascertained by present sampling methods which usually are accurate to only about 5 %. Treatment with insecticides has increased plant production by about one-third. However, this cannot be accounted for only by the reduction in invertebrate stress, since growth rate of plants in treated stands increased by 62 %. This increase results from both the absence of consumers and the stimulative effect of insecticides (Andrzejewska & Wójcik, in Breymeyer, 1971).

In alluvial meadows the role of consumers is even further diminished because of frequent floods which exterminate most small rodents and numerous herbivorous insects. Thus, in these meadows the amount of grazed plant biomass can be ignored. The degradation activities of detritovorous meso- and micro-fauna are included in the decomposer system.

The ecosystem complex in temperate meadows can be illustrated by a hydrosere of semi-natural alluvial meadows in Czechoslovakia (Lanžhot) studied over six years by Rychnovská (1972, 1976) and Úlehlová (1976). There were four ecosystems, all well supplied with minerals, but differing in their water factors and management. Three of them were mown twice a year, while one was left unmown. The 'dry' stand was flooded only exceptionally, the 'moist' and 'damp' ones inundated regularly in spring and occasionally in summer. In the 'wet' stand the water table remained above the soil surface almost all year. The dominant species of plants in these four ecosystems were, respectively, *Festuca sulcata*, *Alopecurus pratensis*, *Phalaris arundinacea* and *Glyceria maxima*. These meadows were never fertilised artificially, the pool of minerals being replenished from flooding and precipitation. Basic climatological data were: mean annual temperature, 9.5 °C; mean annual precipitation, 585 mm; precipitation in the growing season (April–September), 365 mm.

Table 14.1. *Characteristic structural units in seminatural temperate meadow ecosystems (alluvial hydrosere)*

Data set no.	Source of information[b]		*A*	*B*	*C*	*D*
	(*b*)	Type of meadow	Dry	Moist	Damp	Wet
		mown for hay per year	2×	2×	2×	Not
		occurrence of flood (especially in spring)	No	Yes	Yes	permanently
		average soil moisture (%)	25	54	60	74
		Primary production				
		Above-ground plant biomass in g/m²				
1	(*a*), (*b*)	Maximum standing crop (sum of three maximum values in mown)	957	1016	1388	1237
2	(*a*), (*b*)	Permanent pool spring value living	52	81	90	97
3	(*a*), (*b*)	Permanent pool spring value dead	131	65	62	83
4	(*a*), (*b*)	Permanent pool autumnal value (total)	234	149	195	852
5		Biomass sampled above pool per year in *A*, *B*, *C*: 1−(2+3+4); in *D*: 1−3	540	721	1041	1153
6		Harvest taken off as hay per year	291[a]	525[a]	731[a]	—
7	(*c*)	Litter produced per growing season	166	349	556[a]	1277
8		Total biomass produced per year (5+7)	706	1070	1597	2430
9	(*b*)	Leaf area index – LAI	1.3	3.9	4.2	8.7

11	(*b*)	Permanent pool	1556	1221	1664	2640
12		Biomass produced above pool per year	389[a]	407[a]	554[a]	880[a]
13		Above-ground + under-ground production per year (8+12)	1095	1477	2151	3310
		Decomposition chain in g/m²				
14		Above-ground source for decomposition for year 8−(6+2+3)	232	400	714	2250
15	(*c*)	Above-ground litter decomposed per growing season	166	376	581[a]	1477
16		Above-ground dead material decomposed per winter (4−3)	103	84	133	769
17		Under-ground litter decomposed per year (12)	389[a]	407[a]	554[a]	880[a]
		Energy flow in kJ/m²				
18		Input of global radiation for growth (from start to the peak of cumulative growth curve)	2719642	2332823	2339913	1702281
		Energy content of plant biomass produced per year				
19		above-ground (8)	12171	18443	27531	41890
20		under-ground (12)	6284	6586	8954	14226
21		Total (13)	18456	25033	36480	56116

[a] Estimated values.

[b] Data from original IBP studies, except as indicated: (*a*) Jakrlová, 1968, 1971; (*b*) Rychnovská, 1972; (*c*) Tesařová, 1975.

Time indications: interval between start and the peak of cumulative growth curve (number of days): 165 (dry), 143 (moist), 144 (damp), 94 (wet); growing season is considered as 200 days (10 Apr. to 27 Oct.); winter is considered as 165 days (28 Oct. to 9 Apr.).

Terminology used is according to Šesták *et al.* (1971).

Data sets 1–6, 11, 14, 16 and 18–21 are based on six years of measurement (in *A*, *B*, *D*) and three years (in *C*).

Table 14.2. *Processes in seminatural temperate meadow ecosystems (alluvial hydrosere)*

Data set no.	Source of information[b]		*A*	*B*	*C*	*D*
22		Efficiency of radiant energy conversion (per cent) (from start to the peak of cumulative growth curve)	0.68	1.07	1.56	3.30
23	(*a*)	Maximum efficiency of radiant energy conversion in dominant grasses (per cent)	3.1	9.8	7.5	7.6
24		Crop growth rate (CGR) (g/m² daily) (from start to the peak of cumulative growth curve)	3.06	5.44	7.36	10.12
25	(*d*)	Decomposition rate per growing season (g/m² daily)	0.77	1.75	2.78[a]	6.87
26		Decomposition rate in winter (16) (g/m² daily)	0.63	0.51	0.81	4.64
27		Relative growth rate (RGR) (mg/g daily) (from start to the peak of cumulative growth curve)	14.6	13.8	17.9	21.9
28	(*d*)	Litter disappearance rate per growing season (LDR) (mg/g daily)	6.4	10.5	10.6[a]	10.7
29		Litter disappearance rate per winter (LDR) (16) (mg/g daily)	3.5	5.0	7.0	14.1
30	(*c*)	Transpiration - maximum values (g/m² daily)	2008	10540	9584	8307
31	(*b*)	Water consumption per year (kg/m² annually)	155[a]	260[a]	320[a]	480[a]
32	(*c*)	Water turnover rate (g/g water daily)	7.22	6.93	5.18	4.24

[a] Estimated values.

[b] Data from original IBP studies, except as indicated: (*a*) Gloser, 1976; (*b*) Rychnovská, 1972; (*c*) Rychnovská *et al.*, 1972; (*d*) Tesařová, 1975.

Time indications: interval between start and the peak of cumulative growth curve (number of days): 165 (dry), 143 (moist), 144 (damp), 94 (wet); growing season considered as 200 days (10 Apr. to 27 Oct.); winter considered as 165 days (28 Oct. to 9 Apr.).

Terminology used according to Šesták *et al.* (1971).

Data sets 24, 26, 27 and 29 are based on six years of measurement (in *A*, *B*, *D*) and three years (in *C*).

The characteristic structural units, with the biomass budget in meadows under study, are presented in Table 14.1. Rates of processes concerned with primary production, decomposition and water turnover are given in Table 14.2. Some of these values are interpreted further in Fig. 14.1. These measured and estimated values can be considered to constitute a static model of function and turnover of biomass in semi-natural temperate meadows.

In order to determine the extent to which these examples are typical of meadows of similar type elsewhere, let us compare the principal data with those obtained from other projects (Table 14.3). Annual production of stems

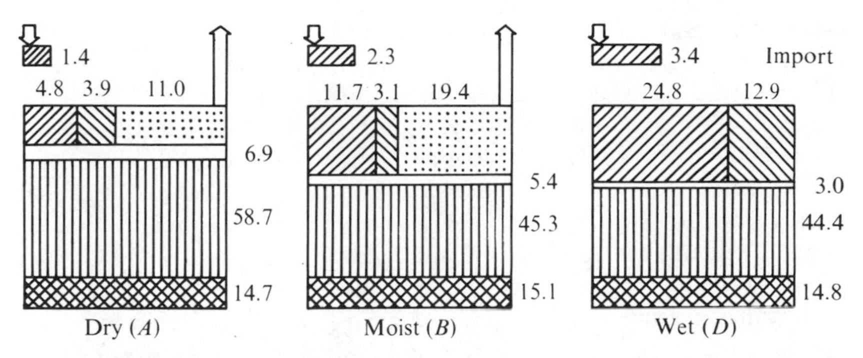

Fig. 14.1. Biomass budget in meadow ecosystems expressed as percentages of total standing crop. *A*, *B* and *D* refer to types of meadow described in Table 14.1. □, permanent pool above ground; ▥, permanent pool under ground; ▤, export (hay). Inputs into decomposer system: ▨, above-ground biomass decomposed per summer; ▧, above-ground biomass decomposed per winter; ▩, under-ground biomass produced and decomposed per year.

Table 14.3. *Comparison of primary production and decomposition in European meadows*

Parameter		Czechoslovakia (Lanžhot)	Poland (Kazuń, Kampinos)	Poland (Ojców)
Annual above-ground production	(g/m²)	706–2430	710–750	874
Crop growth rate (daily)	(g/m²)	3.1–10.1	—	4.7
Under-ground standing crop	(g/m²)	1221–2640	—	1313–4200
Annual under-ground production	(g/m²)	389–880	—	508–2153
Litter disappearance rate (daily)	(mg/g)	6.4–10.7	4.8–14.0	10.6

Data from Poland: Breymeyer, 1971; Kotańska, 1970, 1973; Jankowska, in Medwecka-Kornaś, 1967; Traczyk, in Breymeyer, 1971.

and leaves ranged from 706 to 2430 g/m² in Lanžhot meadows, while values for three Polish meadows ranged from 710 to 874 g/m². Mean daily crop growth rate in Ojców meadows was 4.7 g/m², compared with 3.1–10.1 g/m² at Lanžhot. Mean daily rate of litter disappearance in three Polish meadows ranged from 4.8 to 14.0 mg/g, while it was 6.4–10.7 mg/g in Lanžhot meadows. The standing crop of under-ground plant parts in the Ojców site was 3212 g/m² in the dry meadow, 1313 g/m² in the mesophytic meadow and 4200 g/m² in the wet meadow, compared to Lanžhot values that ranged from 1221 to 2640 g/m². Annual production estimates of under-ground plant parts at Ojców ranged from 508 to 2153 g/m², compared to the Lanžhot values of 389–880 g/m². The turnover rate of above-ground living biomass in Lanžhot meadows is 73–196 % (data set No. 8 divided by data set No. 1 in Table 14.1). In a Hungarian dry grassland the corresponding values range between 50 and 60 % (Précsényi, 1971). Turnover of under-ground parts was estimated to take 2–3 years in the Ojców meadows (Kotańska, 1970, 1973), but 3–4 years at Lanžhot. Few values are available to compare transpiration rates. Rychnovská *et al.* (1972) have previously reported daily transpiration of 4292 g/m^2 in a mesophytic meadow and 15505 g/m^2 in a wet meadow; the Lanžhot IBP values ranged from 2008 to 10 540 g/m^2. These comparisons show clearly that the functioning of meadows under similar ecological conditions and similar management is amenable to comparison and is dependent only to a small degree (if at all) on their floristic composition. On the other hand, among ecological factors, the water factor and availability of mineral elements seem to play the decisive role. The interaction of these two factors results in a characteristic structure and functioning of temperate meadow ecosystems.

Nutrient cycling

A. A. TITLYANOVA & N. I. BAZILEVICH

The large export of biomass in the form of hay from mown meadows leads to a decrease in mineral elements, which is balanced to some extent by such inputs as those carried from other ecosystems by flooding and those contained in precipitation. However, the characteristics of mineral cycling within the ecosystem are determined by productivity factors and by the selective capability of plants in the uptake of nutrients.

Plant species of meadows can be divided into three groups on the basis of their capacity to accumulate various mineral elements. These are: (i) grasses that accumulate mainly silicon and potassium; (ii) legumes that take up mainly nitrogen, calcium and potassium; and (iii) other forbs of intermediate character in this respect.

Above-ground parts of plants usually contain a higher percentage of ash elements than under-ground parts. Concentration of mineral elements increases considerably in grassland plants from northern regions to southern ones (Table 14.4). It increases also in mesohalophilous and, particularly, in halophilous species, first in respect to potassium and sodium and then, with an increase in halophytic stress (soil salt concentration), on account of sulphur, chlorine and sodium. In the above-ground dead biomass (both in the canopy and litter layer) the content of such relatively mobile elements as potassium, chlorine, sodium and sulphur is reduced in comparison to that in green shoots, while silicon, iron and aluminium content is increased.

The pool of mineral nutrients and nitrogen in living and dead biomass of meadow and meadow–steppe ecosystems ranges from 60 to 320 g/m^2, attaining its maximum level in mesohalophytic meadows (Table 14.5). The under-ground plant parts contain 60–80 % of the nitrogen and mineral element pool, because of the greater biomass under ground as compared to above ground. The sum of mineral elements in shoots ranges from 10 to 80 g/m^2, depending mainly on the different chemical composition of the dominant plant species. For example, in steppe dominated by *Bromus riparius* the store of elements has been reported to be 24 g/m^2 (in a maximum standing crop of green shoots of 300 g/m^2), while in hygrohalophytic meadows of the Netherlands the store is 38 g/m^2 (in a standing crop of similar magnitude). The effect of species composition is even more marked in halophytic ecosystems, where the concentration of nitrogen and ash in green shoots sometimes exceeds 49 g/m^2 (in a maximum standing crop of green shoots of only 150 g/m^2).

The average relationship of the amount of minerals contained in plants of halophytic meadows is as follows:

$$\text{above-ground parts}\quad \begin{matrix}\mathrm{N}\\ \mathrm{Si}\\ \mathrm{N}\end{matrix} > \mathrm{Ca} > \mathrm{Mg} > \begin{matrix}\mathrm{P}\\ \mathrm{S}\end{matrix} > \mathrm{Na} > \begin{matrix}\mathrm{Al}\\ \mathrm{Fe}\end{matrix};$$

$$\text{under-ground parts}\quad \begin{matrix}\mathrm{N}\\ \mathrm{Si}\end{matrix} > \mathrm{Ca} > \mathrm{K} > \begin{matrix}\mathrm{Mg}\\ \mathrm{Fe}\\ \mathrm{Al}\end{matrix} > \begin{matrix}\mathrm{P}\\ \mathrm{S}\end{matrix} > \mathrm{Na}.$$

In halophytic ecosystems, the role of K, Na, Mg, Cl increases both above ground and under ground.

The annual uptake of nitrogen by meadow vegetation ranges from 4 to 40 g/m^2 and that of ash elements from 15 to 310 g/m^2. In meadow steppes and mesohalophytic grasslands, the amounts of nitrogen and ash elements are always higher in living roots than in green shoots, while in mesophytic grassland, contents are frequently similar in roots and shoots. This indicates that under mild ecological conditions, the intensity of biogeochemical processes is similar in shoots and roots, while under extreme ecological conditions (drought, salinity) they are more intensive in roots than in shoots.

The biological cycle of chemical elements depends, both geochemically and ecologically, on abiotic exchange processes. In order to illustrate the relationship between biotic and abiotic exchange processes, data are interpreted in Fig. 14.2 that were obtained from a mesohalophytic meadow in western Siberia (Bazilevich *et al.* in Sochava, 1973; Titlyanova, in Sochava, 1973; Tikhomirova, in Sochava, 1973; Titlyanova & Bazilevich, in Rodin & Smirnov, 1975).

Input of mineral elements to the ecosystem occurs mainly in rainfall, surface runoff and under-ground water. Output occurs by the latter two routes, and by leaching when the soil is deeply waterlogged (an exchange of minerals takes place between the soil and under-ground water). The following internal cycles can be found within the ecosystem: (i) uptake of mineral elements by plants; (ii) their release from dead plant biomass; and (iii) mineral cycles linked with formation and decomposition of humus. As fodder for consumers, plants with relatively high contents of nitrogen, potassium, calcium and phosphorus are desirable; quality is reduced if the content of sulphur exceeds 0.25 % and of chlorine exceeds 0.50 %. The consumption of phytomass by herbivores was not considered separately in this study. According to Zlotin & Khodashova (1973), it does not exceed 5 % of primary production.

Table 14.4. *Chemical composition of plant biomass from various meadow ecosystems*

Index of grass-land type	Grassland-type, locality, author	Plant material analysed[a]	Nitrogen (%)	Ash elements (%)								Sum of ash elements
				Silicon	Calcium	Potas-sium	Mag-nesium	Phos-phorus	Sulphur	Sodium	Chlorine	
1.1	Meadow-steppe (*Bromus riparius*, *B. inermis*,	*G*	2.20	1.12	0.43	2.40	0.16	0.23	0.18	0.01	0.36	5.12
	Koeleria gracilis, *Stipa joannis*, *Medicago*	*D*	0.97	3.48	0.46	0.13	0.08	0.07	0.08	0.07	0.18	4.88
	falcata, *Galium verum*, *Filipendula hexapetala*,	*L*	1.62	4.06	0.89	0.18	0.14	0.11	0.15	0.02	0.46	7.32
	etc)	*R+V*	1.25	0.78	1.12	0.68	0.22	0.09	0.13	0.12	0.47	3.92
	USSR Russian Plain (Bazilevich, 1962; Bazilevich, in Rodin & Smirnov, 1975)											
1.2	Meadow-steppe (*Festuca pseudovina*, *Poa*	*G*	1.86	1.38	1.24	1.44	0.20	0.27	0.09	0.01	0.05	4.80
	angustifolia, *Koeleria gracilis*, *Vicia cracca*,	*D+L*	1.40	7.20	2.20	0.64	0.36	0.18	0.09	0.13	0.08	11.90
	Medicago falcata, etc)	*R+V*	0.80	1.34	0.76	0.40	0.19	0.09	0.05	0.08	0.07	3.80
	USSR Novosibirsk region (Bazilevich, in Rodin & Smirnov, 1975)											
1.3	Halophytic meadow-steppe (*Festuca pseudovina*,	*G*	1.36	1.74	0.25	1.39	0.16	0.16	0.07	0.08	0.22	4.20
	Agropyron repens, *Artemisia pontica*)	*D+L*	0.97	7.90	0.73	0.73	0.31	0.15	0.15	0.13	0.19	11.00
	SSR Novosibirsk region (Vagina &	*R+V*	0.71	0.95	0.31	0.29	0.18	0.09	0.09	0.12	0.14	2.60
	Shatokhina, 1971; Bazilevich, in Rodin & Smirnov, 1975)											
2.1	Dry alluvial meadow (*Festuca sulcata*,	*G*	1.64	.	0.85	1.02	0.09	0.21	.	0.06	.	.
	Anthoxanthum odoratum, *Colchicum*	*D+L*	1.37	.	0.63	0.30	0.09	0.14	.	0.05	.	.
	autumnale, *Lathyrus pratensis*)	*R+V*	0.82	.	0.65	0.55	0.04	0.07	.	0.09	.	.
	Czechoslovakia, Southern Moravia (Jakrlová, in Rychnovská, 1972; Petřík, in Rychnovská,											

	angustifolia, *Potentilla humifusa*, etc) USSR Tambovskaya region (Bazilevich, in Rodin & Smirnov, 1975)	*R+V*	2.24	1.73	1.41	0.24	0.28	0.10	0.31	0.11	.	4.18
2.3	Steppe meadow (*Calamagrostis epigeios*, *Poa pratensis*, *Agropyron repens*, *Lathyrus pratensis*, *Vicia cracca*, *Filipendula hexapetala*) USSR Novosibirsk region (Bazilevich, 1970)	*G*	2.28	1.93	0.67	0.91	0.21	0.12	0.16	0.04	0.16	4.20
		D	1.53	3.00	0.83	0.22	0.35	0.12	0.09	0.01	0.04	4.66
		L	1.70	3.36	1.12	0.28	0.45	0.14	0.09	0.02	0.13	5.59
		R+V	1.32	1.36	0.51	0.43	0.11	0.14	0.09	trace	0.10	2.74
3.1	Mesophytic alluvial meadow (*Alopecurus pratensis*, *Agrostis alba*, *Agropyron repens*, *Lathyrus pratensis*) Czechoslovakia, Southern Moravia (Jakrlová, in Rychnovská, 1972; Petřík, in Rychnovská, 1972)	*G*	1.10	.	0.52	2.00	0.07	0.21	.	0.03	.	.
		D+L	1.71	.	0.75	0.64	0.08	0.17	.	0.06	.	.
		R+V	0.87	.	0.98	0.84	0.30	0.06	.	0.09	.	.
3.2	Mesophytic meadow (*Agrostis tenuis*, *Alchemilla monticola*) USSR Leningrad region (Ignatenko & Kirillova, 1970; Druzina, 1972; Miroshnichenko *et al.*, 1973)	*G*	.	0.57	0.93	0.73	0.30	0.09	0.15	0.01	.	2.78
		D+L	.	0.95	1.40	0.33	0.33	0.07	0.11	0.02	.	3.21
		R+V	.	0.20	0.43	0.57	0.33	0.20	0.20	0.10	.	2.03
3.3	Mesophytic alluvial meadow (*Phleum phleoides*, *Festuca rubra*) USSR Moscow region (Evdokimova & Rudina, 1958)	*G*	1.96	1.41	0.26	2.52	0.22	0.36	0.13	.	.	4.90
		R+V	1.51	1.71	0.53	0.79	0.22	0.38	0.14	.	.	3.77
3.4	Mesophytic meadow (*Festuca rubra*, *Anthoxanthum odoratum*, *Trifolium pratense*, etc)	*G*	2.12	.	1.05	1.10	0.28	0.35	.	0.11	.	5.70[b]
		R+V	1.18	—	—	—	—	—	—	—	—	—
3.5[c]	West Germany. Solling plateau (Speidel & Weiss, 1972)	*G*	2.84	.	0.87	2.14	0.19	0.39	.	0.05	.	6.45[b]
		R+V	1.47	.	—	—	—	—	—	—	—	—
3.6	Mesophytic meadow (*Miscanthus sinensis*, *Pteridium aquilinum*, etc) Japan (Koike *et al.*, in Numata, 1973; Yamane & Sato, in Numata, 1973)	*G*	0.70	1.75	0.29	1.20	0.10	0.15	.	0.14	.	3.63
		R	0.16	—	—	—	—	—	—	—	—	3.29[b]
		V	0.40	—	—	—	—	—	—	—	—	8.00[b]

Table 14.4 (*cont.*)

Index of grass-land type	Grassland-type, locality, author	Plant material analysed[a]	Nitrogen (%)	Ash elements (%)								Sum of ash elements[d]
				Silicon	Calcium	Potas-sium	Mag-nesium	Phos-phorus	Sulphur	Sodium	Chlorine	
4.1	Mesohalophytic meadow (*Festuca sulcata, Artemisia monogyna, Limonium tomentellum*)	*G*	2.36	2.14	0.33	1.90	0.46	0.09	0.29	0.17	.	5.38
	USSR Tambovskaya region (Bazilevich, in Rodin & Smirnov, 1975)	*R+V*	2.30	2.71	0.31	0.21	0.46	0.17	0.60	0.21	.	5.67
4.2	Mesohalophytic meadow (*Calamagrostis epigeios, Poa angustifolia, Vicia cracca,*	*G*	2.15	1.00	0.67	1.90	0.40	0.20	0.07	0.23	0.47	5.10
	Artemisia laciniata, Galatella biflora)	*D+L*	1.33	3.05	0.71	0.40	0.40	0.07	0.07	0.07	0.14	5.70
	USSR Novosibirsk region (Titlyanova, 1971, 1972; Bazilevich, in Rodin & Smirnov, 1975)	*R+V*	1.09	2.02	1.04	0.43	0.59	0.12	0.11	0.19	0.20	5.50
5.1	Halophytic meadow (*Puccinellia distans,*	*G*	1.02	0.76	0.17	1.22	0.23	0.17	0.10	0.58	1.01	4.3
	Saussurea salsa)	*D+L*	1.30	3.20	0.45	0.30	0.18	0.07	0.07	0.07	0.14	4.5
	USSR Novosibirsk region (Titlyanova, 1971; Bazilevich, in Rodin & Smirnov, 1975)	*R+V*	0.69	1.85	0.24	0.59	0.15	0.07	0.12	0.62	0.30	4.6
5.2	Wet halophytic meadow (*Salicornia europaea*)	*G*	1.00	0.19	0.43	2.00	0.92	0.24	0.70	9.38	14.42	28.28
	USSR Novosibirsk region (Bazilevich, 1970)	*D*	0.72	0.45	1.02	0.12	0.82	0.05	0.18	0.63	0.51	3.78
		R+V	0.57	0.39	0.35	0.66	0.42	0.26	0.18	2.20	2.38	6.84
6.1	Wet alluvial meadow (*Glyceria maxima,*	*G*	0.75	.	0.53	1.32	0.09	0.29	.	0.08	.	.
	Phalaris arundinacea, Carex gracilis)	*D+L*	1.36	.	0.54	0.82	0.08	0.17	.	0.10	.	.
	Czechoslovakia, Southern Moravia (Jakrlová, in Rychnovska, 1972; Petřík, in Rychnovská, 1972)	*R+V*	1.34	.	0.03	0.94	0.03	0.12	.	0.11	.	.

[a] *G*, living above-ground biomass; *D*, standing dead; *L*, litter; *R+V*, living+dead under-ground biomass.
[b] Calculations based on crude ash values. [c] Fertilisers applied (kg/ha): N, 200; P, 52; K, 200; Ca, 240.
[d] Including titanium and manganese. ., not determined. —, no data.

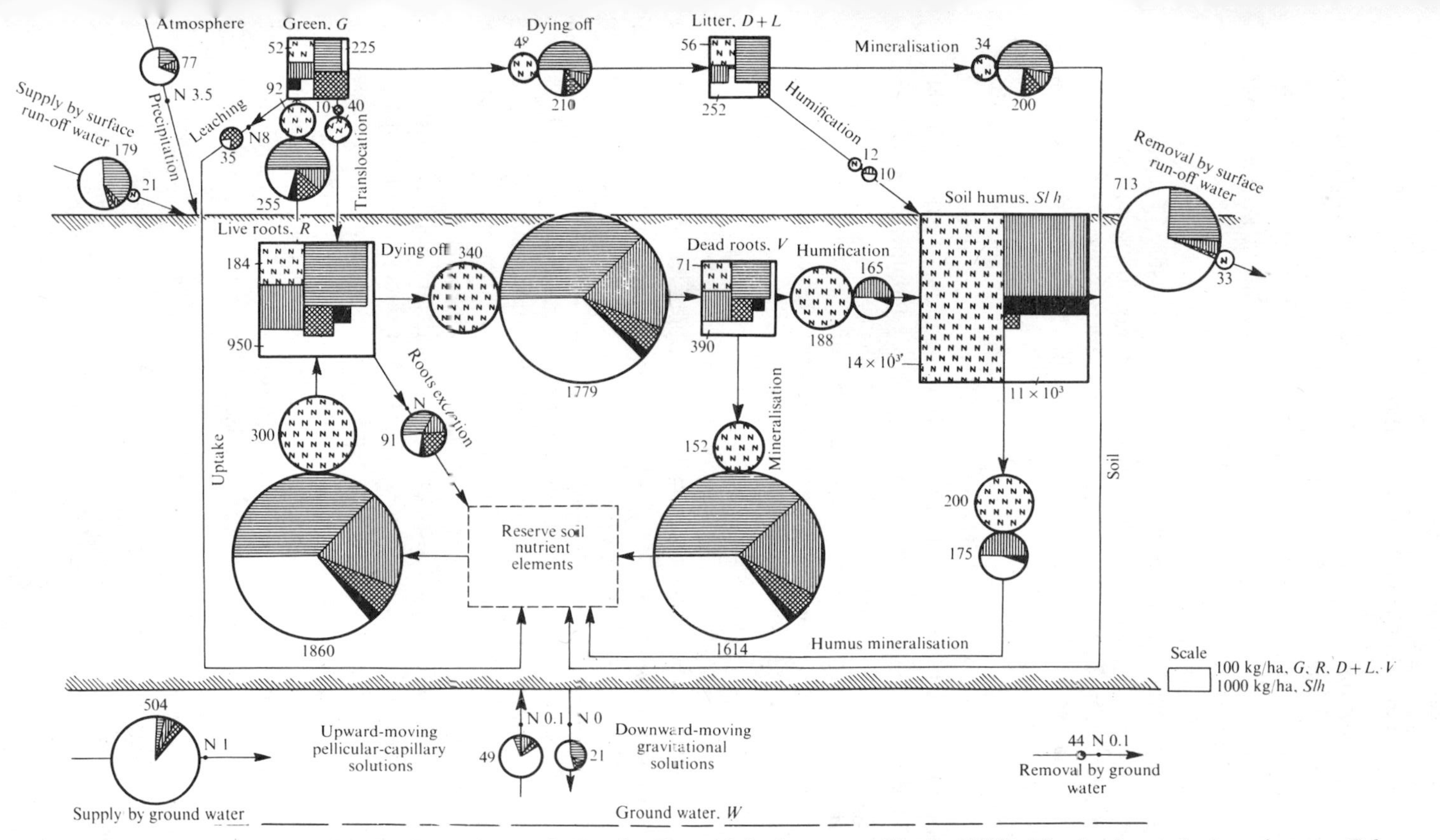

Fig. 14.2. Mineral cycle in a mesohalophytic meadow at the Baraba, Karachi site in western Siberia, USSR. The dominant plant species were *Calamagrostis epigeios*, *Poa angustifolia*, *Vicia cracca*, *Artemisia laciniata* and *Galatella biflora*. Proportional reserves of chemical elements are shown in boxes (in kg/ha). The arrows indicate processes and the circles on them represent the intensity of nutrient flow annually (in kg/ha). ▩, N; ▤, Si; ▥, Ca; ▦, K; ■, P; □, other elements.

The minerals and nitrogen taken from phytomass by herbivores passes through the excreta and dead animals immediately to the litter layer. Their values are included in dead plant material. Nitrogen exchange between the atmosphere and the soil during nitrogen fixation and denitrification also has not been included in this scheme (Fig. 14.2).

Within the course of abiotic processes, easily soluble salts (60–90 %) and some silicon are introduced to and released from the ecosystem. The entrance and release of nitrogen, calcium and, particularly, potassium and phosphorus are of smaller extent. The degree of intensity of biotic processes is considerably greater than that of abiotic processes. Most intense is the process concerned with the uptake of elements by roots from the soil. The principal elements appear to be nitrogen, silicon and calcium, as is reflected in the chemical composition of roots. The differential translocation of elements from roots to shoots results in potassium taking the place of calcium in the shoots as a dominant ion; the concentration of iron and aluminium decreases considerably, while that of phosphorus and sodium increases.

During the growing season, the chemical structure of shoots changes in that the content of potassium, sodium and chlorine decreases and that of calcium and silicon increases. The flow of potassium, nitrogen and sodium is most intensive during the period of rapid increase in shoot biomass in the first half of the growing season. In the second half of the season, the intensity of translocation of these elements from roots to shoots lessens and, sometimes, even the direction of the flow changes. In the autumn, approximately one-half of total nitrogen and 15–20 % of potassium and sodium return from shoots to roots. Simultaneously, leaching occurs of nitrogen, potassium, sodium, chlorine, phosphorus and sulphur from shoots into the soil by rainfall. As a consequence, concentrations of calcium, silicon, iron and aluminium in standing shoot biomass increases and litter becomes more enriched with these elements. The process of leaching continues in litter during its decomposition. While the contents of potassium, sodium, sulphur and also of magnesium and calcium decrease, those of silicon, aluminium and iron increase. During migration of nitrogen in this link of the cycle, its distribution changes; it accumulates in litter due to microbial activity and its concentration is 1.5–2 times as high as in dead shoots. During the decomposition of litter, approximately 25 % of nitrogen and 5 % of minerals are transferred to humus; the remainder is released to replenish the pool of mineral nutrients in the soil (Sln).

Of the total balance of mineral elements transferred by roots from the soil to the biological cycle, approximately 25 % of nitrogen and 10 % of ash elements are utilised in processes associated with the production of above-ground biomass, while the rest is utilised for the growth of the under-ground organs. A small quantity of several elements (K, Na, P, Si, Ca) is released from roots to soil in root excretion; the bulk, however, is released to the soil after the death of the under-ground organs and their decomposition. The latter process is followed by a new arrangement in the distribution of elements: more than half the nitrogen and only 10 % of mineral elements (mainly calcium and aluminium) are fixed in the humus. The remaining mass of chemical elements is released during mineralisation of the root remnants and returned to the mineral pool of the soil.

Úlehlová (1974) and Úlehlová, Klimo & Jakrlová (1976) give data on the mineral element input and distribution in three alluvial meadow ecosystems in Czechoslovakia with different water supply and different plant biomass production. Annual inputs of mineral elements by rain were (in g/m^2): 0.8, N; 0.005, P; 0.8, K; 1.0, Na; 3.8, Ca; 0.0, Mg; and by flood water: 8–13, N; 0.01–0.1, P; 0.1–35, K; 0.2–1.8, Na; 1.2–2.9, Ca; 0.6–2.3, Mg. Distribution of mineral elements in the vegetation are shown in Table 14.6. The yearly budget of the input of mineral elements by rain, floods and their return with litter, and of uptake by plants due to the yearly biomass increment shows that only the unmown wet *Glyceria* site is supplied adequately from the three natural sources. However, the inputs are inadequate in the two mown plant communities, as far as phosphorus and potassium are concerned.

The complex of organic substances of the soil differs considerably from that of its source – the phytomass (shoots and roots). Humus contains more nitrogen than other mineral elements, among which the highest concentration is that of calcium and aluminium and the lowest is potassium. Also, mineralisation of humus replenishes the pool of mineral elements – N, P, Ca. The pool of minerals of soil is used up mainly by plant uptake. Part of the elements may be lost during waterlogging of the soil. The pool of the soil minerals originates from several sources: (i) leaching from shoots and litter; (ii) release by root excretion, during mineralisation of litter and dead roots; (iii) mineralisation of humus; and (iv) inflow of chemical substances from external sources.

The elements essentially limiting plant growth are the most mobile ones (N, P, K). During the same season, they return by all routes to the mineral pool of the soil and may be utilised repeatedly by the plants. Elements, such as iron and aluminium, that are relatively inactive biologically, are taken up to a small extent by the above-ground phytomass and are neither washed out with rain, nor released from the roots. Instead they concentrate in litter and in dead roots and are incorporated into the humus; an intermediate position is held by silicon, calcium and magnesium.

In general, these relationships are characteristic of grassland ecosystems, and they differ only in the intensity of the various processes. In mesohalophytic, halophytic and hygrophytic grassland ecosystems, the role of abiotic processes increases; in steppe (grassland climate) their importance is considerably lower (Bazilevich *et al.*, in Sochava, 1973; Titlyanova & Bazilevich, in Rodin & Smirnov, 1975).

Table 14.5. *Reserves of chemical elements in biomass of certain meadow ecosystems and the amount of their consumption annually in primary production*

	1. Meadow steppe			2. Steppic meadow			3. Mesophytic meadow					
Ecosystem no.[a]	1.1	1.2	1.3	2.1	2.2	2.3	3.1	3.2	3.3	3.4	3.5	3.6
Above-ground maximum living biomass												
Sum of nitrogen and ash elements (g/m²)	24.0	14.9	11.4	9.7[b]	27.2	21.5	15.7[b]	20.9[c]	31.9	17.7	45.8	23.6
Dry weight of standing crop (g/m²)	300	230	180	250	330	330	400	550	450	224	494	546
Sequence of elements	K, N, Si, Ca	N, K, Si, Ca	Si, K, N, Ca	K, N	N, Si, K, Ca	N, Si, K, Ca	K, N	Ca, N, K, Si	K, N, Si	N, Si, K	N, K, Si	Si, K, N
Under-ground biomass												
Sum of N and ash elements (g/m²)	108.0	127.4	99.7	44.8	166.0	35.2	52.5	60.7	135.1	—	—	70.2
Dry weight of standing crop (g/m²)	2250	2170	2760	2020	2200	800	1670	1820	1610	946	738	1188
Sequence of elements	N, Ca, Si, K	Si, N, Ca, K	Si, N, Ca, K	N, K	N, Si, Ca, K	Si, N, Ca	N, K	N, K, Ca, Si	Si, Al, N, Ca, K	—	—	—
Total living biomass												
Sum of N and ash elements (g/m²)	132.0	142.3	111.1	54.5	193.2	56.6	68.2	81.6	167.0	—	—	93.8
Dry weight of total biomass (g/m²)	2550	2400	2940	2270	2530	1130	2070	2370	2060	1170	1232	1734
Sequence of elements	N, Ca, K, Si	Si, N, Ca, K	Si, N, K, Ca	N, K	N, Si, Ca, K	N, Si, K, Ca	K, N	N, K, Ca, Si	Si, Al, N, K	—	—	—
Above-ground standing dead and litter												
Sum of N and ash elements (g/m²)	38.0	25.5	24.3	3.1	—	73.9	3,3	35.5	.	—	—	—
Dry weight of standing dead and litter (g/m²)	460	360	290	120	320	900	96	680	.	95[e]	99[e]	1540
Sequence of elements	Si, Ca, N, K	Si, Ca, N, K	Si, N, Ca, K	N, Ca	—	Si, N, Ca, Al	N, Ca	N, Ca, Si, K	.	—	—	—
Yearly production of above-ground biomass												
Consumption of N (g/m²)	8.0	6.6	3.4	7.5	15.3	12.2	7.9	5.0	10.6[c]	6.6	23.0	6.5
Consumption of ash elements (g/m²)	33.0	19.5	13.5	10.3	34.2	22.2	20.4	17.6	27.4	17.8	52.2	33.6
Primary production of biomass (g/m²)	560	380	270	460	590	530	720	630	540	316	810	925
Yearly production of under-ground biomass												
Consumption of N (g/m²)	12.0	19.4	22.5	9.0	16.5	3.5	8.5	6.2	12.2	5.6	6.5	1.1
Consumption of ash elements (g/m²)	46.0	85.6	92.0	15.3	34.0	8.2	22.0	20.0	55.3	—	—	28.0
Primary production of biomass (g/m²)	1140	1400	3200	1090	730	270	970	960	800	496	440	876
Yearly production of total biomass												
Consumption of N (g/m²)	20.0	26.0	25.9	16.5	31.8	15.7	16.4	11.2	22.8	12.2	29.5	7.6
Consumption of ash elements (g/m²)	79.0	105.0	105.5	25.6	68.2	31.0	42.4	37.6	82.7	—	—	61.6
Primary production of total biomass (g/m²)	1700	1780	3500	1500	1320	800	1690	1590	1340	812	1250	1810
Consumption of N (g/100 g of biomass)	1.2	1.5	0.7	1.1	2.4	2.0	1.0	0.7	1.7	1.5	2.4	0.4
Consumption of ash elements (g/100 g of biomass)	4.7	5.9	3.0	1.7	5.2	3.8	2.5	2.4	6.2	—	—	3.4
Sequence of elements	K, N, Si, Ca	Si, Ca, N, K	Si, N, Ca, K	N, K	N, Si, K, Ca	N, Si, K, Ca	N, K	N, K, Ca, Si	N, Si, K, Al	—	—	—

Table 14.5 (*cont.*)

Ecosystem no.[a]	4. Mesohalophytic meadow		5. Halophytic meadow				6. Wet meadow	7. Wet halophytic meadow
	4.1	4.2	5.1	5.2	5.3[d]	5.4[d]	6.1	7.1[d]
Above-ground maximum living biomass								
Sum of nitrogen and ash elements (g/m²)	14.8	27.7	10.6	49.2	78.6[c]	50.9[c]	31.8[b]	38.0[c]
Dry weight of standing crop (g/m²)	200	240	210	150	438	406	1040	302
Sequence of elements	N, Si, K, Ca	N, K, Si, Ca	K, N, Cl, Si, Na	Cl, Na, K	—	—	K, N, Ca	—
Under-ground biomass								
Sum of N and ash elements (g/m²)	308.0	157.5	137.9	23.2	—	—	74.8	—
Dry weight of standing crop (g/m²)	3300	2330	1730	300	2522	—	2910	1810
Sequence of elements	Si, N, Ca, K	Si, N, Ca, K	Si, N, Na, K, Cl	Cl, Na, K	—	—	N, K	—
Total living biomass								
Sum of N and ash elements (g/m²)	322.8	185.2	148.5	72.4	—	—	106.6	—
Dry weight of total biomass (g/m²)	3500	2570	1940	450	2960	—	3950	2112
Sequence of elements	Si, N, Ca, K	Si, N, Ca, K	Si, N, K, Na	Cl, Na, K	—	—	N, K	—
Above-ground standing dead and litter								
Sum of N and ash elements (g/m²)		30.8	12.4	111.2	—	—	6.9	—
Dry weight of standing dead and litter (g/m²)	370	4150	260	240	208[f]	85[f]	230	129[f]
Sequence of elements		Si, N, Ca, K	Si, N, Ca, K	Ca, N, Na, Mg	—	—	N, Ca	—
Yearly production of above-ground biomass								
Consumption of N (g/m²)	8.1	9.2	3.3	2.2	9.8	9.6	9.5	6.2
Consumption of ash elements (g/m²)	18.0	25.5	13.6	72.1	95.0	54.0	29.3	45.0
Primary production of biomass (g/m²)	360	420	330	220	559	507	1270	406
Yearly production of under-ground biomass								
Consumption of N (g/m²)	25.0	30.0	7.5	1.7	—	—	15.3	—
Consumption of ash elements (g/m²)	68.0	186.0	79.0	21.5	—	—	14.0	—
Primary production of biomass (g/m²)	1100	2680	1080	300	1092	760	1140	578
Yearly production of total biomass								
Consumption of N (g/m²)	33.1	39.2	10.8	3.9	—	—	24.8	—
Consumption of ash elements (g/m²)	86.0	211.5	92.6	93.6	—	—	43.3	—
Primary production of total biomass (g/m²)	1460	3100	1410	520	1651	1267	2410	984
Consumption of N (g/100 g of biomass)	2.3	1.2	0.8	0.8	—	—	1.0	—
Consumption of ash elements (g/100 g of biomass)	5.8	6.8	6.6	18.0	—	—	1.8	—
Sequence of elements	Si, N, Ca, K	Si, N, Ca, K	Si, N, K, Na	Cl, Na, K	—	—	K, N, Ca	—

[a] Explanations for grassland types and authors are in Table 14.4. [b] Values in this column do not include Si, Al, Fe, Mn, Cl, S.
[c] Values in these columns have been estimated only.
[d] Data from Ketner (1972). Types of salt marshes (Netherlands): 5.3, *Plantagini–Limonietum*: *Plantago maritima, Limonium vulgare, Triglochin maritima, etc.*, 5.4, *Puccinellietum maritimae*: *Puccinellia maritima, Plantago maritima, Salicornia europaea*, etc; 7.1, *Junco–Caricetum extensae*: *Juncus gerardii, Plantago maritima, Triglochin maritima, Scirpus rufus, Carex extensa*, etc.
[e] Litter only. [f] Standing dead only. ., not determined. —, no data.

Table 14.6. *Turnover* (g/m^2) *of mineral elements in alluvial meadows* (*Úlehlová* et al., *1976*)

	Nitrogen	Phosphorus	Potassium	Sodium	Calcium	Magnesium
Dry meadow with *Festuca sulcata* (mown)						
Permanent plant pool	12.3	2.7	7.6	1.1	1.3	0.6
Uptake by plants per year	14.2	2.0	10.1	0.7	7.1	0.7
Litter return into the soil per year	6.1	1.1	2.8	0.5	4.6	0.3
Moist meadow with *Alopecurus pratensis* (mown)						
Permanent plant pool	12.2	0.8	4.4	0.8	0.9	0.5
Uptake by plants per year	18.7	1.9	13.5	8.2	8.0	0.8
Litter return into the soil per year	5.9	0.5	2.1	0.5	4.4	0.2
Wet meadow with *Glyceria maxima* (unmown)						
Permanent plant pool	35.9	2.7	25.6	2.4	1.5	0.9
Uptake by plants per year	32.5	3.3	28.1	1.5	7.2	1.3
Litter return into the soil per year	29.7	3.3	28.1	1.5	4.5	1.3

In all grassland ecosystems, the most intensive processes are those exchange processes associated with living organisms. Even when mineral fertilisers are applied artificially, the intensity of mineral cycles influenced by biogenic processes is still high in comparison with that of abiotic processes. This is due to the leading role played by the under-ground organs in the exchange processes occurring in grassland ecosystems.

Clark *et al.* (in Chapter 8 of Breymeyer & Van Dyne, 1979) have considered in detail the principles of nutrient cycling based primarily on studies in natural grasslands in North America.

15. Use and management of meadows

B. SPEIDEL

Some consideration has already been given in Chapter 12 of this volume to the use of semi-natural grasslands for pasture by domesticated animals. Here emphasis will be placed on their use as haylands. In many instances management consists of a combination of these two methods of utilisation. The character of these meadows depends on both the supply of water and the degree of concentration of salts; either or both of these frequently are in excess.

Meadows play a substantial role in water management. They are a means of accumulating water reserves in spring, to be exploited during summer drought. They are capable of filtering out soluble substances, detrimental to human health, that enter surface water through practices followed in intensive agriculture and other activities of man. The relatively high coefficient of plant transpiration in meadows adds to humidity in the air.

In intensively farmed regions, meadows often represent a principal refuge for the natural flora and fauna. Thus, they arc an important means of protecting gene pools. However, expanding arable agriculture constitutes a constant threat to many meadows because of the possibility of drainage. It is to be hoped that greater recognition of the potential productivity of meadows will, in future, prevent unwise decisions from disrupting this natural resource by tillage.

Fresh meadows

The fresh meadows that were subjected to intensive IBP investigation were characterised by one of the following species: *Arrhenatherum elatius* (Arrhenatheretum elatioris), *Trisetum flavenscens* (Trisetetum flavescentis), *Miscanthus sinensis* (Miscanthetum sinensis).

The *Arrhenatherum* ecosystem usually occurs in valleys on the lower part of slopes and terraces where flooding does not occur. It develops in a warm climate and is characterised by fresh soil (perhaps after drainage) and a good supply of nutrients. The moisture in the upper layer of soil is almost entirely dependent on precipitation. Short periods of flooding can be withstood, but more prolonged flooding or waterlogging cannot be tolerated. The sites studied are located in Poland (Ispina and Kazuń in the Vistula Valley, and Saspowka Valley of Ojców National Park). Because of their high productivity and good fodder quality, these meadows are highly rated and usually well managed. According to the level of artificial fertilisation, they can produce two or three crops of hay per year. Instead of a third mowing they are sometimes grazed. In the case of permanent grazing, however, the community that is characterised by *Arrhenatherum elatius* is replaced by a pasture typified by *Lolium* and *Cynosurus* (Lolio-Cynosuretum). Regular organic fertilisation (particularly with liquid manure) results in a considerable increase in certain species of umbellifers, the hard stems of which reduce the quality of the fodder. If the fields are not used, there is a fairly rapid degradation in the vegetation because of accumulation of litter. As a result, both density of shoots and number of species present decline. For example, in a protected area at Ojców, Poland, in which 24 species occurred per 1/16 m^2 in 1964, the mean number decreased to 11 in 1971 and 4.5 in 1973 (Jankowska, 1971 and personal communication). A decrease in production is always associated with this. Since this type of ecosystem is usually suitable for arable agriculture, many potential sites have been under cultivation for a long time.

Especially in the southern part of central Europe, the plant species associated with *Arrhenatherum elatius* are quite varied. When the forbs are in full bloom the landscape is very colourful. Since this community is found commonly, it is not likely to become extinct in the near future. Probably those stands that are protected in nature reserves and in national parks will satisfy the need for protection for some time to come.

The community that is characterised by the occurrence of *Trisetum flavescens* occurs in submontane and montane areas. The vegetative component is largely independent of ground water and receives its moisture principally from precipitation. Where management conditions are not favourable to the survival of a community of *Arrhenatherum elatius* and other tall grasses (*Alopecurus pratensis*, *Dactylis glomerata*, *Festuca pratensis*), there is a type dominated by *Festuca rubra* and other short grasses, such as *Agrostis tenuis* and *Nardus stricta*. The principal IBP study area of this type of ecosystem was the Solling site in the Federal Republic of Germany. In this area the short-grass type prevails (Speidel & Weiss, 1972). Productivity and fodder quality are less than in the types dominated by tall grasses. It occurs principally on the higher sites of the European Central Uplands where the nutrient content of the soil is relatively poor and where there is less fertilisation because of the distance of the sites from villages. Experiments with fertiliser treatments show, however, that the true production potential of these meadows is often not fully exploited. Management is often impeded by the sloping site. Generally, only two crops can be produced, and the second crop is often not satisfactory. This type is partly used for both grazing and mowing. Areas that are situated on steep slopes and those which are too far removed from the farm, or are too stony, are used extensively as permanent pasture for livestock. Shrubs become abundant in the absence of mowing. The sward of the mown areas is generally dense and even on slopes provides good protection from erosion. The grazed slopes are, however, subject to an erosion hazard because of disturbance of the plant cover through treading of livestock. The Solling site lies in a meadow complex at the edge of a wooded area near a holiday resort. The extensive stands of this *Trisetum–Festuca* complex that are situated on slopes and in valleys of the Central Uplands of Europe are attractive to hikers and holiday makers. Potentially these areas could be planted to forest, but the danger of erosion is considerable because of disruption of the grass cover.

The *Miscanthus* grassland of the Kawatabi Station in Japan (Numata, 1975; Numata, Iizumi & Iwaka, 1968; Sato *et al.*, in Numata, 1973) resulted from clearing a beech forest in 1894, since which time it has been used both for hay production and as grazing land. The principal associate of *Miscanthus sinensis* (a tall bunch grass) is bracken (*Pteridium aquilinum*). The best hay crop is produced with one cut per year at heading time in September. Fertilisation with nitrogen increases the productivity of dry matter and crude protein of shoots. However, cutting several times per year reduces the productivity of hay, through a decrease in *Miscanthus* and an increase in undesirable species, particularly bracken. Productivity also declines as a result of grazing, because of selection by cattle of the most palatable species. It is then necessary to mow or burn at least every fourth year to control undesirable species. Under intensive grazing the *Miscanthus* type gives way to a community characterised by *Zoysia japonica* and erosion becomes a danger on slopes where animals form paths. These meadows also deteriorate rapidly when neither mown nor grazed. Fodder production can be improved through cultivation and seeding to a mixture of orchard grass (*Dactylis glomerata*) and ladino clover (*Trifolium repens*).

Inundated meadows

The most intensive studies of inundated meadows during IBP took place in the Lanžhot area in Czechoslovakia and the Ispina site in southern Poland. These types of meadows are subject to wide annual variations in groundwater level. They are regularly inundated in spring. The height and duration of the flooding, as well as the depth of the ground water during the dry period in summer and autumn, determine the composition of the flora in these meadows and also influence management plans.

The Lanžhot study included a series of meadows that can be described as wet, damp, moist and dry, successively, as a result of a decline in the degree of flooding (Balátová-Tuláčková, 1966, 1968 (in Dykyjová, 1970), 1973; Rychnovská, 1972). The wettest community, dominated by *Glyceria maxima*, is usually flooded until the end of June. The community dominated by *Phalaris arundinacea* (Phalaridetum arundinaceae) is subjected to particularly wide variations in the level of the ground water and is usually not flooded after the end of May. In the *Alopecurus pratensis* type (Gratiola officinalis–Carex praecox-suzae-Ass.) water recedes after the middle of April. In the driest type, dominated by *Festuca sulcata* (Serratulo–Festucetum commutatae), flooding occurs only in wet years and ground water reaches the upper layers of soil only early in the growing season. In the wettest type, cutting for hay is restricted to autumn and then only in dry years. The other communities are mown for hay twice annually (in June and August). While no nutrients are applied artificially, there is addition of nutrients to the system by flood water. It would be advantageous, however, to regulate the amount and duration of the inundation. Without regular mowing, trees encroach. If the inundation were eliminated and harvesting of hay continued, the area dominated by *Festuca sulcata* would expand, with a decline in productivity. Under these conditions it would be possible to cultivate the meadow, but crop production would be limited by structural deterioration associated with drying out of the clay soil. This reduction in production potential would be irreversible. These are the last remnants of natural, highly productive inundated meadows in central Europe and should be protected against such deterioration.

In southern Poland an ecosystem dominated by *Carex gracilis* (Caricetum gracilis) occurs in an area with a similar degree of inundation to that of the *Phalaris* type in Czechoslovakia. However, the level of the ground water does not usually fall so much during the dry period, so that soil moisture conditions are good throughout the summer. Mowing for hay takes place twice annually. The hay is of poor quality because of the high content of *Carex gracilis*; it is used partly for fodder and partly for bedding. Improvement of fodder is possible by restricting inundation and lowering the ground-water level; however, opportunities for drainage are usually not present. In areas in the old river bed subjected to flooding, a *Scirpus lacustris–Phragmites communis* (Scirpo–Phragmitetum) swamp occurs with *Equisetum limosum*, *Iris pseudacorus*, *Acorus calamus* and *Carex gracilis* as the chief associates. This area is flooded during most of the growing season, but is mown in late autumn or in winter and used for bedding for animals. *Equisetum* and *Iris* are regarded as poisonous or detrimental to animal health. However, the community is of great ecological importance in protecting the river bank from erosion and raising the level of the soil surface during the silting operation.

Salt marsh communities

The most intensive IBP study in salt marshes was located in a nature reserve on the island of Terschelling in the Netherlands (Ketner, 1972). Salt marshes along the North Sea coast both create land, by retaining silt washed in by the tide, and protect it, by preventing erosion at low tide. The only practical use of such areas is as meadow and grazing lands. The Netherlands study included three plant associations. The area adjacent to the coast and subject to the most flooding and the highest salt content is dominated by *Puccinellia maritima* (Puccinellietum maritimae). This is bordered inland by an area dominated by *Plantago maritima* and *Limonium vulgare* (Plantagini–Limonietum) which in turn gives way to a community of *Juncus gerardii* and *Carex extensa* (Junco–Caricetum extensae). Because of flooding, the hay harvest is never assured; but, when it is possible to mow, the hay is highly valued because of its good quality. This high quality relates to the absence of poisonous plants, of forbs of low nutritive quality, and of animal parasites. This fodder is also rich in minerals, particularly sodium (Wetzel, 1966). Nutrients are only added to this system through sedimentation. This land is of importance for human recreation because of the scarcity of semi-natural landscapes in the Netherlands.

Meadow steppe

Study of meadow steppe was undertaken during IBP at the Karachi Experimental Station at Baraba in the USSR (Bazilevich, 1974). There, the landscape is characterised by low ridges of chernozem soil (rich in nutrients) interlain by depressions with saline and chernozem meadow soil in complexes with solonetz soil. The ridges are cultivated, but the depressions are used as meadow and pasture. While meadow steppe on slightly saline soil produces relatively plentiful hay of good quality, the productivity of solonetsic steppe is much less satisfactory and the quality of the fodder varies with the proportion of halophytes. Swamps can be mown or grazed only in dry years. While their production potential is relatively high, the quality of the fodder is low. Improvement is possible through leaching of the salts, but this is complicated by lack of drainage and the complexity of the relief. The alternative method in solonetsic soils with a thick humic horizon is to plough deeply. Where there is a thin humic horizon, surface loosening with a heavy harrow is sufficient. In addition, it is necessary to add nitrogenous and phosphoric fertilisers together with gypsum. When cultivated in this way meadow–steppe solonetsic soils can be sown with such salt-tolerant species as *Melilotus albus*, *Bromus inermis* and *Medicago sativa*. In meadow solonetsic soils rhizomes and seeds left after tillage produce a natural grass sward with an abundance of *Alopecurus ventricosus*, *Hordeum brevisubulatum*, *Agropyron repens*, *Puccinellia distans* and *P. tenuifolia*.

Part IV. Tropical grasslands

Subeditor: R. MISRA

16. Introduction

K. C. MISRA

Tropical and subtropical grasslands occupy an area of approximately 32000000 km^2 in the plains and mountains located within 30°N and 30°S (Misra, 1974). Detailed knowledge concerning trophic structure and functional relationships of organisms in these ecosystems has mostly been acquired during the last decade and primarily as a result of IBP investigations. The locations in which detailed studies of ecosystem parameters have been made are listed in Table 16.1, together with an indication of the nature of each site and the type of study undertaken. Of the 22 sites listed, only two are south of the equator and only one is in the western hemisphere.

Tropical grasslands are subject to the influence of strong seasonality with respect to precipitation and, at higher latitudes, to temperature. The annual evaporation demand exceeds precipitation at most of the sites and occasional long spells of drought are characteristic. At the Indian sites there are three well-marked seasons, namely, summer (March to mid-June), rainy (mid-June to October) and winter (November to February). Day length ranges from 10 h during winter to 14 h in summer.

The Indian sites are grouped in Table 16.1 into four ecoclimatic types on the basis of the moisture index values of Thornthwaite (1948). The data available to this author did not permit classification, in this manner, of the sites in Africa and Panama.

The Indian sites are located in the monsoon belt. Most of the precipitation occurs in July and August, although the rainy season extends to mid-October. About 80–90 % of the total annual precipitation is received during this period. The relative humidity varies in accordance with temperature and precipitation. Maximum values are recorded during the monsoon, while the minimum occur in the hot, dry summer. Mean annual values for the moist subhumid, dry subhumid and semiarid zones are, respectively, 62, 53 and 48 %, while the moisture index decreases from the humid to semiarid zones (Table 16.2). Mean annual temperature rises as precipitation declines. January and February are the coldest months, while May and June are the hottest. Monthly mean temperature fluctuates from 15 to 30 °C, 16 to 35 °C, 15 to 35 °C and 25 to 43 °C, respectively, in the humid, moist subhumid, dry subhumid and semiarid zones. Solar radiation, temperature, precipitation, relative humidity and soil moisture are the main variables that influence the productivity of tropical grasslands. The amount and period of rainfall influence the structure and composition of the plant communities in the various ecoclimatic zones. Mean averages of these variables are indicated in Table 16.2.

Table 16.1. *List of IBP study sites in tropical grassland, with an indication of the type of studies undertaken*

Country, region and site	Location		Elevation (m) above m.s.l.	Type of data available[a]	Dominant plant taxa	Source of information
India						
Humid region						
Ambikapur	23°08′N	83°15′E	599	1, 3, 4	*Bothriochloa, Digitaria, Eragrostis*	Naik (1973)
Moist subhumid region						
Varanasi	25°20′N	83°0′E	76	1, 3, 4, 5	*Dichanthium annulatum, Heteropogon contortus, Cynodon dactylon, Desmodium triflorum*	Singh (1967), Singh (1972)
Jhansi	25°07′N	78°07′E	251	1, 3, 4, 5	*Heteropogon contortus, Sehima nervosum, Dichanthium annulatum*	Shankar *et al.* (1973)
Sagar	23°50′N	78°40′E	513	1, 3, 4, 5	*Heteropogon contortus, Dichanthium, Cymbopogon*	Jain (1971)
Delhi	28°35′N	77°12′E	216	1	*Heteropogon contortus*	Varshney (1972)
Ratlam	23°19′N	75°03′E	486	1, 3, 4, 5	*Sehima nervosum*	Billore (1973)
Dry subhumid region						
Kurukshetra	29°57′N	76°49′E	251	1, 3, 4, 5	*Bothriochloa pertusa, Cenchrus ciliaris, Cynodon dactylon*	Singh & Yadava (1974)
Ujjain	23°11′N	75°43′E	492	1, 3, 4, 5	*Dichanthium annulatum, Cymbopogon martinii, Sehima nervosum*	Misra (1973)
Rajkot	22°18′N	70°47′E	138	1, 3, 4, 5	*Cenchrus ciliaris, Heteropogon contortus, Aristida depressa*	Pandeya (1974)
Semiarid region						
Jodhpur	26°18′N	73°01′E	224	1, 2, 3, 4, 5	*Lasiurus indicus, Cenchrus, Desmostachya, Zizyphus nummularius*	Gupta *et al.* (1972), Mann & Ahuja (1974)

[illegible]	[illegible]	[illegible]	[illegible]	[illegible]	*Lasiurus indicus*, *Cenchrus*, *Desmostachya*, *Zizyphus nummularius*	Kumar & Joshi (1972)
Udaipur	24°32′N	73°55′E	560	1	*Dichanthium annulatum*, *Apluda mutica*, *Cassia tora*	Vyas *et al.* (1972)
Ethiopia						
Geech Plateau	13°20′N	38°29′E	3660	1	*Danthonia subulata* Rich., *Festuca abyssinica* Hochst. ex Rich.	Klötzli (1975)
Ivory Coast						
Lamto	6°13′N	5°02′E	120	1, 2, 4, 5	Palm savanna – *Borassus aethiopium* Mart., *Loudetia*, *Andropogon schirensis* Hochst., *Hyparrhenia*	Vuattoux (1970), Lamotte *et al.* (1974)
Senegal						
Fété Olé	14°N	15°W	40	1, 2	*Aristida*, *Panicum*, *Guiera senegalensis* Gmel., *Balanites aegyptiaca* Delile, *Acacia senegal* Willd., *Boscia senegalensis* Lam.	Lepage (1974), Poulet (1974), J. C. Bille (IBP data), Gillon & Gillon (1973), Morel & Morel (1974)
Ferlo	16°13′N	15°06′W	40	1, 2		
Nigeria						
Lagos	7°30′N	4°E	Sea level	1, 2	*Andropogon gayanus* Kunth.	Egunjobi (1974)
Uganda						
Mweya Peninsula	3°N	30°30′E	2000	1, 2	—	Edroma (1974)
Ruwenzori Park	0°10′S	29°55′E	922	1	*Hyparrhenia filipendula* Stapf., *Themeda triandra* Forsk.	C. D. Piggott (IBP data)
Zaire (Albert Park)						
Kivu	2°5′S	28°E	2000	1	*Themeda*, *Heteropogon*, *Imperata*	Bourlière & Hadley (1970)
South Africa						
Welgevonden	24°20′S	26°51′E	940	1, 4	Arid open savanna – *Acacia erubescens*, *Eragrostis rigidior*, *Panicum maximum*	J. Van Wyk (IBP data)
Panama	9°5′N	79°30′W	180	1, 2	*Hyparrhenia rufa* Stapf.	Breymeyer (1974)

[a] Type of data available: 1, primary producers; 2, secondary producers; 3, edaphic and nutrient status, decomposers; 4, climatic parameters; 5, energy status.

Table 16.2. *Typical climatic data for Indian grasslands. The values are annual means computed from 15 to 20 yr of recent meteorological records*

Parameter	Ecoclimatic zone: Humid	Moist subhumid	Dry subhumid	Semiarid
Precipitation (mm)	1381	1190	942	492
Temperature (°C)	24	26	25	33
Relative humidity (%)[a]	—	62	53	48
Aridity index[a]	24	25	47	55
Moisture index[a]	−22	—	−27	−33

[a] After Pandeya (1974).

The soils of the Indian sites are mostly of loam texture, with a marked increase in sand in semiarid regions (Table 16.3). They are mostly of alluvial and lateritic origin; however, the sandy soils of the semiarid zone are of aeolian origin. The soil profile is poorly developed; nevertheless, at certain places in the dry subhumid and subarid zones a hard pan of calcium carbonate and bicarbonate is formed at a depth of 1 to 2 m. Soil colour varies from pale brown to deep black, with intergrading shades of dark brown and reddish brown. Bulk density of soil increases with increasing aridity. Rooting depth of the grasses and forbs extends to 1 m below the soil surface. Soil moisture

Table 16.3. *Edaphic characteristics of typical tropical grassland soils in the sites studied in India (compiled from sources indicated in Table 16.1)*

Soil characteristics	Ecoclimatic zone: Humid	Moist subhumid	Dry subhumid	Semiarid
Genesis	Alluvial	Alluvial	Alluvial	Aeolian
Colour	Red to yellow	Pale brown to dark brown, black	Red to deep black	Pale brown to dark reddish brown
Mechanical composition (%)				
Sand	32–34	45	23–77	83–89
Silt	42–43	29	15–35	4–9
Clay	20–23	26	2–42	4–9
Textural class	Loam	Sandy loam	Sandy loam	Sandy loam
Bulk density (%)	—	1.35	1.40	1.60
Field capacity	22–27	17	19	12–14
pH	7.0	6.0–7.3	6.0–8.5	7.6–9.2
Organic matter (%)	—	3.7–3.9	1.8–4.0	0.15
Total nitrogen (%)	—	0.08–1.10	0.03–0.69	0.024–0.029
Total phosphorus (%)	—	0.063	0.023–0.028	0.040

Plate 16.1. The humid tropical grassland at Ambikapur, India, dominated by *Bothriochloa pertusa*. (Reprinted from Naik, 1973.)

Plate 16.2. The moist subhumid tropical grassland at Ratlam, India, dominated by *Sehima nervosum*, as it appears in September. (Reprinted from Billore, 1973).

Plate 16.3. Tropical *Andropogon–Hyparrhenia* grassland associated with *Borassus aethiopium* in the palm savanna on the Lamto site in the Ivory Coast, as it appears in late December. (Photograph by R. T. Coupland.)

shows wide seasonal fluctuation, according to the distribution of precipitation. Texture is another important factor that affects soil moisture. Field capacity ranges from 12 to 19 %. Depth of the water table fluctuates from 8 to 10 m in the rainy season to 15 to 20 m below the soil surface in the dry season. The soils are slightly alkaline, rich in calcium and poor in nitrogen and phosphorus. Soil organic matter decreases with increasing aridity.

Tropical grassland communities are rich in species. All the sites studied have the character of savannas with widely scattered trees (Plates 16.1, 16.2, 16.3), but annual and perennial graminoids occupy the major portion of the landscape. The Indian sites are dominated by perennial grasses of the following genera: *Dichanthium*, *Cynodon*, *Heteropogon*, *Aristida*, *Cenchrus*, *Cymbopogon* and *Chrysopogon*; in the African sites species of *Andropogon*, *Danthonia*, *Festuca* and *Guiera* dominate. The trees are mostly species of *Acacia*, *Prosopis*, *Zizyphus*, *Boscia*, *Combretum*, *Euphorbia* and *Albizia* (Table 16.1). Biological spectra of the grasslands located in the moist subhumid and dry subhumid zones show preponderance of therophytes (more than 60 %) and of cryptophytes (more than 6 %), while hemicryptophytes and phanerophytes are poorly represented.

Vertical distribution of the photosynthetic structure is in three strata: the tree, shrub and grass layers. In addition, a crust of green and blue-green algae is formed on the soil surface during the wet period of the year. The trees are umbrella shaped and generally 10 m high, while height of shrubs is 2–3 m and that of grasses ranges from about 2 cm in dry areas to approximately 2 m in the humid areas.

With the onset of rains these grasslands rapidly turn green and remain productive during the wet part of the year, while in the hot and dry season the graminoids become dormant. The annual ephemerals complete their life cycle during the wet period and perennate through seeds or spores in the soil. The perennials survive in a dormant condition.

These grasslands provide an ideal habitat for a large variety of herbivores and carnivores. These include large conspicuous mammals, as well as many small vertebrates (birds and reptiles) and invertebrates (termites, flies, molluscs and oligochaetes).

Almost all of the sites under study were subject to biotic interference, especially grazing. Grazing has been so severe in India that it is difficult to locate a virgin grassland (Puri, 1960). Studies were made at several sites to determine the effects of such treatments as burning, scraping, clipping and exclusion from grazing. These provided valuable information regarding structure and function in these ecosystems.

Scientific names of most Indian plants used in Chapters 16 to 21 of this volume are according to Bor (1960) for grasses and to Duthie (1903–20) for non-grasses. Authorities are given for species not listed in, and for nomenclature which deviates from, these mannuals.

17. Primary production

J. S. SINGH & M. C. JOSHI

The tropical grasslands are seral in nature, owing their origin to deforestation or abandoned cultivation. According to Whyte (1974), these grasslands, occurring in a forest climate, may be considered as tertiary communities, to distinguish them from secondary communities constituted by shrub and by shrub–tree complexes. They are maintained at various successional levels, mainly due to such biotic factors as grazing and harvesting and to recurrent fire. Differences in the habitat complex (both edaphic and climatic), age and mode of origin, and intensity of biotic effects result in an array of very diverse grassland communities. Since these communities represent different seral stages, each having its unique biotic complement, they differ widely in their efficiency of energy capture and water use.

Despite the extensive geographical development of tropical monsoonal grasslands, as discussed in the preceding chapter, relatively few sites were studied under the IBP and few data sets are available. Further, the various aspects of the primary producer subsystem have not been studied to the same extent and precision in all sites. Because of these limitations, it is difficult to make a large number of worthwhile generalisations. We have attempted here, however, to deal with as many aspects of the structural and functional attributes of the tropical grassland vegetation as possible using the cross-section of available results. In some instances we have referred to pre-IBP and IBP-related studies.

Species structure and phenological diversity

On the basis of a country-wide survey of grasslands in India, Dabadghao & Shankarnarayan (1973) have recognised five broad grass-cover types: *Sehima–Dichanthium*, *Dichanthium–Cenchrus–Lasiurus*, *Phragmites–Saccharum–Imperata*, *Themeda–Arundinella* and temperate–alpine. In these grassland types 56 species of legumes are associated with *Sehima–Dichanthium*, 19 with *Dichanthium–Cenchrus–Lasiurus*, 16 with *Phragmites–Saccharum–Imperata*, nine with *Themeda–Arundinella* and six with the temperate-alpine type. According to Beard (1967), the characteristic grass genera of monsoonal affinities common to tropical northeastern Australia and Africa are *Dichanthium*, *Bothriochloa*, *Chrysopogon*, *Eragrostis*, *Sorghum*, *Heteropogon*, *Themeda*, *Xerochloa*, *Ereapogon*, *Sporobolus*, *Triraphis*, *Chloris*, *Panicum* and *Cenchrus*.

Within the Indian subcontinent, the grass-cover types recognised by Dabadghao & Shankarnarayan (1973) are not homogeneous. Detailed studies within any one type reveal the occurrence of several grassland associations due to differential intensity of biotic influences and local variations in topography and soil depth. Thus, Pandeya (1964) was able to delimit eight grassland associations, namely, *Aristida–Melanocenchris*, *Heteropogon–Andropogon*, *Cymbopogon–Eulalia*, *Themeda–Coix–Ischaemum*, *Bothriochloa–Dichanthium*, *Sehima–Chrysopogon*, *Tripogon* and *Cynodon–Bothriochloa–Dichanthium* at Sagar in central India which would otherwise come under the *Sehima–Dichanthium* cover type of Dabadghao & Shankarnarayan (1973). Different grassland associations within a single climatic area perhaps represent different stages in local seres.

The flora of the tropical grasslands, in general, is dominated by therophytes and cryptophytes because of the strong periodicity in climate and intense biotic pressure (Pandeya, 1964; J. S. Singh, 1967; Singh & Yadava, 1974). Therophytes are abundant during the rainy season and are able to survive the hostile dry and hot periods as seeds. Cryptophytes, on the other hand, are able to withstand severe grazing pressure because of the hidden, subsurface position of the buds. Thus, cryptophytic species dominate the vegetation during the winter and summer seasons. However, because of their origin from degeneration of forests, these grasslands do contain variable amounts of phanerophytic elements. Thus, scattered trees and shrubs of *Prosopis cineraria* McBride (khejra), *Zizyphus nummularia* Wt. & Arn. (beri), *Balanites aegyptiaca* Mart. (hingot) and *Capparis decidua* Edgew. (karil) are common in semiarid grassland (Kumar & Joshi, 1972; Gupta, Saxena & Sharma, 1972). For a more humid *Sehima–Dichanthium* grassland, Dutta & Pandey (1971) have reported heavy infestations of such shrubs as *Zizyphus nummularia*, *Acacia catechu* (kher), *Acacia leucophloea* (reonj), *Butea monosperma* Taub. (dhak), *Mimosa rubicaulis* (shiah kanta), *Carissa spinarum* (karaunda) and *Lantana camara* (kuri), accounting for up to 30 % of the canopy cover. In the savanna vegetation of Africa, the woody components are more important. *Guiera senegalensis* Gmel., *Balanites aegyptiaca* Mart. (hingot), *Grewia bicolor* Juss., *Commiphora africana* Engl., *Acacia senegal* Willd. (gum arabic) and *Boscia senegalensis* Lam. have been recorded to be most important ligneous species in the savanna grassland at Fété Olé in Senegal where phanerophytes average 133 individuals/ha (J. C. Bille, IBP data). Woody components of the savanna at Lamto (Ivory Coast) have been described by Vuattoux (1970).

The impact of strong seasonality of climate manifests itself in marked phenological diversity. Detailed phenological observations on individual species have been made in India at Varanasi (Singh, 1967), at Kurukshetra (Singh & Yadava, 1974) and at Pilani (Kumar, 1971; Gill, 1975). A maximum number of species produce seedlings and vegetative sprouts just after the first few showers of monsoon rains, starting usually in the later part of June each year. The majority of species flower, fruit and produce

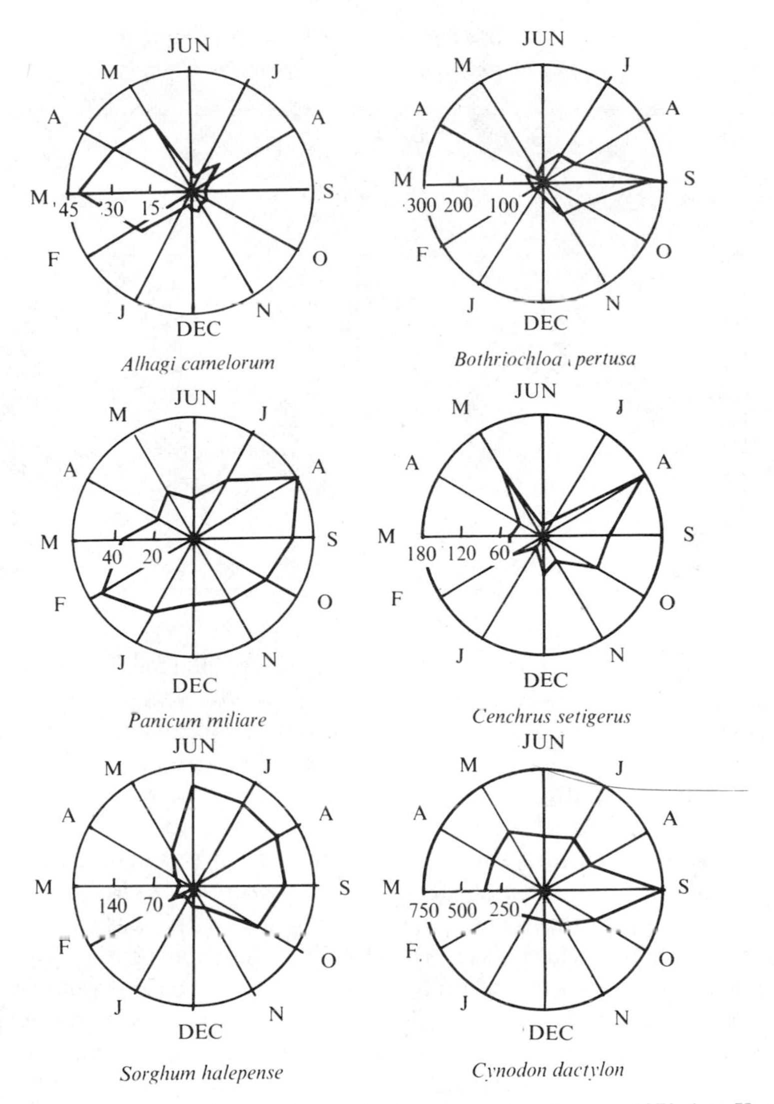

Fig. 17.1. Variation in the density of grassland species during the year 1970–1 at Kurukshetra (tillers/m^2).

mature seeds during the rainy season. While sprouting, flowering and fruiting in a number of perennial species are activated during winter, others are favoured by longer photo-periods and high temperatures of summer. On the basis of phenology, J. S. Singh (1967) grouped various species in the *Dichanthium* grassland at Varanasi into five categories: (i) those that complete their life cycle in one season, e.g. *Alloteropsis cimicina* and *Lindernia ciliata* Pennel;

Table 17.1. *Seasonal variations in total plant density (shoots/m²) in four tropical grasslands in India*

Month	Kurukshetra 1970–1 Singh & Yadava (1974)	Varanasi 1964–5 Singh (1967)	Ujjain 1971–2 Misra (1973)	Ratlam 1971–2 Billore (1973)
May	931	—	—	—
June	1490	4024	142	1213
July	1552	8214	344	2506
August	2041	9874	572	3934
September	2143	11055	822	4700
October	1408	9312	862	1271
November	830	1997	376	1112
December	471	2857	104	983
January	564	3642	367	460
February	780	2196	555	89
March	817	3242	419	55
April	950	2623	267	44
May	1217	1963	255	21

(ii) those that complete their life cycle in one season, but recur in the following season also, e.g. *Paspalum royleanum* and *Panicum humile*; (iii) those that germinate, flower and fruit in one season, but complete seed maturation in the next season, e.g. *Volutarella ramosa* Santapau (rissa) and *Trichodesma indicum* (chota kulfa); (iv) those with two or more flowering and fruiting flushes in a year, e.g. *Desmodium triflorum* (kudaliya) and *Dichanthium annulatum* (marvel grass); and (v) those with one flowering and fruiting flush each year, although the period of flowering may cover two seasons, e.g. *Indigofera linifolia* (torki) and *Evolvulus alsinoides* (sankhapushpi).

Total vegetation density, as well as density of individual species, fluctuates throughout the year in response to changes in environmental variables. The pattern of changes in density differs from species to species and exhibits differential adaptations to growing conditions. For example, in a study of grassland vegetation at Kurukshetra, Singh & Yadava (1974) noted that some species exhibit maximum density during the rainy season (e.g. *Bothriochloa pertusa* (sandhur) and *Sorghum halepense* (Johnson grass)) and others during the summer season (e.g. *Alhagi camelorum* (camel thorn) and *Evolvulus alsinoides*) or winter season (e.g. *Panicum miliare* (little millet)). In the same grassland, *Cenchrus setigerus* (dhaman grass) shows two peaks in a year, while *Cynodon dactylon* (Bermuda grass) is represented almost evenly throughout the year (Fig. 17.1). The ratio of the total number of species to the average total number of individuals/m² (variety ratio) is at a maximum following the maturity of the monsoon vegetation (Kumar & Joshi, 1972).

The total density of vegetation varies from site to site, probably in response to the botanical composition and available water. However, the values are highest for the rainy season (Table 17.1). The more arid grasslands, such as those at Pilani (Kumar & Joshi, 1972) and at Fété Olé (76 to 229 plants/m^2, J. C. Bille, IBP data), support fewer shoots. Singh & Yadava (1974) have reported a significant positive linear relationship ($r = 0.92$, $P < 0.01$) between total vegetation density and above-ground live biomass for the dry subhumid mixed grassland at Kurukshetra. Mall & Billore (1974) have also reported a positive linear relation ($r = 0.86$, $P < 0.05$) between total density and net above-ground primary production for the *Sehima* grassland. These studies indicate that the vegetation in these multiple-species communities is not density saturated; this is in contrast to single-species populations that rapidly colonise ruderal habitats during the rainy season and where productivity may decline after an optimal density is reached (Singh, 1969*a*).

Distribution of plant biomass, growth strategy and canopy architecture

Shankarnarayan, Sreenath & Dabadghao (1969) have studied the vertical distribution of above-ground biomass in six major grasses of the *Sehima–Dichanthium* grassland. These authors found a positive quadratic relationship ($Y = a+bx+cx^2$; where Y is the percentage dry weight and x is percentage height from the top) between height and weight which explains more than 98 % of the variability in all six species. A significant positive relationship between absolute height and dry weight has also been reported in grassland species by Mall, Billore & Misra (1973).

Information on the distribution of biomass of individual species in different above-ground strata of a mixed grassland led Singh & Yadava (1974) to group various grassland species into five categories. In the case of grasses, the biomass was more concentrated towards the base to give the appearance of an abruptly tapering upright pyramid, the extent of which depended on the height of the plants (e.g. *Cynodon dactylon* and *Dichanthium annulatum*). In some forbs, on the other hand, biomass was concentrated at mid-height (e.g. *Kochia indica* (bui)) or both at the base and again at some distance upwards (e.g. *Chenopodium album* (white goose-foot)). In certain other plants, biomass was almost uniformly distributed to a considerable height (e.g. *Alhagi camelorum*). However, when data for total vegetation were plotted this way a gradually tapering upright pyramidal shape resulted, indicating that these differing canopies of individual species are so oriented as to form a multi-layered, close canopy. In such a canopy each quantum of incident solar radiation has a greater probability of being intercepted and used than in a single-layered canopy.

In a strongly seasonal climate where growing conditions vary markedly through the year, species with a variety of production strategies are needed if the vegetation is to continue producing in all seasons. That this is so is illustrated by differential growth behaviour of perennial species. Thus, Singh (1968) grouped the various important perennials of the moist subhumid grassland at Varanasi into five categories on the basis of their growth pattern: (i) those with peak growth during the rainy season and a steep fall in biomass thereafter, e.g. *Bothriochloa pertusa*; (ii) those with maximum growth during the rainy season and a steady but slow decline during the winter season, e.g. *Indigofera linifolia*; (iii) those with continued growth to a maximum in winter and a steady decline thereafter, e.g. *Desmodium triflorum*; (iv) those with two peaks of growth, one in the rainy season and the other at the winter–summer interphase, e.g. *Volvulopsis numularia* G. Roberty; and (v) those with maximum growth in summer, e.g. *Convolvulus pluricaulis* (dodak). Growth forms and growth behaviour of several grasses for the moist subhumid savanna at Lamto, Ivory Coast, have been described in detail by Monnier (1968).

When species with differential vertical distribution patterns of biomass and with differential temporal growth behaviour are combined into a community, a canopy of highly dynamic, layered structure results. Such a dynamic multi-layered canopy has been described for the dry subhumid mixed grassland at Kurukshetra (Singh & Yadava, 1974). In this grassland different layers of vegetation are dominated in different months by different species according to their light- and water-adaptation characteristics. This permits the community to maintain year-long productivity. Further, the height of the canopy adjusts itself to variations in growing conditions. For example, canopy height is at a maximum during the wet period when both the number of species and density of plants are at their maximum (Mall, Misra & Billore, 1973; Pandeya, 1974; Singh & Yadava, 1974). Increases in the height of plants may be induced through decreased light intensities in close canopies (Singh, 1969*b*). Through this well-developed canopy, the vegetation makes best use of the opportunities offered by the environment in the form of an abundant water supply at that time.

Besides layered structure, the chlorophyll content and the leaf area index are important features of community architecture influencing productivity. Grassland communities in the same region show marked differences in these attributes, depending upon botanical composition. Year-long variation in leaf area index and chlorophyll content of a *Dichanthium* grassland and a *Sehima* (sena) grassland from central India are compared in Table 17.2 on the basis of data reported by Mall, Misra & Billore (1973). While the *Dichanthium* community shows a maximum leaf area index of 7.9 (in September), the maximum for the *Sehima* community is only 3.4 (in August). Chlorophyll content is also greater in the *Dichanthium* community. For three semiarid grassland communities at Pilani, Kumar & Joshi (1972) have reported

Table 17.2. *Leaf area index and chlorophyll content of two grassland communities in Central India in 1971–2 (after Mall, Misra & Billore, 1973)*

Month	*Dichanthium* community (Ujjain)		*Sehima* community (Ratlam)	
	Leaf area index (m^2/m^2)	Total chlorophyll (mg/m^2)	Leaf area index (m^2/m^2)	Total chlorophyll (mg/m^2)
June	0.15	141	0.60	100
July	1.47	367	2.23	199
August	5.10	761	3.44	524
September	7.90	1053	2.24	481
October	6.77	1370	0.56	197
November	3.81	272	0.46	96
December	0.50	47	0.22	22
January	0.29	67	0.04	6.3
February	0.77	124	0.03	2.9
March	0.39	85	0.02	2.9
April	0.16	27	0.01	1.3
May	0.10	18	0.001	0.6

0.11–1.47 g/m^2 total chlorophyll in the above-ground herbage. In all the cases, however, both leaf area index and chlorophyll content are highest for the rainy season when vegetation makes its best growth. The leaf area index of an ungrazed grassland in Ruwenzori Park, Uganda, dominated by *Hyparrhenia filipendula* Stapf. and *Themeda triandra* Forsk., ranged from 0.71 (in November) to 2.49 (in April) (C. D. Pigott, IBP data).

Working on the *Sehima* grassland, Billore (1973) found that leaf area index was positively related to chlorophyll content ($r = 0.97$, $P < 0.05$). Further, net above-ground productivity was related positively to both chlorophyll content ($r = 0.98$, $P < 0.05$) and leaf area index ($r = 0.94$, $P < 0.05$). These relationships emphasise the interdependence of the architectural properties of the community and net productivity. The relationship may be linear, as was found by Billore (1972), or curvilinear, depending upon the degree of mutual shading and the amount of structural, non-photosynthetic tissue which may increase with increase in leaf area index.

Standing crop of plant biomass

There is a marked variation in the standing crop of plant biomass from site to site. Minimum and maximum values of above-ground live biomass for 17 grassland types are included in Table 17.3 for comparison. The minimum values range from 0 to 871 g/m^2, while the maximum values range from 76 to about 3000 g/m^2. In general, the subhumid grasslands support greater biomass, than in both the semiarid to arid and in humid types. The general

Table 17.3. *Maximum and minimum values of above-ground live biomass (g/m²) for certain tropical grasslands*

Site	Grassland type	Minimum biomass	Maximum biomass	Source of data
Rajkot	*Cenchrus ciliaris* Linn.	0	228	Pandeya (1974)
Jodhpur[a]	Mixed grass	7	164	Gupta *et al.* (1972)
Pilani	Mixed grass	35	76	Kumar & Joshi (1972)
Kurukshetra	Mixed grass	105	1974	Singh & Yadava (1974)
Delhi[a]	*Heteropogon*	0	771	Varshney (1972)
Varanasi[a]	*Eragrostis*	871	3296	Singh (1972)
	Desmostachya	573	2360	Singh (1972)
Jhansi[a]	*Sehima–Heteropogon*	496	1408	Shanker *et al.* (1973)
Ujjain	*Dichanthium*	24	457	Misra (1973)
Ratlam	*Sehima*	1	363	Billore (1973)
Ambikapur[a]	Mixed grass	123	423	Naik (1973)
Sagar	*Heteropogon*	14	572	Jain (1971)
	Dichanthium	11	337	Jain (1971)
Lamto (Ivory Coast)	Palm savanna	0	690	Data bank[b]
Ruwenzori Park (Uganda)	*Hyparrhenia* and *Themeda*	59	405	Data bank[b]
Welgevonden (South Africa)	Drysweet bush veld, open savanna	8	753	Data bank[b]
Panama	*Hyparrhenia*	501	2088	Breymeyer (1974)

[a] Includes standing dead shoots also.
[b] Lamto data were submitted to the Fort Collins data bank by M. Lamotte; South African data were submitted by J. Van Wyk; Ruwenzori Park data were submitted by C. D. Pigott.

pattern of biomass changes during the year is similar in all grasslands. However, examples of year-long variations in live biomass from several grassland types of India are given in Table 17.4.

Active growth of vegetation in these Indian sites is triggered by the advent of monsoon rains, and the biomass grows to a peak value in about September. Following the maturity of rainy season vegetation, biomass declines (depending upon the botanical composition), either rapidly (Singh, 1968; Choudhary, 1972) or slowly (Varshney, 1972) because of the transfer of material to the standing dead and litter compartments. This period of decline may be followed by a second or third spurt of growth during winter or summer, depending upon the availability of soil water. In many tropical grasslands, however, water stress becomes too severe during the dry season to permit any recognisable spurt of growth during the post-monsoon period.

The data from Welgevonden (South Africa), on a drysweet bushveld open savanna, indicate a maximum live biomass of 753 g/m² in February and a minimum biomass of 8 g/m² in August. The active growth on this site starts from October onwards and the decline in biomass following maturity begins

Table 17.4. *Variations in the above-ground live biomass* (g/m^2) *in selected grasslands throughout the year*

Month	Rajkot 1973	Jodhpur[a] 1970–1	Pilani 1969–70	Kurukshetra 1970–1	Delhi[a] 1969–70	Varanasi[a] Lowland	Varanasi[a] Upland	Jhansi[a] 1970–1	Ujjain 1971–2	Ratlam 1971–2	Sagar 1969–70
June	15	7	42	773	106	871	573	—	36	25	—
July	161	29	56	1103	128	1092	908	568	126	60	14
August	228	93	72	1526	436	2051	1596	850	252	186	86
September	207	146	76	1974	597	2571	1725	898	435	363	212
October	49[b]	164	39	1707	771	3169	1976	1408	457	219	572
November	6[b]	161	39	486	507	3296	2037	952	313	158	488
December	2[b]	115	35	105	465	2588	2360	795	51	64	299
January	—	144	38	187	400	2219	1845	1132	66	15	147
February	—	145	53	261	200	2828	1720	1106	70	6	158
March	—	135	71	445	—	2750	2250	547	53	4	171
April	—	127	71	560	9	2331	2062	1187	28	3	48
May	—	149	64	800	—	2695	1491	496	24	1	38

[a] Includes standing dead also. [b] These values are for 1972.
Grassland types and sources of data are given in Table 17.3, except that the information for Jodhpur is from S. K. Saxena, S. K. Sharma & B. Ram (personal communication).

Table 17.5. *Variation in the standing dead shoot material (g/m²) throughout the year in selected grasslands*

	Rajkot 1972–3	Pilani 1969–70	Kuruk-shetra 1970–1	Varanasi 1968–9	Ujjain 1971–2	Ratlam 1971–2	Sagar 1969–70
June	3[a]	4	203	—	37	148	—
July	44[a]	—	196	68	40	96	359
August	95[a]	0	318	35	53	86	—
September	210	10	142	61	85	105	—
October	88	21	400	322	127	229	399
November	290	20	1268	205	220	242	518
December	430	17	845	210	422	316	383
January	337	27	782	110	290	302	247
February	256	20	631	—	227	283	239
March	—	1	553	48	185	275	360
April	—	0	466	—	167	241	253
May	—	2	471	102	112	209	283

[a] Values for 1973; the rest of the values for this site are for 1972.
Grassland types and sources of data are given in Table 17.3, except that the information for Varanasi is from Choudhary (1972).

in March (J. Van Wyk, IBP data). The above-ground live biomass in the Ruwenzori Park grassland increases from a minimum of 59 g/m² in November to a maximum of 405 g/m² in early June (C. D. Piggot, IBP data).

The maximum amount of standing dead material in the South African site occurs in June (421 g/m²) and the minimum (63 g/m²) in January. The standing crop of dead shoots in the Uganda grassland, on the other hand, continues to increase from December (28 g/m²) to the end of the sampling period in August (reaching 364 g/m²). On the Indian sites, however, the material in the standing dead compartment increases considerably in the post-monsoon period and the peaks lie within the period from October to January (Table 17.5). The dynamics of the standing dead compartment, thus, follows the dynamics of the live compartment, and changes in the litter compartment follow the changes in the standing dead compartment (Singh, 1973).

The under-ground plant biomass (roots plus rhizomes) also varies significantly through the year, with most of the biomass being concentrated in the uppermost 30–40 cm of soil. Among the Indian sites, the lowest amount of root biomass occurs in dry, semiarid grasslands, such as at Pilani (16–86 g/m²), while the maximum (705–1381 g/m²) is recorded for moist subhumid localities (Table 17.6). The moist subhumid palm savanna dominated by *Loudetia simplex* (Nees) C. E. Hubbard (fritibé) at Lamto, Ivory Coast, supports 1400 to 2800 g/m² of root biomass (M. Lamotte, IBP data), while that of the more arid open savanna at Welgevonden (South Africa) is in the

Table 17.6. *Variation in the under-ground biomass (g/m²) in selected grasslands throughout the year*

Month	Jodhpur 1970–1	Pilani 1969–70	Kuruk-shetra 1970–1	Varanasi Low-land 1970–1	Varanasi Upland 1970–1	Jhansi 1970–1	Ujjain 1971–2	Ratlam 1971–2	Sagar 1969–70
June	—	42	955	165	200	—	675	575	788
July	280	—	1000	534	224	268	650	474	1020
August	470	86	671	812	431	204	600	576	1272
September	670	53	611	1282	788	140	550	723	1381
October	780	16	675	1282	443	333	610	845	1221
November	540	18	1167	838	667	298	715	873	1115
December	540	27	643	901	443	191	925	765	871
January	320	44	871	925	430	204	775	749	1037
February	370	48	768	833	687	206	750	767	994
March	240	55	802	783	638	170	765	735	705
April	260	65	634	648	556	251	800	668	763
May	240	43	809	668	410	233	790	627	883

Grassland types are given in Table 17.3: sources of data are given in Table 17.4.

range of only 184–410 g/m². The *Hyparrhenia–Themeda* grassland in Ruwenzori Park (Uganda), with an annual rainfall (900 mm) a little lower than that at Lamto, supports 512 to 2007 g/m² of under-ground biomass. The green shoot:root ratio varies from 0.34 to 3.21 at the time of peak live standing crop, thus exhibiting a very wide range among the sites.

Turnover of under-ground plant material in tropical grasslands seems to be quite rapid. Thus, Singh & Yadava (1974) have reported an annual rate of 97 % for the dry subhumid grassland at Kurukshetra, and calculations based on the results reported by Kumar & Joshi (1972) indicate an average annual turnover rate of 93 % for the semiarid grassland at Pilani. Presence of a large number of rainy season ephemerals, strong periodicity of climate and high temperatures appear to be responsible for this high rate of turnover.

Energy content of plant biomass

Singh & Yadava (1973) have recently reported on energy values of several species in the mixed grassland at Kurukshetra. In annual species, maximum energy concentration in shoots occurred at maturity during August and September. The perennial species, on the other hand, exhibited more than one peak of energy concentration coincident with flushes in flowering and fruiting. Energy values of various compartments (above-ground live, standing dead, litter and roots) were significantly different from each other, and also there was a significant variation in the energy concentration during the year.

Table 17.7. *Seasonal averages of energy values (kJ/g) of different primary producer compartments in certain grasslands*

Site and grassland type	Live above ground			Standing dead			Litter			Below ground		
	Rainy	Winter	Summer	Rainy	Winter	Summer	Rainy	Winter	Summer	Rainy	Winter	Summer
Kurukshetra mixed grass Singh & Yadava (1973)	17.1	17.2	17.7	16.2	17.9	19.1	16.2	15.9	16.7	15.7	16.2	16.6
Ujjain *Dichanthium* Misra (1973)	15.1	15.5	14.1	13.7	14.1	13.6	12.6	12.1	11.0	13.9	15.4	15.5
Ratlam *Sehima* Billore (1973)	15.9	16.2	14.9	14.0	15.2	14.2	12.7	13.6	13.1	13.7	14.2	15.0
Sagar *Heteropogon* Jain (1971)	14.6	15.1	13.9	11.7	14.2	14.4	9.0	11.3	11.8	8.8	9.2	11.5

Seasonal averages in energy values of four grasslands have been calculated in Table 17.7. In a majority of cases the plant material in all the compartments exhibits greater energy concentration during the post-monsoon period, particularly in the winter season. For all four grasslands, the energy concentration in roots increases as the season progresses from rainy to winter to summer.

It may be of interest to note that among the four sites given in Table 17.7, the energy concentration in plant biomass of all compartments is lowest in the wettest site (Sagar, 1410 mm annual rainfall) and highest in the driest site (Kurukshetra, 790 mm annual rainfall). Unfortunately detailed chemical analysis of biomass from these four sites is not available, so that it is not possible to explain this apparent inverse relationship of energy concentration to rainfall.

Standing state of nutrients

Along with the growth resulting from carbon and energy fixation, plants absorb a host of nutrients from the soil and incorporate these in their biomass. Whyte (1964) has reviewed earlier work on the nutrient content of major grassland species of India.

Nutrient content of various compartments for a semiarid (Pilani) and two subhumid (Ratlam, annual rainfall 1257 mm, and Sagar, annual rainfall 1410 mm) grasslands is given in Table 17.8. Nitrogen seems to be the most abundant nutrient in plant material with maximum values for the more humid Sagar grassland and minimum values for the semiarid grassland. Misra (1972) and Billore (1973) have reported on the nitrogen and the phosphorus contents in vegetation and soil for the grasslands at Varanasi and Ratlam, respectively. It is evident from their data that only approximately 2 % of the total nitrogen residing in the system is contained in vegetation; hence, only a minor fraction of the total is being circulated through organisms at any one time. On the other hand, in the case of phosphorus, while at Varanasi only 4 % of the total was found in the plant biomass, at Ratlam up to 30 % of the available phosphorus resided in the vegetation. It also appears that the active zone of absorption of soil nutrients in these grasslands is rather shallow. Thus, Pandey & Kothari (1972) reported that active absorption of ^{32}P in *Dichanthium* grassland was confined to a depth of 15 cm only.

Net primary production

Bourlière & Hadley (1970) have reported minimum estimates of net aboveground primary production from some 22 tropical savanna and related communities as determined from peak standing crop data. These communities

Table 17.8. *Standing state of certain nutrients (g/m²) in tropical grassland (based on average annual standing crop)*

Plant compartment	Pilani (mixed grass) (range for 3 sites) J. S. Gill (personal communication)	Ratlam (*Sehima*) Mall, Misra & Billore (1973)	Sagar (*Heteropogon*) S. K. Jain (personal communication)
Above-ground live			
Nitrogen	0.20–0.68	1.041	3.484
Phosphorus	0.04–0.13	0.250	0.139
Potassium	0.14–0.39	0.343	1.431
Calcium	—	0.468	—
Standing dead			
Nitrogen	0.03–0.18	1.066	5.881
Phosphorus	0.01–0.02	0.247	0.101
Potassium	0.01–0.03	0.304	1.059
Calcium	—	1.104	—
Litter			
Nitrogen	0.11–0.30	1.092	2.383
Phosphorus	0.02–0.07	0.168	0.067
Potassium	0.05–0.14	0.285	0.443
Calcium	—	1.025	—
Under ground			
Nitrogen	0.29–0.98	3.498	0.379
Phosphorus	0.07–0.20	1.276	0.364
Potassium	0.24–0.69	0.537	—
Calcium	—	2.216	—

occur in areas experiencing from about 200 to 1660 mm annual rainfall and they range in net annual aerial productivity from 37 to 1750 g/m². It is apparent from these values that productivity depends more on botanical composition than on amount of rainfall. For example, in *Themeda–Heteropogon* grassland the annual productivity ranged from 32 to 630 g/m², while in the *Imperata* grassland it was 1750 g/m²; both grasslands occur in Albert Park at Kivu with an annual rainfall of 860 mm.

The data available to the present authors also indicate that precipitation is not the most important variable influencing net aerial production in tropical grasslands. Calculated values for annual net above-ground production (based on harvest data) are presented for 21 tropical grassland communities in Table 17.9. For most sites these values were calculated using trough–peak analysis of above-ground live and standing dead shoots (Singh, Lauenroth & Steinhorst, 1975). These grasslands occur in areas receiving from 209 to 1410 mm annual rainfall. The values for above-ground net production range from 82 g/m² for a semiarid grassland in Senegal to 3396 g/m² for *Eragrostis* lowland subhumid grassland in India. Contrary to expectations, Fig. 17.2 indicates no significant relationship between the above-ground net production and annual rainfall. As we have discussed earlier, the tropical grasslands may

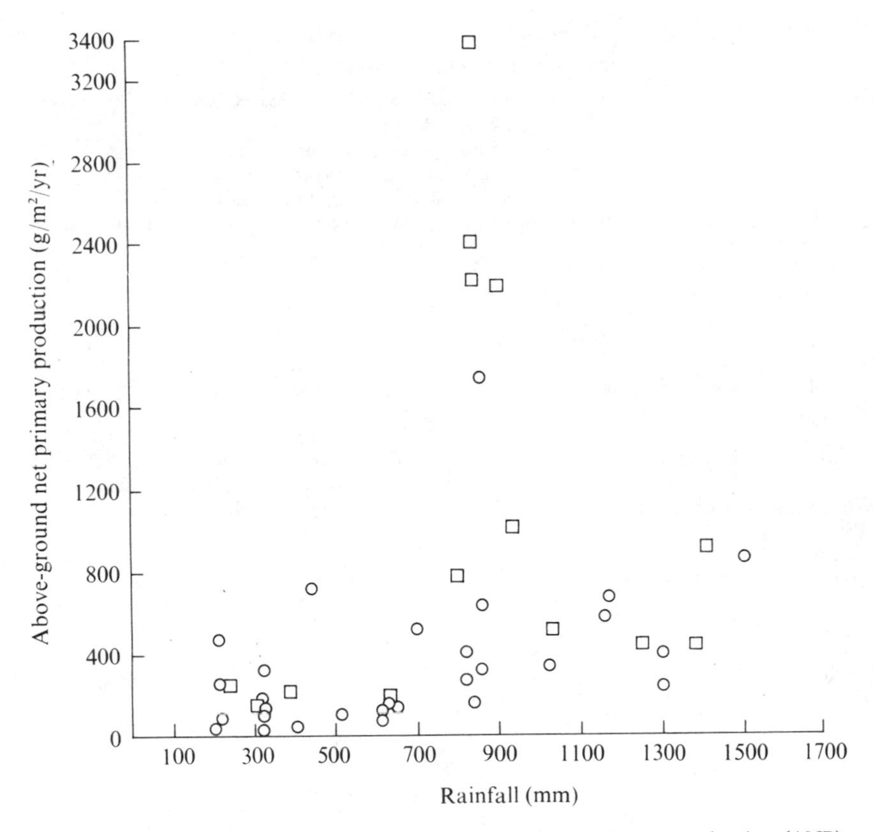

Fig. 17.2. Relationship between above-ground annual net primary production (ANP) and annual rainfall. Circles represent values reported by Bourlière & Hadley (1970) and squares represent values from Table 17.9.

represent different successional stages in local seres, depending upon the mode and age of origin, and the kind and intensity of biotic influences. Although a majority of studies reported in Table 17.9 were made in protected grasslands, the period of protection varied from one to many years.

Root biomass data are available from comparatively fewer sites. The values of under-ground net production reported in Table 17.9 were calculated through trough–peak analysis on the total under-ground biomass data. Similar to above-ground net production, there is a marked variation in net under-ground production in these grasslands. The values range from 61 g/m² for semiarid grassland to 1495 g/m² for *Hyparrhenia–Themeda* grassland. The total annual net production ranges from 278 to 4558 g/m². Grasslands occurring in the subhumid region tend to show a greater annual net production than do semiarid or humid grasslands. In fact, maximum values of produc-

Table 17.9. *Net annual primary production and annual rainfall values for certain tropical grasslands*

Site	Grassland type	Above-ground net production (g/m²)	Below-ground net production (g/m²)	Total net production (g/m²)	Rain fall (mm)	Source of data
Fété Olé (Senegal)	*Aristida* dominated	82	—	—	209	J. C. Bille[a]
	Panicum and *Chloris* dominated	256	—	—	209	
	Panicum and *Zornia* dominated	476	—	—	209	
Rajkot	*Cenchrus ciliaris*	244	—	—	242	Pandeya (1974)
Jodhpur	Mixed grass	164	—	—	311	Gupta *et al.* (1972)
Pilani	Mixed grass	217	61	278	391	Kumar & Joshi (1972)
Welgevonden (South Africa)	Bushveld open savanna	710	—	—	436	J. Van Wyk[a]
Udaipur	Mixed grass	184	—	—	627	Vyas *et al.* (1972)
Kurukshetra	Mixed grass	2407	1131	3538	790	Singh & Yadava (1974)
Delhi	*Heteropogon*	798	—	—	800	Varshney (1972)
Varanasi	*Heteropogon* (Vindhyan Plateau)	—	—	4200	843[b]	Ambasht *et al.* (1972)
	Dichanthium (alluvial)	—	—	2200	843	Ambasht *et al.* (1972)
	Eragrostis (lowland)	3396	1161	4558	843	Singh (1972)
	Desmostachya (upland)	2218	1377	3595	843	Singh (1972)
Ruwenzori Park (Uganda)	*Hyparrhenia–Themeda*	701	1495	2196	900	C. D. Pigott[a]
Jhansi	*Sehima–Heteropogon*	1019	497	1516	936	Shanker *et al.* (1973)
Ujjain	*Dichanthium*	520	464	984	1030	Misra (1973)
Lamto (Ivory Coast)	Palm savanna	498	—	—	1158	M. Lamotte[a]
Ratlam	*Sehima*	433	399	832	1257	Billore (1973)
Ambikapur	Mixed grass	436	563	999	1379	Naik (1973)
Sagar	*Heteropogon–Apluda–Cymbopogon*	914	937	1851	1410	Jain (1971)

[a] Data submitted to IBP grassland data bank.
[b] Mean precipitation during years of study; long-term annual mean is 1134 mm.

tion (3000 g/m² annually) are recorded for grasslands located in areas receiving 790–843 mm of rainfall.

The strong seasonality of growing conditions and of growth pattern of major species is reflected in marked variations in rate of production in different seasons and in partitioning of production above ground and under

Table 17.10. *Seasonal net primary productivity (g/m^2) in three tropical grasslands of India. (ANP, above-ground net production; BNP, under-ground net production; TNP, total net production)*

Site and grassland type		Season: Rainy	Winter	Summer	Source of data
Kurukshetra					
Mixed grass	ANP	1706	156	544	Singh & Yadava (1974)
	BNP	137	785	209	
	TNP	1843	941	753	
Varanasi					
Desmostachya (upland)	ANP	1052	636	530	Singh (1972)
	BNP	588	424	365	
	TNP	1640	1060	895	
Eragrostis (lowland)	ANP	1699	1333	364	Singh (1972)
	BNP	1117	24	20	
	TNP	2816	1357	384	

ground. To illustrate this seasonality the values of net production for three grassland types have been calculated in Table 17.10 for different seasons. These values indicate maximum production during the rainy season and minimum production during winter or summer seasons. Thus, in the grassland at Kurukshetra dry matter is accumulated at a daily rate of 15.1 g/m^2 during the rainy season and at 6.2–8.2 g/m^2 during the post-monsoon season. Further, in the mixed grassland at Kurukshetra, while in the rainy season only 7 % of the total net production was translocated under ground, in winter and summer seasons 83 and 28 % respectively of total net production was used for biomass accumulation in under-ground parts. Thus, the wet period seems to be more congenial for above-ground growth, resulting into an extensive development of photosynthetic canopy during this period, while the dry period is more favourable for biomass accumulation under ground. However, root growth during the dry period may be inhibited in lowland sites, perhaps because of clay soil which becomes too hard when relatively dry. This is exemplified by *Eragrostis* grassland in which only 2–5 % of total production was accounted for by root growth during the dry season.

Efficiency of energy capture

Values of efficiency of energy capture (proportion of half of incident total solar radiation fixed in net primary production (Singh & Misra, 1968)) have been calculated for 13 tropical grasslands in Table 17.11. A perusal of this table indicates a wide range of values, with a minimum of 0.23 % (for the growing season only) for the semiarid grassland at Pilani to a high of 1.66 % for the dry subhumid grassland at Kurukshetra. With the exception of the

Table 17.11. *Efficiency of energy capture (%) in certain tropical grasslands*

Site	Grassland type	Above-ground net production	Below-ground net production	Total net production
Rajkot	*Cenchrus ciliaris*	0.09	—	—
Pilani[a]	Mixed grassland	0.09	0.14	0.23
Kurukshetra	Mixed grassland	1.13	0.53	1.66
Delhi	*Heteropogon*	0.39	—	—
Varanasi	*Heteropogon* (Vindhyan Plateau)	—	—	1.28
	Dichanthium (alluvial)	—	—	0.67
	Eragrostis (lowland)	1.07	0.34	1.41
	Desmostachya (upland)	0.70	0.40	1.10
Jhansi	*Sehima–Heteropogon*	0.52	0.24	0.76
Ujjain	*Dichanthium*	0.21	0.18	0.39
Ratlam	*Sehima*	0.18	0.14	0.32
Ambikapur	Mixed grass	0.22	0.27	0.49
Sagar	*Heteropogon–Apluda–Cymbopogon*	0.45	0.45	0.90

[a] For growing season only.

moist subhumid grassland at Sagar and the humid grassland at Ambikapur the amount of energy accumulated in above-ground parts is greater than in under-ground parts. Most of the tropical grasslands exhibit higher efficiency of energy capture (around 1 %) as compared to temperate grasslands (Sims & Singh, 1971). This higher efficiency may be due to several reasons. A majority of plants in tropical environments possess the C_4-pathway of carbon dioxide assimilation and have higher water-use efficiency, little or no light saturation, a higher range of optimal temperatures for photosynthesis, and no photo-respiration. These characteristics result in higher growth rates of tropical plants. Many of the grasslands possess dynamic multi-layered canopies efficient in production throughout the year. Temperature never being limiting, year-long growth is possible depending upon the availability of soil water.

Water-use efficiency

Since there are no data concerning actual amounts of water used by the tropical grasslands, a crude index of water-use efficiency has been calculated as a ratio of annual above-ground net production (g/m^2) to annual rainfall (mm) in Fig. 17.3. This ratio ranges from 0.1 to 4.0 for these grasslands. There is apparently no marked relationship between annual rainfall and the efficiency of water use, except for the fact that more communities with higher water-use efficiency tend to occur in areas receiving 200–900 mm of rain each year. Thus, there seems to be a tendency for water-use efficiency to decline in areas of high rainfall.

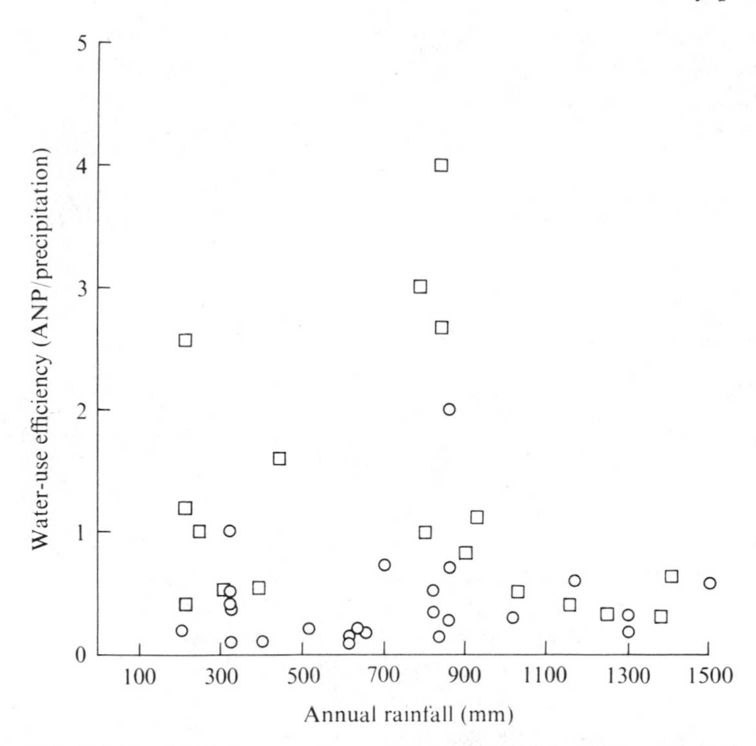

Fig. 17.3. Relationship between water-use efficiency (ANP in g/m²: precipitation in mm) and rainfall in tropical grasslands. Circles represent calculations based on data reported by Bourlière & Hadley (1970) and squares are based on data reported in Table 17.9.

Compartmental transfers

Rates of production of dry matter and its transfer through various compartments and its ultimate disappearance from the system, are shown in Table 17.12 for one grassland from each of the semiarid, dry subhumid and moist subhumid types. While in the semiarid grassland at Pilani the net accumulation in the root compartment is twice as great as in the shoot compartment, the reverse is true for the dry subhumid grassland at Kurukshetra. In the moist subhumid grassland, on the other hand, the share of the roots and shoots is almost equal. While the transfer from standing dead to litter compartments in both semiarid and moist subhumid grasslands is faster than, or equal to the accumulation of the standing dead material, the reverse is true for the dry subhumid grassland. Thus, in a subhumid climate, grassland supports a comparatively greater amount of standing dead shoots throughout the year. The greater output as compared to input in the litter compartment at Pilani and Sagar, and the root compartment at Kurukshetra, may be due to excess carryover material in these compartments from

Table 17.12. *Annual net flow rates into, and losses from, compartments in terms of organic matter (in g/m²) for three grasslands in India*

Parameter	Pilani[a] (semiarid zone)	Kurukshetra (dry subhumid zone)	Sagar (moist subhumid zone)
Net primary production	239	3538	1851
Live shoots	84	2407	914
Dead shoots	84	2044	914
Litter	94	1728	990
Under-ground parts	155	1131	937
Disappearance from			
Above-ground litter	116	1666	594
Under-ground parts	155	1185	841
Total	271	2851	1435

[a] Data from Gill (1975).

previous years. Except for semiarid grassland, there is a net surplus of material, as the rate of disappearance is slower than the rate of accumulation of dry matter. Breymeyer (1974), for the *Hyparrhenia rufa* Stapf. (afoutré) grassland in Panama, has reported daily rates of net primary production of 19 g/m² and decomposition of 7.8 g/m². She has suggested that perhaps intensive decomposition of grass occurs before parts of shoots fall to the litter layer. However, since the calculations in Table 17.12 have been made on the basis of net movement of dry matter (Singh & Yadava, 1974), any decomposition of standing dead matter would be accounted for. Perhaps it is because of this surplus production that grazing or burning becomes important for the maintenance of these grasslands at a given level of succession. How-

Table 17.13. *System transfer functions for certain tropical grasslands (ANP, above-ground net production; BNP, under-ground net production; TNP, total net production)*

Site	Grassland type	Transfer functions: TNP to ANP	Transfer functions: TNP to BNP	Source of data
Pilani	Mixed grass	0.35	0.65	Gill (1975)
Rajkot	*Cenchrus ciliaris*	0.03	0.97	Pandeya (1974)
Kurukshetra	Mixed grass	0.68	0.32	Singh & Yadava (1974)
Varanasi	*Desmostachya* (upland)	0.62	0.38	Singh (1972)
	Eragrostis (lowland)	0.74	0.26	Singh (1972)
Ujjain	*Dichanthium*	0.53	0.47	Misra (1973)
Ratlam	*Sehima*	0.55	0.45	Billore (1973)
		0.52	0.48	

ever, since each set of data represents only one year of observation, the validity of this statement needs to be checked through long-term studies.

System transfer functions calculated on the basis of annual production values for certain sites are presented in Table 17.13. While in comparatively dry grasslands at Pilani and Rajkot 65–97 % of total net production is channelled under ground, in grasslands experiencing comparatively higher rainfall (Table 17.9), only 26–48 % of total production is used for root growth.

Effect of system perturbations

Grazing

It is perhaps not justifiable to regard grazing and burning as system perturbations for the tropical grasslands because both are causal to the origin and maintenance of these systems. It is difficult to visualise a tropical grassland ecosystem which would be free of grazing and periodic burning. However, grazing does influence the structure and function of these grasslands in several ways, depending upon the vegetation type, rainfall, and the period and intensity of grazing. In general, grazing creates relatively open canopies, thus, making the invasion by annuals and other aliens possible and increasing plant diversity (Singh, 1968; Singh & Misra, 1969). Under heavy use, unpalatable, coarse plants of low successional order often invade the grasslands (Dabadghao & Shankarnarayan, 1973). Continued heavy use decreases the basal cover of plants (Chakravarty, Ram & Singh, 1970) and may trigger soil erosion. The intensity of grazing may also influence the inter-specific association among plant species through changes in habitat conditions and differential response of species populations (Singh, 1969*c*). Moderate grazing in dry subhumid to humid climates often increases the above-ground net primary production through enhanced tillering and increased diversity (Singh, 1968). In semiarid climates the same intensity of grazing also increases the production. However, at the same time heavy grazing in semiarid grasslands leads to decrease in net primary production through the deterioration of the vegetation (Kumar & Joshi, 1972; Gill, 1975).

Besides grazing, harvesting of the biomass close to the soil surface is a common practice in the tropics. In a comparative study in the *Dichanthium* grassland at Varanasi, Singh (1968) found that, while harvesting removes 87–89 % of biomass, cattle grazing removes only from 15 to 51 %. Grazing efficiency (in terms of percentage of biomass removed) was found to be better for pastures which were only under grazing use as compared to those from which herbage was both grazed and harvested during different periods. Grazing efficiency was also higher for the period immediately following the rainy season as compared to winter or summer seasons when plant growth was comparatively depauperate.

Burning

The role of fire in maintaining tropical grasslands at a given seral stage is well realised. Recently 20 research and review papers documenting the effect of fire on the structure and succession of savannas in Africa have been published (Kumarek, 1972). However, not one of these was devoted to the effect of fire on plant productivity.

Egunjobi (1974) has reported that, in savanna grassland in Nigeria dominated by *Andropogon gayanus* Stapf., protection from fire for one year slightly depressed current production and led to a decrease in the crude protein content of new herbage. The lower production in the protected plot was thought to be due to the accumulation of litter which weathered slowly. Pandey (1971) reported that root and shoot production in *Dichanthium* grassland at Varanasi increased considerably subsequent to burning. He recorded total annual net production of 2480 g/m^2 on plots burned twice in a year as compared to 1325 g/m^2 for unburned plots. The same study has revealed that the nitrogen content of soil in burned grassland increases rapidly, partly due to rapid and abundant growth of the legume, *Desmodium triflorum*, the germination inhibitive seed coat of which is charred as a consequence of burning. Stimulation of tillering, acceleration of nutrient cycling and elimination of seedlings and saplings of plants of higher successional order seem also to be responsible for the favourable influence of fire.

Fertiliser applications

Dabadghao & Shankarnarayan (1970) have reported on the effect of applications of nitrogen and phosphorus on the productivity of three moist subhumid grassland communities of *Sehima–Dichanthium* grass cover at Jhansi. Nitrogen fertilisation appeared to be most effective in increasing the productivity of these grasslands. A 40 kg/ha application of nitrogen increased the net above-ground herbage yield of the *Iseilema* community by 42 % and that of the *Heteropogon* community by 66 %, while a 60 kg/ha application increased the yield of the *Sehima* community by 83 %. Increases due to phosphate applications were also significant, but were not as remarkable as nitrogen application. No study seems to have been made of the long-term effects of fertilisers on the structure and function of tropical grasslands.

18. Consumers

M. C. DASH

This discussion of the numbers, diversity, biomass and energetics of consumers of tropical grasslands is based on data available from one site in Panama, four sites in Africa (Lamto in the Ivory coast, Ferlo and Fété Olé in Senegal and Mweya Peninsula in Uganda), and two sites (Berhampur and Jodhpur) in India. Not all aspects of the consumers subsystem have been studied at any site, but most of the data available have been used to prepare this chapter. Considerably more detailed discussions of consumers and their activities in tropical grasslands and savannas have been prepared for the companion volume (Breymeyer & Van Dyne, 1979; see particularly Chapter 6 by Harris & Bowman and Chapter 7 by Coleman & Sasson).

Populations

A list of consumers that have been studied at various tropical sites is presented in Table 18.1. Among vertebrates, domestic cattle, goats and rodents at the Indian sites and buffalo, elephants and lions in the African sites are the dominant consumers. Among invertebrates, beetles (above ground) and earthworms and termites (under ground) are the principal consumers.

Some 15000 and 2769 elephants were counted, respectively, in Kabalega Park and in Ruwenzori Park of Mweya Penninsula of Uganda in September, 1973 (Edroma, 1974). Herds of 50 elephants are common. Besides the elephants, a large number of buffalo, reedbuck, bushbuck and rodents are also found. Carnivores, like hyaena and lion, have considerable impact on their prey species in African savannas. It has been possible to recognise individual lions by their whisker patterns and photographs. In the study area of 60 km^2 in Ishasha areas of Uganda it has been observed that 22 lions predate on 13 % of topi, 8 % of kob and 6 % of buffaloes annually (Edroma, 1974). Rodents are the common granivorous consumers in many grassland sites. Poulet (1974) has shown that desert mammalian species like *Desmodillisa braneri* and *Gazella clama* were observed more often than usual in drought periods at Fété Olé, Senegal. During the drought period, although the numbers of *Taterillus pygarus* (the most common granivorous rodent) decreased sharply, the species was able to reproduce despite almost complete lack of grass cover. In the state of Rajasthan, India, it has been observed that the annual food requirement of a population of 477 rodents is 1044 kg of grass when the annual net production of perennials is estimated to be between 800 to 1400 kg/ha (Gupta, 1974). Besides these mammals, birds are found to

Table 18.1. *Consumers that are considered to be important in seven tropical grassland sites in which they were studied*

Taxa	Panama	Africa: Lamto	Africa: Ferlo	Africa: Fété Olé	Africa: Mweya Pen.	India: Ber-hampur	India: Jodhpur
Vertebrates							
Domestic cattle	.	.	.	.	.	+	+
Elephants	.	.	.	.	+	.	.
Buffaloes	.	.	.	.	+	.	.
Rodents	.	.	.	.	+	+	+
Reedbuck	.	.	.	.	+	.	.
Bushbuck	.	.	.	.	+	.	.
Lion	.	.	.	.	+	.	.
Hyaena	.	.	.	.	+	.	.
Birds	.	+	+	+	+	+	+
Reptiles	.	+	.	.	+	+	.
Amphibians	.	.	.	.	.	+	.
Invertebrates							
Arachnids and Myriapods	.	.	+	.	.	.	.
Termites	.	.	.	+	.	+	.
Beetles	.	.	+	+	.	+	.
Oligochaetes	.	+	.	.	.	+	.
Nematodes	.	.	.	.	.	+	.
Testate protozoa	.	.	.	.	.	+	.

\+ Data available or occurrence noted.

remove a substantial quantity of leaves of grasses in Ugandan sites for building nests. The annual average number of birds in Senegal sites was 6.3/ha during the normal rainy periods and 2.9/ha in dry periods (Morel & Morel, 1974). In Indian sites, the ecological equivalents of African buffalo are chiefly the domestic cattle.

Analysis of data on invertebrate consumers from sites in Senegal indicates that bettles, arachnids, myriapods and termites are the chief consumers (Gillon & Gillon, 1973, 1974). Estimates of the numbers of arthropods living in the grass layer of the dry thornbush savanna of Ferlo, Senegal have been made during the end of the dry season (July), during the period of peak grass production (September and January), and the beginning of the next season (Gillon & Gillon, 1973). The number of arachnids and myriapods reaches a maximum of 18000/ha in September and a minimum of 13000/ha in January. The arachnids and myriapods accounted for 49 % of the total arthropods in January and 68 % in July. Tenebrionid beetles were less abundant than arachnids and myriapods. These beetles are five to six times as abundant in this Sahelian savanna as in the Guinean savanna of the Ivory coast (Gillon & Gillon, 1974). Termites are found quite deep in the soil, although their nests are located in tree stumps at Fété Olé, Senegal (Lepage, 1974). Drought causes a decrease in termite numbers, and the termites prefer dead wood over dry grass as food. The consumption of dry grass decreased significantly, but that of dead wood increased, during the drought period.

Studies in Australia and Africa have shown that qualitative and quantitative changes in the termite fauna occur when tropical woodlands and savannas are either completely or partially cleared for pastoral and agricultural purposes. Grazing animals affect the termite fauna by direct or indirect competition for food with grass-harvesting termites, the disruption of surface runways by hoof action, and the provision of a new food source in the form of dung (Wood, 1974).

In a dry subhumid grassland site at Berhampur, India, oligochaete numbers reached a maximum of $7800/m^2$ during the rainy season (late August to October). The minimum density reached some $560/m^2$ during summer (May to June). Regression analysis shows a significant positive correlation between percentage of soil moisture and oligochaete number ($r = 0.83$, $n = 16$, $y = 2.62x - 17.33$), indicating the importance of moisture for survival and growth of oligochaete populations (Thambi & Dash, 1973; Dash, Patra & Thambi, 1974; Dash & Patra, 1977). Summer drought during March to June was one of the main causes of reduction in the population to a minimum. Immature oligochaetes dominated the population during the rainy season and in winter (July to mid-October and October to February), while adults were proportionally more abundant during the summer (March to June). Soil testate protozoans reached a maximum number of about 2500/g of moist soil during October and November, and the minimum density was about 250/g in May. A significant positive correlation was found between soil testate amoebae ($r = 0.95$ at the 1 % level) and the percentage of soil moisture (Dash, 1975). Soil temperature and moisture are important factors controlling the population dynamics of soil oligochaetes, protozoans and other invertebrates in tropical soils.

The density of oligochaetes and testate protozoans in tropical grasslands is compared in Table 18.2 with that in temperate grasslands. Among oligochaetes, the density of enchytraeids (500 to $7000/m^2$) in Indian dry subhumid grasslands is 5 to 50 % of the numbers obtained for temperate grasslands and pastures in Europe. The density of Megascolecidae and Ocnerodrilidae (64 to $800/m^2$) in Indian sites is comparable with the density of Lumbricidae found in temperate European grassland and arable soils (25 to $850/m^2$). The testate protozoan population density in the Indian site is much more than the numbers (3–46/g of moist soil) obtained for New Zealand grazed pastures (Stout, 1962), but the values for the Indian site are about 10–15 % of those given for English and sub-antarctic grassland habitats (Heal, 1964, 1965).

Diversity

Consumers show low diversity in grasslands when compared with other biomes; for example, in the Indian sites, the occurrence of only two species of lizards (*Varanus* sp. and *Calotes versicolor* Daud) and in the Ivory Coast site (Lamotte *et al.* 1974) of only one species (*Mabuya buettneri*) have been reported.

Gillon & Gillon (1974) have shown that species with the highest populations do not necessarily have the highest biomass. Although species of beetles belonging to the genus *Zophosis* are abundant in numbers (63 % of the total number of beetles) in certain African sites, two other beetles (*Pimelia senegalensis* and *Vieta senegalensis*) have higher biomass. Different species of beetles have different ecological niche requirements in the same savanna habitat. *Zophosis* occurs on bare ground; *Pimelia senegalensis* and *Stortho-cnemis abyssinicus* are the dry season forms.

Oligochaete consumers comprise three families in dry subhumid grassland in Berhampur, India, i.e. Enchytraeidae, Megascolecidae and Ocnerodrilidae. They include six genera and seven species. Among the oligochaetes, Megascolecidae was found to contribute 90 % of the total annual production of oligochaete biomass. The oligochaete, *Millsonia anomala* Beddard, was dominant, comprising as much as 99 % of the total production of consumer biomass in the Ivory Coast site.

Beetles, among the above-ground consumers, and oligochaetes, among the under-ground consumers, contribute significantly to the species diversity in tropical grasslands. However, we do not have much information on nematodes which may be important consumer elements in tropical grasslands. Although there is no information on the role of termites in the sites sum marised in this chapter, their importance is well established in tropical grasslands (Brian, 1977).

Biomass

In terms of biomass, domestic cattle are the most important group of above-ground consumers in tropical grasslands. However, quantitative data on mammalian biomass values are not available. The annual average biomass of birds has been estimated to be 40 mg/m at Fété Olé, Senegal, but in drought years this was reduced to 19 mg/m^2. Morel & Morel (1974) attribute the decrease in bird population in drought periods to food shortage. The average biomass of lizards was 6.6 mg/m^2 in the Ivory Coast site (Lamotte *et al.*, 1974).

Beetles above ground and oligochaetes under ground dominate the invertebrate biomass in tropical grasslands. Nematodes and termites are found in very large numbers, although their biomass may be small; but considering

Table 18.2. *Population, biomass and annual production per square metre of oligochaetes and testate protozoa in various grassland sites*

Habitat and location	Taxon	Population	Biomass (g) (wet weight)	Production (g) (wet weight)	Source
Sandy permanent pasture, Denmark	Enchytraeidae	44000	2.97		Nielsen (1955)
		30000	3.03		
		74000	10.50		
Alluvial grassland and *Nardus* grassland, moorland, England	Enchytraeidae	10000–25000	15		Peachey (1963)
		37000–200000	35		
Base-rich grassland, Moor House, England	Lumbricidae	389–470	52–110		Quoted in Satchell (1967)
Base-rich grassland, Bangor, N. Wales	Lumbricidae	481–524	112–120		Quoted in Satchell (1967)
Arable land, Germany	Lumbricidae	220	48		Quoted in Satchell (1967)
Arable land, Bardsey Island	Lumbricidae	287	76		Quoted in Satchell (1967)
Apple orchard under grass, Wisbech, England	Lumbricidae	848	287		Quoted in Satchell (1967)
Apple orchard under grass, Holland	Lumbricidae	25–500	11–122		Quoted in Satchell (1967)
Merlewood, England	Lumbricidae			100	Satchell (1970)
Andropogon, Tennessee, USA	Lumbricidae	13–41	3–8		Reynolds (1973)
Lamto, Ivory Coast	*Millsonia anomala*		6.5	46	Lamotte *et al.* (1974)
Tropical grassland, Orissa, India	Megascolecidae, Ocnerodrilidae	64–800	30.25	140	Dash *et al.* (1974), Dash & Patra (1977)
	Enchytraeidae	500–7000	0.14	0.25	Thambi & Dash (1973)
Grazed pasture, New Zealand	Testate protozoa	3–46			Stout (1962)
English grassland, moorland, deciduous woodland, sub-antarctic grassland	Testate protozoa	20000–70000			Heal (1964, 1965)
Tropical grassland, Orissa, India	Testate protozoa	250–2500			Dash (1975)

their large numbers and feeding activities, they may be important consumer elements.

Tenebrionid beetles represent the most important family of the above-ground arthropods in the dry thornbush savanna at Ferlo, Senegal. They contribute 19 % (2.4 mg/m²) of the biomass in January, 58 % (8 mg/m²) in July and 45 % (40 mg/m²) in September. Arachnids and myriapods contribute 18 % (2.3 mg/m²) of the total biomass of the arthropods in January and 6 % (4.6 mg/m²) in September. The maximum biomass of consumers coincides with the peak grass production in the site in September. Gillon & Gillon (1973) observe that there is a broad correlation between the biomass of grass in a given area and the biomass of the arthropods living in that area. During the rains, the arthropod biomass is four times as great as during dry periods.

The quantity of under-ground biomass varies greatly among the tropical grasslands and perhaps is limited by the amount of primary production. It is evident from Table 18.2 that the mean annual biomass turnover of oligochaetes is as high as five to seven times the mean biomass; this high turnover rate, along with the occurrence of large biomass, suggests that they may be important in tropical grasslands. The biomass value of earthworms for tropical grasslands is lower than in many European grasslands, but it is higher than the values given by Reynolds (1973) for Tennessee. Lakhani & Satchell (1970) and Satchell (1970) report that estimates of earthworm biomass are 100 to 287 g/m² (wet weight) in several European sites. The values for the Berhampur site in India range from 6 to 60 g/m² (wet weight).

Production

Production values for vertebrates in tropical grasslands are scarce, and chiefly related to domesticated species. For example, Mann & Ahuja (1974) found that in the grasslands of Rajasthan (India) yearling heifers gained more weight per unit area than did yearling ram lambs. Supplemental feeding of yearling heifers from December to June doubled the rate of gain in biomass. Increase in biomass of yearling lambs was not affected by different deferred and rotational grazing treatments. Gains in biomass differed significantly between breeds. Increase in biomass of the Jaisalmeri breed was 27.45 kg/animal annually. The growth rate is highest during July to October (17.7 kg/lamb) and minimal (0.62 kg/lamb) during the dry period of March and April (Gupta, 1974).

The annual secondary production of lizards in the Ivory Coast site (Lamotte *et al.*, 1974) was 20 mg/m², with annual turnover of three times the average biomass.

Secondary production data for soil oligochaetes are available from the Ivory Coast site, the Berhampur site (India) and from European sites. Satchell (1970) quotes some 100 g/m² (wet weight) of secondary production for

Lumbricus terrestris L. in European sites. Similar estimates have been made for *Lampito mauritii* Kinberg in Indian sites, but values in the Ivory Coast are lower.

Energetics

The development of a consumer energy-flow model requires information of four types: (i) amounts and kinds of food available; (ii) energy intake rates; (iii) energy partitioning within consumers; and (iv) changes in (i), (ii) and (iii) that result from changes in the state of consumers or of the environment. Since available data on tropical vertebrate consumers are scanty and superficial in nature, this discussion of the energy flow in tropical grasslands will involve invertebrate consumers; it is based on the model for soil oligochaetes developed by Dash & Patra (1977). Evaluation of food habits of invertebrate fauna is difficult because of diversification and specialisation of taxa lower than family level. Dash & Cragg (1972) have demonstrated that some members of the soil fauna graze on micro-flora and ingest dead plant materials.

Information is available concerning energetics of oligochaete populations in a dry subhumid grassland in India where the annual net energy input by producers was 32133 kJ/m², not allowing for consumption of plant material by invertebrates and birds (Dash *et al.*, 1974; Dash & Patra, 1977). Oxygen consumption by earthworms amounted to about 60 l/m² annually, indicating the expenditure of 1205 kJ of energy; annual energy output in mucus production (142 g/m²) was 2377 kJ/m²; and 678 kJ/m² of energy was utilised in the production of oligochaete tissue (35 g/m²). Thus, annual energy utilisation for metabolism, growth and reproduction of these invertebrates amounted to an estimated 4260 kJ, which approximates to 13 % of net primary production. However, this estimate of energy flow is minimal, because nephridial excretion has not been measured, so perhaps the proportion used by earthworms was nearer 14 %. This annual level of energy utilisation by earthworms represents only assimilated energy; actual consumption is greater.

The index of utilisation (ratio of annual mean plant biomass to mean consumer biomass) can be used to compare the relative standing crops of producers and consumers in different sites. Breymeyer (1974) reported a value of this index in relation to invertebrates of 819 for the site studied in Panama and compared it to a value of 163 for the temperate region. In the Indian site (Berhampur) being discussed, the index of utilisation for oligochaetes is 309, based on a mean monthly plant biomass of 2472 g/m² and mean oligochaete biomass of 8 g/m².

19. Micro-organisms

R. S. DWIVEDI

Very few studies have been made of the micro-organisms mediating decomposition in tropical grasslands. This account is based on investigations reported during the past two decades in India that are known to the author. A summary of the nature of microbial populations in the Lamto savanna of the Ivory Coast has been prepared for the companion volume (Coleman & Sasson, Chapter 7 in Breymeyer & Van Dyne, 1979).

Populations

The number of species of micro-organisms that become established on recently dead portions of plants is few. However, as decomposition progresses in the litter and soil layers, the number increases rapidly. Decomposition is aided by invertebrates, which break detritus into small pieces and thus expose a larger surface area on which the micro-organisms can operate. Populations of micro-arthropods in one tropical grassland are summarised in Table 19.1.

Population studies of micro-organisms have involved both direct observation of plant material collected in the field and culturing of field collections in the laboratory. Wide fluctuations have been recorded in the number of fungal species occurring on shoots of grass at different times of the year (Table 19.2). The number of species found on stems was less than that on blades and sheaths.

Details of the successional pattern of populations of micro-fungi from early senescence until complete decomposition of litter at Varanasi have been studied in respect to several species of grasses: *Bothriochloa pertusa*, *Cynodon dactylon* and *Dichanthium annulatum* (Khanna, 1964); *Setaria glauca* Beauv. (Sharma, 1967); and *Saccharum munja* Roxb. (Rai, 1968). Populations of micro-organisms appear on all the substrates studied, in the following sequence, although numbers are variable. The genera of fungi that first appear are *Alternaria*, *Curvularia*, *Helminthosporium*, *Nigrospora*, *Phoma*, *Pyrenochaeta* and *Stemphylium*. Nearly all of these readily use cellulose and showed rapid production in laboratory cultures. The dominant genera that follow primary colonisation are *Bispora*, *Annellophora*, *Tetraploa*, *Volutella*, *Cladosporium*, *Melanospora*, *Periconia*, *Spegazzinia*, *Stagonospora*, *Humicola*, *Myrothecium* and *Stachybotrys*. These are less competent colonisers than the primary invaders. They are, in turn, followed by a third group of fungi, of which the dominant genera are *Humicola*, *Hormiscium*, *Lacellina*, *Papulospora*, *Spondylocladium*, *Fumago*, *Endrophragmia*, *Pithomyces*, *Leptothyrium*,

Table 19.1. *Ranges of monthly estimates of populations of micro-arthropods (in thousands/m²) in soil from grassland at Varanasi, India (J. S. Singh, personal communication)*

Taxon	1972–3	1973–4
Acarina	9.3–19.9	9.0–21.3
Collembola	1.8–6.1	2.3–6.1
Other arthropods	1.2–4.5	1.5–3.4
Total	13.5–28.8	13.7–32.3

Table 19.2. *Comparative numbers of fungal species recorded on stem samples of various grasses collected monthly*

Species of grass	Year	No. of species of fungi
Cynodon dactylon (Khanna, 1964)	1961–2	4 (Sept.)–24 (Jan.–Feb.)
	1962–3	6 (June, Sept.)–21 (Jan.)
Saccharum munja Roxb. (Rai, 1968)	1966–7	8 (Jan.)–45 (Aug.)
	1967–8	8 (Jan.)–48 (Aug.)
Bothriochloa pertusa (Khanna, 1964)	1961–2	4 (Aug.–Sept.)–15 (Jan.)
	1962–3	2 (Sept.)–11 (Jan.)
Setaria glauca Beauv. (Sharma, 1967)	1964–5	3 (Sept.)–20 (Nov., Feb.)
	1965–6	4 (Sept.)–26 (Feb.)

Table 19.3. *Populations of soil micro-fungi by plate counts (in thousands/g of dry soil) at different depths in various grassland ecosystems at Varanasi, India (after Dwivedi, 1965)*

	Soil depth (cm)		
Dominant plants	0–15	15–30	30–45
Dichanthium annulatum–Cynodon dactylon	70	76	27
Vetiveria zizanioides–Dichanthium annulatum	30	24	8
Saccharum spontaneum L.	20	10	9
Setaria glauca Beauv.–*Dichanthium annulatum–Cynodon*	44	23	9
Pure stands of *Vetiveria zizanioides*	16	18	6

Bahusandhica, *Hendersonia* and *Stigmella*. These forms survive for only a short period, which indicates that they have the least competitive saprophytic ability and are less able to withstand competition with the other fungi.

Populations of fungi in decomposing litter have also been made in *Dichanthium annulatum* grasslands (Dwivedi, 1959; Khanna, 1964; Mishra, 1964). The commonest colonists are species of *Alternaria*, *Curvularia*, *Helminthosporium*, *Nigrospora*, *Phoma*, *Pyrenochaeta*, *Stemphylium*, *Bispora*, *Humicola*,

Table 19.4. *Microbial populations by plate counts (in thousands/g of dry soil) at different depths in soil of* Heteropogon contortus–Dicanthium annulatum *grassland at Sagar, India (G. P. Misra, personal communication)*

Depth (cm)	Fungi	Actinomycetes	Bacteria
0–7.5	22.0	1524	610
7.5–15	27.6	1839	690
15–30	13.0	4000	375
30–45	3.9	151	150
45–60	3.5	125	1171

Volutella, *Cladosporium*, *Spegazzinia*, *Stachybotrys*, *Hormiscium*, *Lacellina* and *Periconia*. The relative capacity for cellulose decomposition of some dominant fungi has been estimated by laboratory culture methods.

The dynamics of populations of soil micro-organisms have also been investigated (Tables 19.3, 19.4). Population levels are generally highest in the uppermost 15–30 cm of soil, but decline rapidly at greater depths. Population data have not been interpreted in terms of microbial biomass.

Activities

Decomposition

Litter decomposition has been studied in the Varanasi, Rajkot and Kurukshetra grasslands of India. The rate of decomposition is found to be at a maximum during the rainy season of July to September. The process slows down in the drier season. In the *Dichanthium annulatum* grassland at Varanasi 17 % of shoot production was estimated to be decomposed, while the balance was consumed through herbivory. Singh & Yadava (1971) have estimated that 60–80 % of net primary production is processed through the decomposition chain. According to Singh (1972), 430 g/m^2 of litter disappears annually from the *Eragrostis nutans* community in the Vindhyan Hills, while the value for the *Dichanthium annulatum* type, as given by Rao (1970), averages 855 g/m^2.

An approximation of the relative rates of decomposition at various levels below the soil surface has been obtained by burying nylon net bags containing litter (shoots) (Tables 19.5, 19.6). The rate of disappearance declines rapidly below a depth of 15 to 30 cm.

Soil respiration data are available from the work of R. Snehi Dwivedi (1970). By using the cuvette method, he has shown it to be of the order of 30 % of gross photosynthesis in a *Dichanthium annulatum* stand.

Table 19.5. *Rates of shoot litter decomposition* (%) *at different depths in soil, as modified by the activities of ants for two weeks in dry weather* (*summer*), *in grassland at Varanasi, India* (*after R. Shankar Dwivedi, 1970*)

Treatment	Soil depth (cm) 15	30	45
Ants absent	39	16	6
Ants present	70	25	90

Table 19.6. *Rates of decomposition at different depths in soil in grassland at Varanasi, India, based on exposure of shoot material in nylon bags for two weeks in dry weather* (*summer*)

	Soil depth (cm) 15	30	45
Decomposition per day (%)	0.95	0.83	0.52
Time required in days for			
20 % decomposition	21	24	38
100 % decomposition	105	120	190

Nitrogen fixation

Symbiotic nitrogen fixation is of significance in the *Dichanthium* grassland at Varanasi (Rao, 1970). The interaction of *Rhizobium* with the wild legume, *Desmodium triflorum*, results in such an increase in amount of available nitrogen for growth of grasses that shoot production increases by 25 to 50 %.

Nitrogen content of soil increases rapidly following burning of the vegetation in Indian grassland (Pandey, 1971). This is apparently due to rapid, abundant growth of legume seedlings (especially *Desmodium trifolium* due, in turn, to the breaking of seed dormancy through charring of the seed coat.

20. Ecosystem synthesis

J. S. SINGH, K. P. SINGH & P. S. YADAVA

In this chapter we explore the pattern of energy flow and nutrient (nitrogen and phosphorus) cycling in the tropical grasslands occurring in different ecoclimatic zones in India. Because coordinated data for higher trophic levels are not available, our discussion in this section is limited to the primary producer level. The principles of cycling of nitrogen and phosphorus in grassland have been reviewed by Clark *et al.* (in Chapter 8 of Breymeyer & Van Dyne, 1979) with reference to studies in natural temperate grassland in North America.

An ecological system may be visualised as a set of compartments interconnected by flows of energy, matter or information. The selection of the set of compartments to be used in a particular situation is usually made on the basis of the existing knowledge on the structural or functional relationships among the morphologically, physiologically or conceptually recognisable 'sub-units' of the system. In the present case we divide the total mass of plant organic matter, present in the grassland community at any one point in time, into the following compartments (after Singh, 1973): live shoots, standing dead shoots, litter and roots. The compartment designated as 'roots' consists of all under-ground plant parts, both live and dead. The flows between the compartments are represented by net rates of transfer of energy or nutrients from one compartment to another. In energy flow models the incident solar radiation fixed via photosynthesis constitutes the 'source' of energy, while in the case of nutrient cycling models, nutrients present in the soil constitute the source. The net energy that passes through the various compartments is ultimately dissipated from a 'sink' consisting of decomposing dead shoots and roots, but the nutrients are recycled to the soil where they can be reabsorbed by roots.

Time-series data on the biomass, energy content and nutrient concentration (nitrogen and phosphorus) in various producer compartments were available for one year at the following Indian sites: Pilani (Gill, 1975), Kurukshetra (Yadava, 1972), Ujjain (Misra, 1973), Ratlam (Billore, 1973), Sagar (Jain, 1971), Varanasi (Choudhary, 1967) and Ambikapur (Naik, 1973). The semiarid ecoclimate is represented by three sites at Pilani, the dry subhumid by Kurukshetra and Ujjain, the moist subhumid by Ratlam, Sagar and Varanasi, and the humid environment by the grassland at Ambikapur. To obtain the mean compartment values, the time-series data on dry matter in the respective compartments were averaged for each season. This value then was multiplied by the respective season-long average energy content or nutrient content to yield the mean standing crop of energy or nutrient in that compartment. Such values were calculated for each grassland separately and then the values for different grasslands of a given ecoclimate were averaged so as to obtain a single mean value for each compartment for each season. Annual values are weighted daily averages of the three seasonal values. The Indian seasons are: rainy (June–September: 122 days), winter (October–February: 151 days) and summer (March–May: 92 days).

The net fluxes of energy and nutrients were calculated for each season by multiplying the net rates of transfer of dry matter from one compartment to another with the respective energy or nutrient content. A similar procedure was used by Bokhari & Singh (1975) with respect to cycling of nitrogen. The energy flux values were then divided by the number of days in the season to yield daily flow rates. The sum of these values throughout the year represents the annual flow rate.

Estimates of above-ground net production (ANP) were derived through trough–peak analysis of the time-series biomass values of live and standing dead shoots (Singh, Lauenroth & Steinhorst, 1975). Under-ground net production (BNP) was calculated by summing all increments in under-ground biomass through the year.

Rates of transfer of dry matter to the standing dead compartment (STD) and to the litter compartment (L) and rates of litter disappearance (LD) and root disappearance (RD) were calculated using the following expressions:

$$\mathrm{STD} = \begin{pmatrix}\text{Standing crop of live}\\ \text{shoots in the beginning}\\ \text{of the season}\end{pmatrix} + \mathrm{ANP} - \begin{pmatrix}\text{Standing crop of live}\\ \text{shoots at the end of}\\ \text{the season}\end{pmatrix}$$

$$\mathrm{L} = \begin{pmatrix}\text{Standing crop of standing}\\ \text{dead shoots in the be-}\\ \text{ginning of the season}\end{pmatrix} + \mathrm{STD} - \begin{pmatrix}\text{Standing crop of standing}\\ \text{dead shoots at the end of}\\ \text{the season}\end{pmatrix}$$

$$\mathrm{LD} = \begin{pmatrix}\text{Standing crop of litter in}\\ \text{the beginning of the}\\ \text{season}\end{pmatrix} + \mathrm{L} - \begin{pmatrix}\text{Standing crop of litter at}\\ \text{the end of the season}\end{pmatrix}$$

$$\mathrm{RD} = \begin{pmatrix}\text{Standing crop of below-}\\ \text{ground plant parts in the}\\ \text{beginning of the season}\end{pmatrix} + \mathrm{BNP} - \begin{pmatrix}\text{Standing crop of below-}\\ \text{ground plant parts at the}\\ \text{end of the season}\end{pmatrix}$$

The sum of LD and RD represents total disappearance (TD) of dry matter from the system. Basically, the balance-sheet approach of Singh & Yadava (1974) was used in the above calculations.

Energy flow

Energy fixation is at a minimum (3871 kJ/m^2 annually) in the strongly water-limited grasslands of the semiarid zone and reaches a maximum (36213 kJ/m^2) in the dry subhumid zone, while moist subhumid and humid grasslands have intermediate values (Table 20.1). The proportion of incident solar radiation fixed varies from 0.5 % in the dry subhumid zone to 0.05 in the semiarid zone (Table 20.2). The net energy fixed is partitioned equally between roots and shoots in the moist subhumid grasslands, while roots receive a larger proportion both under semiarid (64 %) and humid (67 %) conditions. In the dry subhumid grasslands about 66 % of the net energy fixed is used for shoot growth, the balance being deposited in the under-ground parts. The partitioning of energy above and under ground is markedly influenced by the seasonality of the climate. During the wet rainy season, when the rate of energy capture is at a maximum, a greater proportion of the energy is retained above ground for shoot growth, while in the relatively dry post-monsoon period, greater amounts of energy are channelled under ground for storage

Table 20.1. *Mean energy values (in kJ/m^2) by season for incident solar radiation, net primary production, and plant biomass in each compartment in grasslands of four ecoclimatic zones of India*

Zone, season	Incident solar radiation ($\times 10^4$)	Net energy fixed	Live shoots	Standing dead	Litter	Under ground parts	Disappearance in decomposition chain
Semiarid							
Rainy	173	2363	573	128	297	963	1388
Winter	244	758	425	130	424	858	1678
Summer	293	750	211	90	387	835	393
Annual	710	3871	421	119	372	887	3459
Dry subhumid							
Rainy	253	18085	13062	2103	1668	10673	4707
Winter	237	11192	6199	8868	3370	12492	15376
Summer	218	6936	7797	5791	4075	12311	4220
Annual	708	36213	8607	5656	2987	11844	24303
Moist subhumid							
Rainy	228	9401	3739	1553	948	9958	4562
Winter	267	3253	2276	3164	1594	10135	6910
Summer	228	3002	853	3354	2258	10945	2093
Annual	723	15656	2242	2833	1600	10395	13565
Humid							
Rainy	200	6924	4474	1770	1371	10972	4328
Winter	238	7964	2364	4685	1778	10575	8327
Summer	199	4918	1857	5719	1812	9141	3196
Annual	637	19806	3804	3805	1702	10384	15851

Table 20.2. *Seasonal and annual energy flow ratios (input to output) for tropical grasslands occurring in various ecoclimatic zones in India*

	Semiarid				Dry subhumid				Moist subhumid				Humid			
Transfer	Rainy	Winter	Summer	Annual	Rainy	Winter	Summer	Annual	Rainy	Winter	Summer	Annual	Rainy	Winter	Summer	Annual
ISR–TNP	0.001	0.0003	0.0002	0.0005	0.007	0.005	0.003	0.005	0.004	0.001	0.001	0.002	0.003	0.003	0.002	0.0026
TNP–ANP	0.51	0.14	0.16	0.36	0.94	0.17	0.70	0.66	0.68	0.36	0.08	0.50	0.77	0.10	0.10	0.33
TNP–BNP	0.49	0.86	0.84	0.64	0.06	0.83	0.30	0.34	0.32	0.64	0.92	0.50	0.23	0.90	0.90	0.67
ANP–SD	0.30	1.27	1.00	0.92	0.04	9.30	0.07	0.81	0.11	5.70	3.24	1.00	0.18	6.61	1.23	1.01
SD–L	1.29	0.88	1.31	1.03	0.96	0.72	7.21	0.85	0.85	0.77	0.80	0.78	0.38	0.45	3.78	0.74
ANP–L	0.39	0.69	1.31	0.96	0.04	6.72	0.56	0.69	0.09	4.78	2.61	0.88	0.08	2.98	4.68	0.75
L–LD	1.21	1.07	1.00	1.12	0.98	0.68	1.03	0.75	3.12	0.61	0.20	0.81	2.17	0.15	0.75	0.58
BNP–RD	0.68	1.29	0.41	0.77	3.65	0.69	0.69	0.95	0.88	1.78	0.71	1.06	2.17	1.11	0.33	0.98
TNP–TD	0.59	2.21	0.52	0.89	0.26	1.37	0.61	0.67	0.49	2.12	0.70	0.86	0.63	1.05	0.65	0.80

ISR, incident total solar radiation; TNP, energy captured in total net production; ANP, energy used in above-ground net production; BNP, energy used in below-ground net production; SD, energy transferred from live shoots to standing dead shoots; L, energy transferred from standing dead shoots to litter; LD, energy dissipated via litter disappearance; RD, energy dissipated via root disappearance; TD, total energy dissipated from the system.

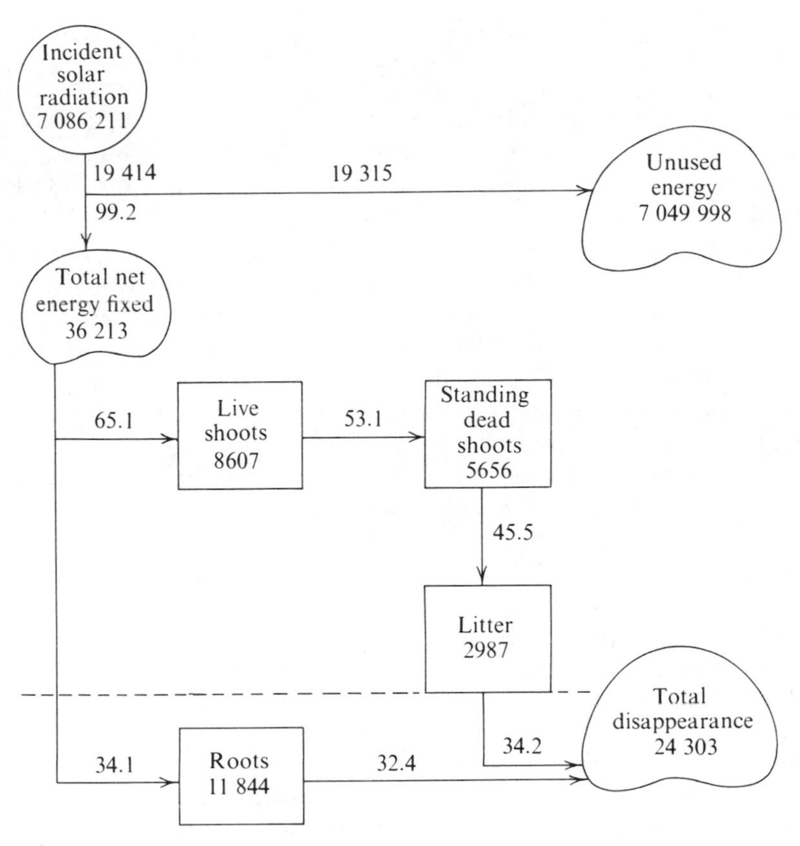

Fig. 20.1. Annual energy flow (kJ/m²) through the primary producer compartments of an average dry subhumid grassland in India. Values in the compartments are mean standing crops of energy. Values on arrows represent net mean daily flux rates. Annual values are given for incident solar radiation, net energy fixed, unused energy and total disappearance.

in roots and rhizomes. The year-long standing crop of energy in roots exceeds that of live shoots in all the grasslands. Most of the energy transferred from live shoots to the dead-shoots compartment finds its way to the litter compartment each year. The net fluxes of energy are greatest in the dry subhumid zone. With the exception of dry subhumid grasslands, root disappearance rate distinctly exceeds the rate of litter disappearance as a consequence, perhaps, of more rapid turnover of the greater root biomass. Approximately 80–89 % of the total net energy fixed by the system is dissipated each year, except in the grasslands of the dry subhumid zone which show a loss of only 67 % of energy fixed. It would appear from this that in these grasslands 11–33 % of net production can be removed without affecting the amount of energy needed for the maintenance of the system in non-drought years.

Maximum rates of energy capture occur in the rainy season in all zones. In the dry subhumid grasslands the mean rate during this season averages 148 kJ/m^2 daily. The lowest average daily seasonal rate (5 kJ/m^2) occurs during winter in the semiarid zone.

Considerable variation exists between sites in the same zone. For example, the annual average daily rate of net production ranged from 6 to 13 kJ/m^2 in three semiarid grassland sites at Pilani. Similarly, in the dry subhumid zone the mixed grassland at Kurukshetra fixed about four times as much energy annually as the *Dichanthium* grassland at Ujjain. The effect of grazing on energy capture is evident from data available from the hill grassland at Varanasi. In stands dominated by *Heteropogon* that had been protected from grazing for two years 1.15 % of incident solar radiation was fixed on an annual basis, while in adjoining areas subjected to grazing by livestock only 0.1–0.3 % of radiation was fixed (Singh & Ambasht, 1975).

A diagrammatic representation of energy flow through an average grassland in the dry subhumid zone is presented in Fig. 20.1.

Nutrient cycling

The strong seasonality in energy fixation and partitioning due to differential growing conditions and growth pattern of major species is also reflected in the uptake, transfer and release of nutrients in the grasslands of the various ecoclimatic zones.

The annual uptake of nitrogen varies from one ecoclimatic zone to another, being maximum, on an average, in dry subhumid grasslands (25.58 g/m^2) and minimum in semiarid grasslands (2.93 g/m^2) (Table 20.3). In semiarid and humid grasslands, respectively, 53–66 % uptake of nitrogen is retained in the under-ground net production, while in the moist subhumid and dry subhumid grasslands 53–73 % of the total uptake is associated with the above-ground net production. Thus, the semiarid and humid climates promote recycling of nitrogen through under-ground parts, while dry and moist subhumid climates favour recycling through above-ground parts. The proportion of nitrogen absorbed by vegetation that is subsequently returned to soil ranges from 63 to 78 %, being higher in humid and semiarid grasslands, than in the zones of intermediate moisture supply.

The rate of nitrogen uptake by plants is highest in the rainy season in all except the humid zone, ranging from 56 % of the annual transfer in dry subhumid grasslands to 71 % in the semiarid zone. However, the rate of release back to the soil is greatest during winter, when from 47 to 64 % of this transfer takes place.

The annual uptake of phosphorus is maximum in dry subhumid grassland (4.96 g/m^2) and minimum in the semiarid zone (0.49 g/m^2) (Table 20.4). The main fraction of phosphorus uptake is associated with the above-ground net

Table 20.3. *Mean seasonal and annual rates of flow of nitrogen (g/m²) between soil and plants and between plants and soil, with mean nitrogen contents (g/m²) of soil and plant compartments, for grasslands in various ecoclimatic zones of India*

								Totals		
Zone, season	In soil	Soil to shoots	In shoots[a] and litter	Litter to soil	Soil to roots	In roots	Roots to soil	Soil to plants	In plants and litter	Plants to soil
Semiarid										
Rainy		1.21	0.80	0.38	0.86	0.70	0.59	2.07	1.50	0.97
Winter	63.9	0.08	0.65	0.50	0.32	0.60	0.58	0.40	1.25	1.08
Summer		0.08	0.41	0.10	0.38	0.51	0.15	0.46	0.92	0.25
Annual		1.37	0.64	0.98	1.56	0.61	1.32	2.93	1.25	2.30
Dry subhumid										
Rainy		13.50	12.13	0.97	0.71	5.84	2.36	14.21	17.97	3.33
Winter	452.0	1.45	11.74	6.56	5.08	6.28	3.74	6.53	18.02	10.30
Summer		3.62	9.49	1.68	1.22	5.95	0.90	4.84	15.44	2.58
Annual		18.57	11.31	9.21	7.01	6.05	7.00	25.58	17.36	16.21
Moist subhumid										
Rainy		6.44	5.26	0.87	3.24	9.52	0.83	9.68	14.78	1.70
Winter	451.2	1.52	4.41	2.08	1.84	4.62	3.38	3.36	9.03	5.91
Summer		0.20	4.53	0.08	2.04	8.80	2.12	2.24	13.33	2.20
Annual		8.16	4.72	3.03	7.12	7.31	6.78	15.28	12.03	9.81
Humid										
Rainy		5.86	7.45	0.41	1.74	11.80	3.78	7.60	19.25	4.19
Winter	398.4	0.80	10.27	0.65	7.50	11.03	8.31	8.30	21.30	8.96
Summer		0.48	8.05	0.82	4.51	9.31	1.50	4.99	17.36	2.32
Annual		7.14	8.76	1.88	13.75	10.85	13.59	20.89	19.61	15.47

[a] Includes live and standing dead shoots.

production in moist subhumid, dry subhumid and humid zones, the proportion of total uptake reflected in net shoot production being, respectively, 67, 70 and 85 %. In contrast, in the semiarid grasslands 65 % of phosphorus uptake is directed to the under-ground net production. The annual release of phosphorus through litter decomposition varies from 0.01 g/m² (humid) to 1.11 g/m² (dry subhumid), while the annual release through root decomposition ranges from 0.26 g/m² (semiarid) to 1.92 g/m² (dry subhumid). The proportion of phosphorus uptake by vegetation that is subsequently recycled to soil is 15, 44, 61 and 98 % for humid, moist subhumid, dry subhumid and semiarid zones, respectively. Thus, in more humid climates phosphorus has a tendency to get locked up in dead plant material.

Like nitrogen, the rate of phosphorus uptake is maximum in the rainy season in all zones (54–84 % of annual uptake). The rate of release from plants to soil is highest during winter, accounting for 43–69 % of annual release.

Table 20.4. *Mean seasonal and annual rates of flow of phosphorus (g/m²) between soil and plants and between plants and soil, with phosphorus contents (g/m²) of soil and plant compartments, for grasslands of various ecoclimatic zones of India*

Zone, season	In soil	Soil to shoots	In shoots[a] and litter	Litter to soil	Soil to roots	In roots	Roots to soil	Totals: Soil to plants	Totals: In plants and litter	Totals: Plants to soil
Semiarid										
Rainy	54.4	0.14	0.20	0.07	0.16	0.13	0.11	0.30	0.33	0.18
Winter		0.01	0.14	0.12	0.07	0.12	0.12	0.08	0.26	0.24
Summer		0.02	0.11	0.03	0.09	0.12	0.03	0.11	0.23	0.06
Annual		0.17	0.15	0.22	0.32	0.12	0.26	0.49	0.27	0.48
Dry subhumid										
Rainy		2.56	2.15	0.09	0.13	1.36	0.49	2.69	3.51	0.58
Winter		0.28	1.93	0.84	1.10	1.50	1.26	1.38	3.43	2.10
Summer		0.65	1.47	0.18	0.24	1.45	0.17	0.89	2.92	0.35
Annual		3.49	1.88	1.11	1.47	1.44	1.92	4.96	3.32	3.03
Moist subhumid										
Rainy	22.9	0.78	0.42	0.02	0.22	0.79	0.16	1.00	1.21	0.18
Winter		0.04	0.45	0.02	0.02	0.84	0.22	0.06	1.29	0.24
Summer		0.02	0.35	[b]	0.18	0.83	0.13	0.20	1.18	0.13
Annual		0.84	0.41	0.04	0.42	0.82	0.51	1.26	1.23	0.55
Humid										
Rainy	40.8	1.67	0.17	[b]	0.04	0.27	0.09	1.71	0.44	0.09
Winter		0.23	0.22	[b]	0.17	0.25	0.19	0.40	0.47	0.19
Summer		0.01	0.15	[b]	0.10	0.21	0.03	0.11	0.36	0.03
Annual		1.91	0.18	0.01	0.31	0.24	0.31	2.22	0.42	0.32

[a] Includes live and standing dead shoots. [b] Less than 0.005.

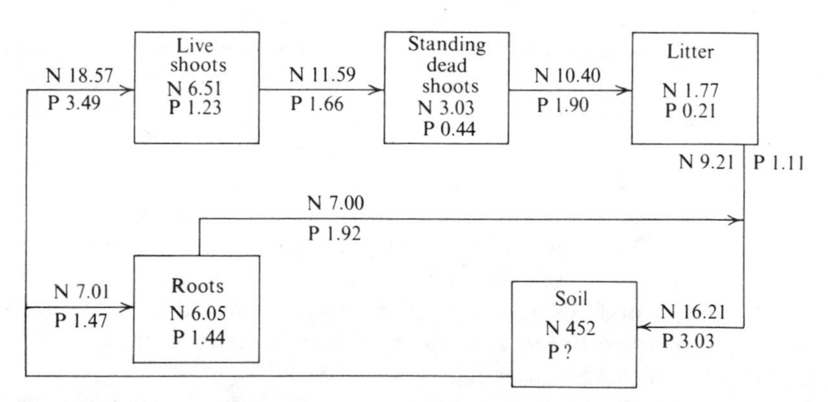

Fig. 20.2. Uptake, transfer and release of nitrogen (N) and phosphorus (P) in an average dry subhumid grassland in India. Values in compartments are mean standing crop of nutrient in g/m², and those on arrows are net annual flux rates in g/m².

A diagrammatic representation of nutrient cycling in an average grassland in the dry subhumid zone is presented in Fig. 20.2.

Relationship between energy flow and nutrient cycling

A comparison of nutrient cycling per unit of energy flow on an annual basis in grasslands of the various ecoclimatic zones of India is made in Table 20.5. Our computation of nutrient uptake: energy capture ratios takes into account both above-ground and under-ground net production, and nutrient release: energy dissipation ratios are based on total disappearance (both from litter and roots). Nitrogen and phosphorus exhibit contrasting behaviour with respect to energy flow. The grasslands of dry regions (semiarid and dry subhumid) appear to need less nitrogen to support the same magnitude of energy flow than the grasslands in relatively humid regions (moist subhumid and humid). As a corollary, more energy is needed in the drier zones to pump similar amounts of nitrogen in the biological system as compared to relatively humid zones. In the case of phosphorus the situation is reversed, as the grasslands of humid regions cycle lower quantities of this nutrient in effecting fixation or dissipation of comparable amounts of energy. It is significant that the amount of nutrients absorbed per unit of energy captured is always greater (except for phosphorus in the semiarid zone) than the amount released from equivalent energy dissipated, indicating a distinct short-term trend for conservation of nutrients in the carry-over herbage, as a result of protection from grazing. However, this conservation trend does not take into account possible nutrient losses through other avenues, e.g. the movement of phosphorus from dying leaves in rainfall.

Table 20.5. *Relationship between annual nutrient uptake (mg/m^2) and energy flow (in thousands of kJ/m^2) in various ecoclimatic zones of India*

	Dry region		Humid region	
Nutrient ratio	Semiarid	Dry subhumid	Moist subhumid	Humid
Nitrogen				
Nutrient uptake: energy capture	0.76	0.71	0.98	1.06
Nutrient release: energy dissipated	0.66	0.67	0.72	0.98
Phosphorus				
Nutrient uptake: energy capture	0.13	0.14	0.08	0.10
Nutrient release: energy dissipated	0.14	0.12	0.04	0.02

21. Use and management

R. K. GUPTA & R. S. AMBASHT

Grasslands in tropical countries are mostly in an advanced stage of degradation due to high animal populations and absence of management practices. 'Fair' and 'good' (Bhimaya & Ahuja, 1969) range condition classes in India are sometimes present due to lower pressure of the stock, but more than 80 % of the grasslands are in the 'poor' range condition class. In some tropical countries politico-religious considerations interfere with the reduction of high animal populations, thus making the task of conservation and management challenging to range ecologists. Common land is subject to particularly heavy use in the tropics.

Improvement of degraded grassland through control of grazing load is an expensive procedure in the tropics because wood fence posts have a very short useful period of service. Various types of fences have been tried in India (Bhimaya *et al.*, 1966). Fencing with angle iron posts and five strands of barbed wire, although initially costly, proved to be the most efficient, economical and durable. Controlled grazing results in an approximate doubling of herbage production after two years of average rainfall in all classes of range. For example, in one study (Ahuja, 1972) the increases reported were 148 % in a 'poor' range, 91 % in a 'fair' range and 116 % in a 'good' range after two years of fencing. However, after two years of protection there was a slight decrease in forage production in all three conditions.

A wider spectrum of large herbivorous mammals was indigenous to the tropical grasslands, as compared with the temperate zone. Raising domesticated livestock in the tropics has certain limitations that are related to their reduced ability to adapt to adverse environments compared with native species. The health of exotic species is poor, with the result that yields of milk, meat and wool are low. The domestication of indigenous species such as various antelopes and zebra has been suggested, partly because they do not damage the habitat to the same extent as domesticated animals.

Range animals make their fastest gains during the period when green herbage is abundant. This is the time when the nutritive value of shoots is highest. Protein content in the vegetative stage of three tropical grasses (11–16 %) averages two to three times that at maturity (3–8 %) (R. K. Gupta *et al.*, unpublished) and drops even further (to 2–3 %) as the shoots die in the dry season (Ahuja, Bhimaya & Prajapati, 1967).

The performance of domesticated animals in tropical grasslands depends in a major way on the nutritive quality of forage available during the dry season (Ahuja *et al.*, 1967, 1974). Where only dead grass herbage is available, and the drought is long, loss in body weight during the annual dry periods may be so great that a beef animal on range may not reach a marketable condition until it is six years old. Where some green herbage is available in the diet (Ganguly, Kaul & Nambiar, 1964), the digestibility of the dry forage is increased (Ray & Nudgul, 1962). Crop residues and tree 'loppings' also are used to supplement the dry grass herbage during drought. Supplemental feeding with oil-cake concentrate has doubled the body weight of yearling heifers in a controlled continuous grazing system (Ahuja *et al.*, 1967). Trees in a savanna provide shade as well as reserve fodder. A density of 20 to 25 plants of *Zizyphus nummularia* per hectare has been recommended by Ahuja & Bhimaya (1966) in the semiarid areas of India.

In some areas of the arid tropics sheep are grazed on range only during the wet season and are held on irrigated, seeded grassland during the dry season. This results in more efficient use of the range and in a highly increased production per animal.

Continuous controlled grazing results in higher gains in body weight of domesticated animals than do various deferred and rotational grazing systems (Ahuja, 1964). However, some deferred and rotational systems have increased the grazing capacity (by 39 %: Soil and Water Conservation Centre, 1971), have reduced fluctuations in gains from year to year, and have favoured regeneration of range plants by seed (Prajapati, 1970).

Clipping experiments have shown that maximum yield of herbage is obtained in tropical grasslands when they are defoliated to a height of 15 cm above the soil surface at intervals of one month (Dabadghao & Das, 1960–63) to two months (Indian Grassland and Fodder Research Institute, 1971). The vertical distribution of biomass in the canopy is variable among species (Das, Dabadghao & Debroy, 1964; Shankarnarayan, Sreenath & Dabadghao, 1969). The relationship between percentage height clipped and percentage weight removed is quadratic. The difference in the distribution pattern of height becomes considerably less when about 70 % of the height is clipped. This indicates proper utilisation of grasslands at the safest limit.

Fire, as a management tool, produces different effects in different grassland types. Egunjobi (1974) considers a programme of controlled burning to be necessary in grazed tall-grass savannas in Africa. Pandey (1972) nearly doubled herbage production in *Dichanthium* grassland in India by annual and semi-annual burning. However, no such effect on vigour or yield of grasses was found in *Sehima* grassland in India, whether grazed or not (Indian Grassland and Fodder Research Institute, 1971). Protection from fire for only one year caused a measurable decline in both herbage production and protein content of new herbage. Time of burning had no noticeable effect on quality of herbage (Egunjobi, 1974). Changes in floristic composition may result from burning. For example, *Heteropogon contortus* increases proportionally to other species in burned areas, while *Sehima nervosum* decreases (Indian Grassland and Fodder Research Institute, 1971). Fire is used as a regular management practice to control woody plants in savannas. However, in open grassland burning more commonly occurs as a result of fires that are set accidentally.

Contour furrowing has been used as a means of reducing the amount of run-off water from ranges. This practice has resulted in herbage yields as much as three to five times that of the untreated control (Erasmus, 1969; Ullah *et al.*, 1972; Soil and Water Conservation Centre, 1971). Generally, perennial grasses are favoured by this treatment, while annuals decline proportionally.

Economic returns from animals often are not commensurate with the high cost of forage production. Improvements that are necessary include: the use of higher-yielding animals; better care of animals; and the provision of better marketing facilities.

Perennial grasses have an important role as soil conservation agents. Their influence on soil structure is such as to reduce the hazard of water erosion to a minimum.

Abundant roots near the surface of the soil and the production of root exudates result in an increase in the proportion of water-stable aggregates and in binding of root particles together. For example, in one study (Chatterjee & Sen, 1964) the proportion of water-stable soil aggregates in annual crops of maize and wheat ranged from 16 to 29 %, compared with 23 to 46 % under four perennial species of tropical grasses. The extensive root systems of perennial grasses also add more organic matter to the soil than do annual crops.

Infiltration of water occurs at a faster rate under perennial grasses, with a reduction in the amounts of run-off and soil erosion. Measurements in tropical grasslands indicate that the rate of water infiltration is 50–100 % greater in areas of permanent grassland, compared to cropland (Mistry & Chatterjee, 1965; Vasudewaiah, Singh-Teotia & Gupta, 1965; Tejwani, Gupta & Mathur, 1975). The proportion of water lost in run-off from perennial grassland is often 1 % or less in situations where losses are 20 % or more from adjacent cropland (Tejwani & Mathur, 1972). There is a considerable variation in the effectiveness of different species of tropical grasses in modifying run-off. In one study (Tejwani *et al.* 1975) the mean (six-year) range in four species tested under the same conditions was from 1.8 to 4.2 % of natural rainfall. The degree of protection against soil erosion that is provided by perennial grasses varies even more between species (Ambasht, 1970; Prajapati, Phadke & Agarwal, 1973), as does their soil-binding ability (Bhimaya, Rege & Srinivasan, 1956; Bhaskaran & Chakravarty, 1965).

Artificial seeding and renovation are range improvement practices that are receiving the attention of experimentalists in tropical grasslands. Seeding in range with woody cover generally leads to failure because of severe competition by established weedy plants. Thorough tillage is required prior to seeding.

Seeding during the wet season has resulted in the best stands, while dry season seedings lead to failure (Dabadghao, 1959). The use of pelleted seeds has given variable results in comparison with unpelleted seeds (Chakravarty & Bhati, 1968; Chakravarty & Verma, 1966, 1968). Other matters that have been considered are depth and rate of seeding (Chakravarty *et al.*, 1966; Chakravarty & Verma, 1960).

Experiments with herbicides and fertilisers in tropical grasslands have illustrated that herbage production can be increased considerably through their use. However, care must be exercised in adjusting the treatment to the particular situation. Marked changes in floristic structure result from both herbicides and fertilisers. In some instances they favour desirable range species, but sometimes they cause undesirable species to increase. Herbicides show particular promise in range improvement through their use in controlling shrubs. Applications of lime to acid soils and of trace elements in deficient areas have markedly increased herbage yield. The limiting factor with these additives is the economic feasibility of their use.

Through nomadism, man has provided a means whereby his domesticated animals can migrate according to the seasonally changing supply of herbage in different habitats, as is done by wild herbivores. Pastoral nomads in India are largely concentrated in arid and semiarid regions near the boundary with Pakistan. However, they are common also in all dry regions of south-western Asia, Asia Minor and northern Africa, and in some countries they form the largest proportion of the human population. Nomads have a permanent home, but move out in search of forage and water for their livestock during periods of scarcity. They earn their livelihood through casual labour, traditional caste occupations and agriculture. Their movements are not chaotic, but follow a regular cycle and annual rhythmic pattern (Malhotra, 1968; Gupta, 1975). Because of recent socio-economic changes, the sedentary population is much less dependent on the produce from this system.

Grasslands are important to man in some tropical areas as sources of household fuel and as a means of moderating the environment. The woody plant resources of the range are used for the production of charcoal. The high albedo values of certain grasses, as much as 65 % in one species of *Aristida*, create oasis conditions in arid areas. Extensive parks and lawns are maintained for this purpose.

Changes in grassland ecosystems under various management practices have been monitored for several years in several tropical sites. It is important that these sites be conserved for studies in UNESCO's Man and the Biosphere Programme.

Part V. Arable grasslands

Subeditor: W. M. WILLOUGHBY

22. Introduction

W. M. WILLOUGHBY & P. J. VICKERY

Increasing world demand for low-cost animal products has stimulated the need to augment output from various grazing resources. This has been accomplished by modifying the nature of the vegetative cover to provide a larger area of grassland as well as a higher yield of forage per unit area. These improvements in forage supply have permitted an increase in numbers of livestock supported, greater production per head and better quality of product.

Expansion of the area of grassland has been achieved mainly by clearing forests and woodlands. In some situations croplands are also converted to grasslands, either to provide a permanent and stable community or on a temporary basis to increase the fertility of the soil for future crops. In the reverse direction, the area of grassland is sometimes reduced by permanent or temporary conversion to croplands.

Grassland productivity is usually improved by replacing natural grassland, or the volunteer species that follow cropping or the clearing of forests, with sown species of better quantity or quality. This is commonly referred to as 'pasture improvement'.

Establishment of arable grassland

Arable grassland is the outcome of the pasture improvement process in which cultivable soils are tilled and sown with improved species, usually with the addition of fertiliser. However, many soils which are non-cultivable because of shallowness, stoniness or steepness, or which currently are considered non-arable because of infertility, may have relatively similar sown grasslands established upon them. Considerable experimentation has been done in techniques for such establishment. These include zero or minimal cultivation, chemical suppression of competing plants, oversowing with pasture species and fertiliser from ground level or from aircraft, and the use of grazing animals to reduce competition from native species.

The number of plant species used throughout the world for grassland sowing is relatively few (Hartley & Williams, 1956). Commonly two or more species are sown, and one at least is usually a legume. Legumes provide the system with access to atmospheric nitrogen, obviating the need for fertiliser nitrogen. In addition, production per animal is generally higher than from non-legumes (Reed, 1972). The new species often require an improved soil environment. Elements such as phosphorus, potassium, sulphur, zinc, molybdenum, boron or copper may be deficient and require correction through fertiliser application and the legumes may need appropriate nodulating bacteria.

Grasslands and animal production

In most areas of the world, regardless of the species grown or fertiliser applied, the nature of the climate is such that during the year the grassland has peaks and troughs in quantity and quality of herbage produced and available to the animals grazing thereon. Indeed, most grassland species used for sowing appear to have been selected for maximum production at the peaks with little regard for the distribution of production throughout the year, let alone for such distribution under grazing conditions.

The troughs in herbage availability either in quantity or quality impose the major restraints on animal productivity from an area of grassland. The greater the severity and duration of the troughs the lower the level of food that the animals can harvest during these periods and the more days that reduced levels of consumption or intake persist. Periods of low quality of intake restrict production per head. Periods of low quantity of intake restrict production per head and the number of livestock that the herbage can support during these periods and, thus, throughout the whole year or grazing season.

The number of animals and the amount and composition of herbage that each consumes each day, as dictated by their type, size and physiological status, and by the climate, influence the total amount and composition of the herbage consumed on that day. The unconsumed herbage together with any new growth generated therefrom, less losses in quantity and quality due to ageing, decomposition, and consumption by insects, constitute the amount and composition of the herbage on the following day; and so on. Thus, the grazing animal exerts a constant influence on the herbage, the most critical effects being those influencing the severity and duration of the periods which restrict animal numbers and production per head.

The restriction in animal numbers imposed by the troughs results in a surplus of herbage in the peaks. Among the attempts man may make to use this surplus are to graze more animals and to hand-feed them in the trough periods either with excess herbage cut and conserved from earlier peak periods, or with food produced on other areas. The total efficacy and profitability of these procedures has rarely been experimentally investigated (see for example, Hutchinson, 1971).

IBP investigations

Studies of the interactions between the processes of herbage production and consumption in management systems, aimed at optimising animal production from grassland, were made during IBP at Armidale in Australia (Plate 22.1) and similar studies were started at Migda in Israel. Other studies of this type may have been conducted elsewhere, but if so, the data obtained were not presented for consideration in this international synthesis.

The investigations at the Armidale site have been concerned with the structure and function of animal productivity from arable grasslands grazed year-long. They have included studies of primary production, herbage consumption, conversion of intake within the animals, the importance of other consumers and nutrient cycling. In Israel the investigations have similar aims, but because they started later the studies did not reach the same level of detail during the IBP years.

Armidale is located on the eastern highlands of New South Wales, Australia, at an altitude of 1046 m, latitude 30°31′ S, longitude 151°39′ E.

Plate 22.1. Sown *Phalaris–Trifolium* grassland at Armidale, Australia, in early summer (December), grazed continuously by 10 sheep/ha. Vacuum collection of herbage invertebrates under quick traps (Turnball & Nichols, 1966) is shown in the background. (Photograph supplied by P. J. Vickery.)

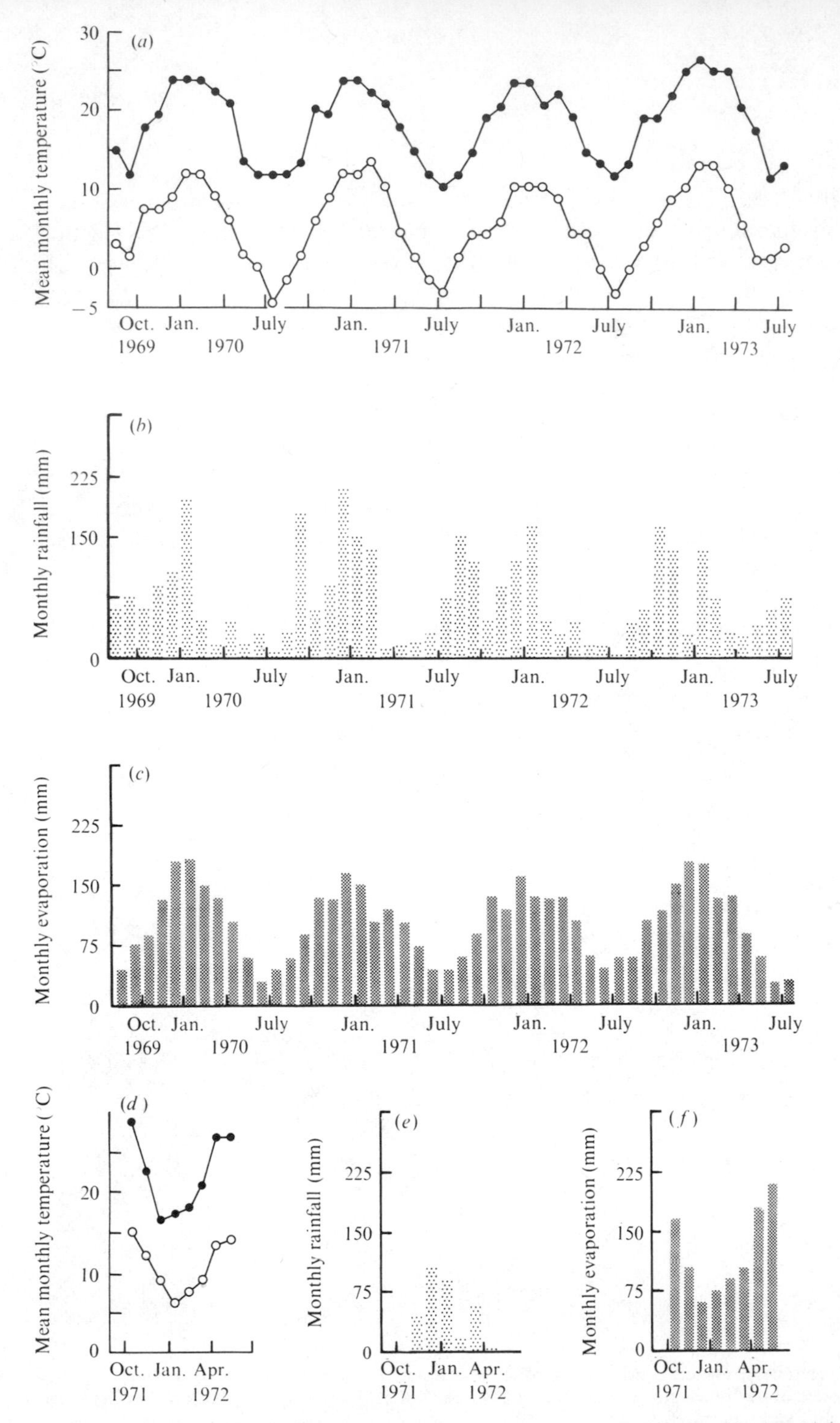

Fig. 22.1. Meteorological data summarised for years reported. (*a*), (*b*), (*c*), Armidale; (*d*), (*e*), (*f*), Migda. ●——●, maximum temperatures; ○—○, minimum temperatures.

Migda is located in the northern Negev desert of Israel at an altitude of 100 m, latitude 31°22′ N, longitude 34°25′ E.

The Armidale site has a cool temperate climate, with a mean annual rainfall of 750 mm; 60 % falls between September and February, and potential evaporation parallels rainfall so the effective rainfall is not seasonal (Hilder, 1963). Winter temperatures are low (July mean minimum 0.4 °C, July mean maximum 11.7 °C) and summer temperatures between 27 °C and 32 °C commonly occur (January mean minimum 12.8 °C, January mean maximum 26.7 °C) (Roe, 1947). Frosts occur on approximately 120 occasions per year.

The Migda site is in a winter rainfall area with a Mediterranean climate, rain occurring between October and April; 60 % of the annual precipitation is concentrated in December and January. There are extreme rainfall fluctuations between years (42 mm in 1962–3, 412 mm in 1964–5), but without discernible periodicity. The pattern of rainfall distribution within a single season greatly influences vegetation development. Temperatures are mild in winter. The coldest month is January with mean minimum and mean maximum temperatures of 7.6 °C and 18.1 °C. There are 20 to 30 nights of frost, with the soil surface temperature dropping to −1 to −4 °C. Temperatures are high in summer. The hottest month is August with mean minimum and mean maximum temperatures of 20.2 °C and 33.5 °C (Tadmor, Eyal & Benjamin, 1974).

Meteorological records from the two sites are presented in Fig. 22.1 to summarise the environmental conditions during the periods of the IBP studies.

The soil at the Armidale site is a grey-brown podzolic loam overlying an orstein layer at a depth of 40 to 70 cm. Field capacity is approximately 30 % (w/w) and wilting point 10 %. The bulk density varies with stocking rate, being 1.27 g/cm^3 for stocking rates in the vicinity of 20 sheep/ha. This allows storage of up to 100 mm of available moisture in the 40 cm above the orstein layer.

The Migda site soil is a structureless sandy loam loess, 10–20 m deep. 'Field capacity' is 16 % (w/w) and 'wilting point' 5 %. Bulk density is 1.35 g/cm^3. This gives an available moisture range of 11 % per weight or 15 % per volume, or 150 mm of available water temporarily stored per metre of depth. Thus, in this 250-mm rainfall area, wetting is usually not deeper than 150 cm and soil moisture seldom exceeds the rooting depth of the annual vegetation. On impact, rain drops form a crust and infiltration decreases rapidly to 3.4 mm per hour. This causes run-off even on slight slopes.

Originally the native vegetation of the Armidale site was a temperate woodland. However, it has changed to a large extent to semi-natural grassland as a result of the activities of man during 150 yr of European settlement. These activities included clearing trees, reduction or exclusion of native animals, invasion by exotic animals (e.g. rabbits, hares) and plants, year-round confinement of sheep or cattle to the grasslands and cropping in some areas. Soil fertility was lowered, the natural grassland changed in composition and its productivity was reduced. The natural grassland communities have been replaced by pastures based on white clover (*Trifolium repens* L.), together with phalaris (*Phalaris aquatica* L.), perennial rye grass (*Lolium perenne* L.) or tall fescue (*Festuca arundinacea* L.). These pastures often require substantial quantities of superphosphate fertiliser to maintain the sown species, but they can support six times the sheep population of semi-natural grasslands, resulting in at least a proportional increase in animal production per unit area (Willoughby, 1966; Langlands & Bowles, 1974).

In the Migda site the original vegetation was semi-desert grassland; however, it now consists of a regeneration of herbaceous annual species which have developed after the abandonment of cropping (Tadmor *et al.*, 1974). Further details of the vegetation on these two sites is given in the chapter on primary productivity which follows.

Chapters 23 to 27 are based principally on research carried on at the Armidale site. Additional information concerning these investigations is provided by Hutchinson & King (in Chapter 12 of Breymeyer & Van Dyne, 1979), while a simulation model of the grazed ecosystem is described by Seligman & Arnold (in Chapter 13 of Breymeyer & Van Dyne, 1979).

23. Producers

P. J. VICKERY

The grassland community at the Amidale site consisted of two sown species, *Phalaris aquatica* L. (phalaris) and *Trifolium repens* L. (white clover), which contributed the major portion of the vegetative biomass. Non-sown species were also present and contained representatives from the genera *Bothriochloa*, *Bromus*, *Danthonia*, *Hordeum*, *Panicum* and *Vulpia*, as well as a wide range of annual diotyledonous species common to cultivated land. However, the relative species composition of the community varied both with season and number of grazing ruminants per unit area (stocking rate); details of such changes are reported by Hutchinson & King (Breymeyer & Van Dyne, 1979, Chapter 12). These authors classified the non-sown graminaceous species as annuals, because this reflects the behaviour of these plants in this environment, even though in other grasslands similar plants may behave as perennials. At low to medium stocking rates of sheep (10 and 20/ha) *P. aquatica* and *T. repens* were dominant, while at higher stocking rates (30/ha) the forbs and annual grasses were a significant component of the vegetation in the period August, 1969–August, 1971 for which net primary productivity (NPP) will be reported.

Details of the vegetation at the Migda site in Israel have been reported by Tadmor *et al.* (1974). It consists of abandoned cropland with a volunteer cover of herbaceous annuals, the major grass genera being *Brachypodium*, *Elymus*, *Hordeum*, *Phalaris* and *Stipa*. Legumes are represented by *Medicago* and *Trigonella*, while other dicotyledons belong to the genera *Athemis*, *Centaurea*, *Erucaria* and *Reboudia*. Seasonal development of the vegetation is wholly dependent on rain and before the rainy season grazed areas are often completely bare. After the onset of rain, between October and November, germination and emergence takes place within 5 to 15 days. Germination, emergence and vegetation development are all restricted by low temperatures in the early part of the rainy season and significant development is often delayed until February. The vegetation matures and dries to a standing hay crop between March and April. Net primary production is reported for the 1971–2 growing season, which had a good intraseasonal distribution of rain and near optimal growth of vegetation.

Both the Armidale and Migda sites were fertilised with sufficient major nutrients to ensure that these were not limiting. However, at Armidale nitrogen was not used, because this was considered unnecessary in the legume-based grassland.

Seasonal dynamics

The seasonal patterns of green and dead herbage and plant roots at the Armidale site are shown in Table 23.1. Data are presented for two stocking rates which produced light and heavy utilisation of herbage during two years having different distributions and total amounts of rainfall. The year has been divided into four periods, each of 12 weeks, which conveniently divides the year into four growth phases. The remaining 29 days at the end of the winter period have been omitted from this analysis, but measurements taken at this time have been used in the calculation of annual production. The trends in productivity over the two years sampled, as estimated from carbon dioxide exchange measurements (Vickery, 1972), are shown in Fig. 23.1.

Table 23.2 shows the seasonal trend in herbage biomass at the ungrazed Migda site, together with production estimates for the sampling periods.

Table 23.1. *Seasonal mean herbage biomass (g/m^2) at Armidale for two stocking rates*

Year and S.E.[a]	Stocking rate[b]	Period[c]				Annual mean
		Spring	Summer	Autumn	Winter	
		(*a*) Green herbage				
1969–70	L	104	257	257	94	183
27	H	105	111	52	17	71
1970–1	L	59	282	436	220	249
23	H	20	58	63	40	45
		(*b*) Dead herbage				
1969–70	L	—	183[d]	266	249	233
25	H	—	22[d]	20	16	19
1970–1	L	164	134	305	547	288
13	H	1	9	21	43	18
		(*c*) Roots				
1969–70	L	1104	839	655	782	845
84	H	741	468	435	423	517
1970–1	L	741	793	970	809	828
81	H	521	516	453	527	504
		(*d*) Precipitation (mm)[e]				Annual totals[f]
1969–70		201	322	133	78	734
1970–1		265	320	288	56	984

[a] Standard error for comparing period by stocking rate means.
[b] L, light (10 sheep/ha); H, heavy (30 sheep/ha).
[c] Periods are 12 weeks in length and began on 8 Aug. 1969 and 31 July 1970, respectively, in the two years sampled.
[d] Based on 2 samplings in the last six weeks of the period.
[e] Precipitation data are included for comparison.
[f] Including rain in four weeks outside sampling period.

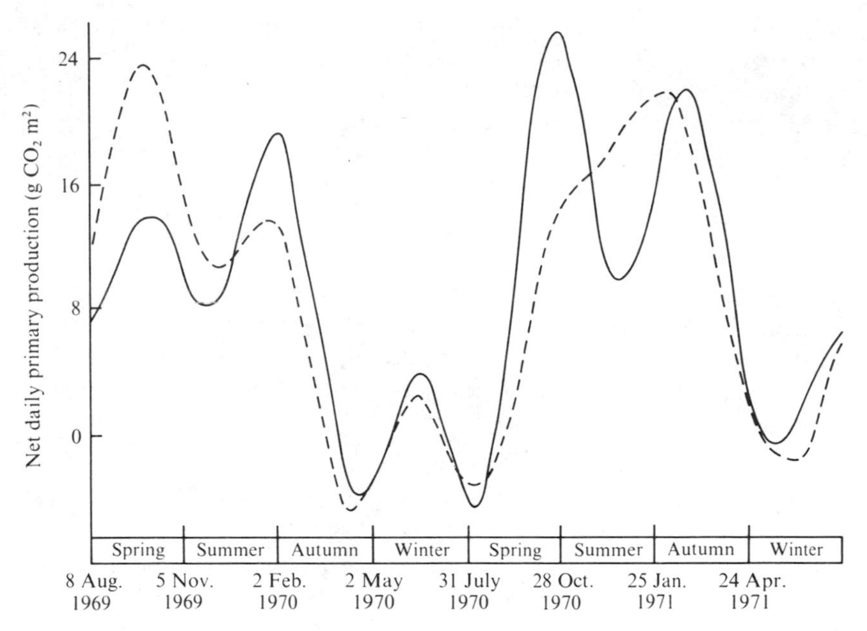

Fig. 23.1. Seasonal trends in the rate of above-ground and under-ground NPP for two years at 10 sheep per ha (——) and 30 sheep per ha (- - -); the curves are sixth harmonic Fouriers.

Table 23.2. *Biomass, precipitation and production* (*in g*/*m*^{2}) *at the Migda site in 1972*

Days from 1 Jan.	Herbage biomass	Production period (days)	Estimated production for period[a]	Rainfall for period (mm)[b]
30	28	30	28	90
67	325	37	297	8
89	578	22	253	55
105	629	16	51	0

[a] Assuming zero biomass at 1 Jan. 1972.
[b] Rainfall in the two months prior to 1 Jan. 1972 was 148.0 mm.

Annual production

For the site at Armidale estimates of NPP have been derived from carbon dioxide exchange data using the procedures outlined by Vickery (1972). Annual NPP of the whole plant system (tops plus roots) was estimated by integrating measurements of daily carbon dioxide balance made once every 21 days, assuming that the same level of productivity was maintained for 10 days before and 10 days after the sampling date. Carbon dioxide balance figures were converted to dry matter, assuming 11.3 kJ/g CO_2 and 18.0 kJ/g of dry matter. By this method the annual total whole plant net primary production in grams dry matter per square metre was:

	Sheep/ha		
	10	30	S.E.
1969–70	1905	2115	422
1970–1	2422	2042	156

Above-ground NPP (net shoot production) was also estimated at Armidale by applying the Wiegert & Evans (1964) analysis as adapted by Hutchinson (1971) to measurements of herbage biomass taken at 42-day intervals. For each of the periods into which the year was divided, the relationship used to calculate shoot production was:

$$\text{Above-ground NPP} = \Delta A + \Delta D + rD + Ia + Id.$$

In this relationship the changes in the amounts of green and dead herbage are represented as ΔA and ΔD, respectively, while the consumption of green and dead herbage by herbivores (in this case sheep) is represented as Ia and Id. The loss of dead and litter material is calculated from an instantaneous disappearance rate (r) and the mean amount of dead and litter material present (D). Estimates of herbage consumption by the sheep were derived from sheep live weights, using the model of Vickery & Hedges (1972), while seasonal instantaneous disappearance rates were obtained from on-site measurements (K. J. Hutchinson & K. L. King, personal communication). Using this method, the annual above-ground net primary productivity (g dry matter/m^2) was estimated to be:

	Sheep/ha	
	10	30
1969–70	697	1033
1970–1	1184	797

By comparing the figures for total plant NPP with those for above-ground NPP it is possible to obtain an estimate of the proportion of above-ground and under-ground annual NPP. A 40 % above-ground and 60 % under-ground NPP is consistent with the results at 10 sheep/ha, while at 30 sheep/ha a 45 % above-ground and 55 % under-ground NPP would be consistent with the results.

The above method of estimating above-ground NPP also enables calculation of the proportion of above-ground NPP consumed by the sheep. For the two years reported, this was 56.9 % at 10 sheep/ha and 91.5 % at 30 sheep/ha. These figures indicate that domesticated livestock were relatively efficient in harvesting the herbage produced by the grassland community at Armidale. This was particularly so at the higher stocking rate where the sheep consumed nearly all of the herbage produced, whereas at the lower stocking rate the proportion of net shoot production consumed was more variable. At 10 sheep/ha in the least productive year (1969–70) it was 70.8 %, while in a more productive year (1970–1) it was only 43.0 %. Further information on herbage consumption by all the major groups of the consumer community is given in Chapter 24 of this volume, where energy expenditures resulting from herbage consumption are examined in detail.

For the Migda site, measurements of peak standing crop on an ungrazed site have been used to estimate production. Because this is an annual grassland, carry-over between growing seasons is negligible, as is herbage senescence and death during the growing season. Here the maximum above-ground herbage biomass recorded in 1972 was 629 g/m^2. This was after 105 days of growth from 1 January; during this period and the preceding two months 301 mm of precipitation was recorded on the experimental site. The herbage biomass above-ground prior to 1 January was less than 28 g/m^2.

The annual level of NPP at the Armidale site responded to the amount of precipitation, a 34 % increase in precipitation in 1970–1 compared with 1969–70, resulted in a 27 % increase in NPP at 10 sheep/ha and a 4 % decrease at 30 sheep/ha, based on the production figures from carbon dioxide exchange. In 1970–1 a 16 % depression in NPP was recorded at a stocking rate of 30 sheep/ha, compared with 10 sheep/ha.

Although the annual above-ground production at Armidale is greater than that at the Migda site, the two levels of production are consistent with the rainfall received; greater than 700 mm at Armidale, compared with 300 mm at Migda. However, it does appear that the efficiency of water use for NPP may be lower at Armidale than at Migda. This could be because a significant amount of the rainfall at Armidale is received during summer (Table 23.1) when evaporative conditions are high and when *Phalaris aquatica* growth may be reduced by high temperatures (Ketellapper, 1960). Despite this difference, the results are consistent with the hypothesis of Whittaker (1970), which suggests that annual NPP is strongly correlated with rainfall.

Efficiency of energy capture

The input of solar energy at Armidale was estimated from hours of bright sunlight recorded with a Campbell–Stokes recorder. This has been converted to energy units using the regression recommended by Hounam (1969) for Australian conditions. The total solar radiation received in 1969–70 was 649.8 kJ/cm^2 and in 1970–1, 629.3 kJ/cm^2. Using the figures for annual NPP from the carbon-balance experiments, this represents a 0.53 % efficiency of energy capture at 10 sheep/ha and a 0.59 % efficiency at 30 sheep/ha in 1969–70. In 1970–1 the efficiencies at the same stocking rates were 0.69 and 0.58, respectively. In terms of above-ground production estimates, assuming an energy content of 18.0 kJ/g, the efficiencies for 10 and 30 sheep/ha were 0.18 % and 0.27 %, respectively, in 1969–70, while in 1970–1 they were 0.30 % and 0.21 %.

The efficiencies of energy capture found are lower than those reported by Black (1964) for subterranean clover (*Trifolium subterraneum* L.) swards maintained at their optimum leaf area index (LAI). In terms of total solar radiation, Black reported annual efficiencies of 1.54 %; howeveı, this figure represents a potential, as both water and nutrients were maintained at non-limiting levels as well as LAI. Thus, the efficiencies reported here of 0.53 to 0.69 % are reasonable for plant communities which were not maintained at optimum levels of moisture, or LAI, throughout the year.

24. Consumers

K. J. HUTCHINSON & K. L. KING

The increase in herbage production that has taken place in the Armidale region, because of the replacement of semi-natural grassland by stands of sown, fertilised, perennial grass-legume mixtures, has resulted in large increases in domestic stock numbers; there is evidence of accompanying increases also in the abundance of invertebrate consumers (Table 24.1). An increase in consumer numbers is a logical consequence of improved primary production and nutritional level. A full understanding of the production changes that follow grassland improvement involves assessment of the total consumer community.

Table 24.1. *Changes in annual shoot production and numbers of selected groups of invertebrates under low intensity of grazing, following grassland improvement in the Armidale region*

Grassland type	Annual shoot production (dry matter/ha (kg))	Regional[a] stocking practice sheep/ha	Number of micro-arthropods (24 pitfall traps)	
			Collembola	Acarina
Semi-natural grassland	1100 to 2800[b]	2.5	668	86
Sown grassland	11000	8.0	4256	1100

[a] Cook & Malecky (1974). [b] Data from Begg (1959).

Abundance and biomass

Census data respecting invertebrates in *Phalaris–Trifolium* (phalaris-white clover) pastures at Armidale are presented in Table 24.2. Scarabaeid larvae of the genera *Anoplognathus*, *Rhopaea* and *Sericesthis*, were the most important group of invertebrate herbivores encountered in sampling. The mean dry biomass of these root-feeding animals was calculated from the data presented in Table 24.2; the values ranged from 28 to 85 kg/ha in different years with the highest mean density of 92 larvae/m^2. There was a consistent seasonal rhythm in larval biomass, with peak values occurring in autumn and winter and the lowest values occurring in spring and summer when the adults emerge above ground.

Meso-fauna, including Collembola, Acarina, Enchytraeidae and Nematoda, were the most important invertebrate saprophages with a mean dry biomass in 1971–2 of 14 kg/ha; the associated density was 456700 individuals/m^2.

Table 24.2. *Mean dry matter content, abundance (No./m²) and fresh biomass (kg/ha) of invertebrates on* Phalaris–Trifolium *grassland sites at Armidale grazed with 10 wether sheep per hectare*

Group	Dry matter (%)	1970–1		1971–2		1972–3	
		No./m²	kg/ha	No./m²	kg/ha	No./m²	kg/ha
Coleoptera (adult)[a, b]	34	23	15	30	10	46	8.7
Coleoptera (larva)[a, b]	28	36	15	41	19	53	23
Scarabaeidae (larva)[b]	23	89	381	41	120	63	192
Formicidae[a, b]	22	43	1.2	110	5.4	96	3.3
Lepidoptera (adult)[a]	10	0.1	0.03	0.2	0.2	0.2	0.1
Lepidoptera (larva)[a, b]	12	3.5	1.9	4.6	2.6	11	3.9
Orthoptera[a, b]	19	1.7	4.3	2.4	4.9	7.1	7.5
Diptera (adult)[a]	13	0.7	0.1	1.0	0.2	1.2	0.4
Diptera (larva)[a, b]	23	15	1.4	16	2	37	9.6
Hemiptera[a]	23	2.5	0.2	2.1	0.04	2.8	0.1
Diplopoda[a, b]	41	460	49	263	44	183	25
Oligochaeta (large)[b]	28	59	113	48	61	82	124
Enchytraeidae	15	—	—	6900	8.0	—	—
Nematoda	15	—	—	298000	1.5	—	—
Acarina	45	25600	2.8	49183	5.3	—	—
Collembola	45	24600	5.6	100279	22.8	—	—
Dermaptera[a, b]	23	6.5	2.6	5.9	2.6	7.1	2.4
Araneae[a, b]	19	2.2	0.4	8.2	1.3	12	2.5
Chilopoda[a, b]	19	0.9	1.2	0.8	0.1	2.2	4.7
Invertebrate total			595		311		407
		No./ha	kg/ha	No./ha	kg/ha	No./ha	kg/ha
Sheep		10	458	10	532	10	550

Above-ground macro-arthropods[a] were sampled on 26 occasions (1970–3) using a quick-trap method (Turnbull & Nicholls, 1966). Collembola and Acarina (19 samplings), and enchytraeids and nematodes (4 samplings) were extracted from soil cores using Tullgren and wet extraction techniques (King & Hutchinson, 1976). Large soil invertebrates[b] were sampled on 26 occasions by hand sorting soil cores taken to a depth of 25 cm.

Collembola was the most important of these groups of meso-fauna; the hemi-edaphic species, *Hypogastrura communis* (Folsom), *Brachystomella parvula* (Schäffer) and *Cryptopygus thermophilus* (Axelson) were the most abundant. The total for the large saprophages, Oligochaeta, Diplopoda and Dermaptera, ranged in annual dry biomass from 3.9 to 5.3 g/m².

While invertebrates have been classed as phytophages, saprophages and predators on the basis of their principal feeding behaviour, many of the groups have polyphagous species; herbivores commonly ingest both living and dead plant material and associated micro-organisms.

The effect of management on populations of invertebrates in seeded grasslands has been discussed by us (in Chapter 12 of Breymeyer & Van Dyne, 1979).

Energetics

Energy budget calculations for sown pastures at Armidale are presented in Table 24.3; the transactions are summarised in Fig. 24.1; the notation given follows the recommendation of Petrusewicz & Macfadyen (1970). Energy budgets were determined for each time period of 42 days from

$$I = P + R + FU \pm \Delta B,$$

with kJ/m² as the common unit. Hemmingsen's (1960) equation for poikilotherms was used to calculate respiration (R) with the Armidale census data and site measurements for aerial, litter and soil temperatures used as inputs. A Q_{10} of 2 was assumed. The equation of Young & Corbett (1968) was used to estimate the maintenance energy expenditure of grazing sheep and the sheep's production was estimated from the calorific values of wool and live-weight gain, with adjustment for conversion inefficiency (Agricultural Research Council, 1965). Energy conversion parameters for invertebrates were derived from appropriate feeding and production studies (Table 24.4).

Table 24.3. *Mean annual energy budgets (kJ/m²) calculated for sown grasslands at Armidale, grazed year-long with 10 wether sheep per hectare*

Group	Respiration	Production	Egestion	Ingestion
Phytophages				
Coleoptera (adult)	38	10	52	100
Coleoptera (larva)	67	51	464	582
Scarabaeidae (larva)	525	400	2075	3000
Formicidae	31	2.2	25	58
Lepidoptera (adult)	0.3	0.2	0.4	0.9
Lepidoptera (larva)	10	7.9	32	50
Orthoptera	11	4.0	28	43
Diptera (adult)	0.4	0.3	0.4	1.1
Diptera (larva)	16	9.1	18	43
Hemiptera	0.7	0.5	1.7	2.9
Saprophages				
Diplopoda	208	159	683	1050
Oligochaeta (large)	272	103	2125	2500
Enchytraeidae	143	109	1428	1680
Nematoda	102	62	109	273
Acarina	212	46	295	553
Collembola	558	479	1553	2590
Dermaptera	8.7	6.7	88	103
Predators				
Araneae	11	4.5	1.5	17
Chilopoda	4.2	1.6	2.9	8.7
Sub-total	2218	1456	8982	12656
Sheep (wethers 2 to 5 yr)	4180	260	3630	8070
Total	6398	1716	12612	20726

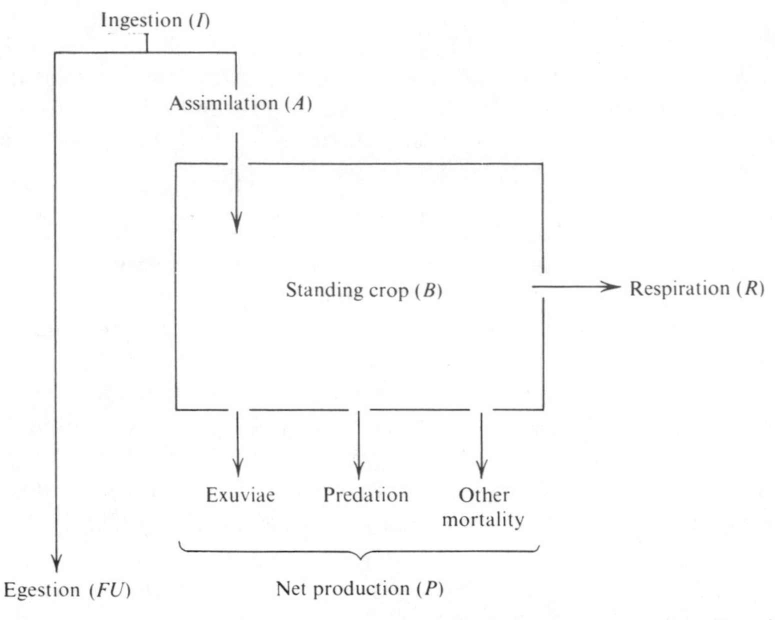

Fig. 24.1. Energy calculations following the notation of Petrusewicz & Macfadyen (1970).

Laboratory determinations of respiration do not include the energy costs of food search and ingestion. Van Hook (1971) has pointed out that metabolic rates determined in a respirometer may be considerably lower than the field values for some free-living invertebrates. Hence, the estimates of respiration given in Table 24.3 for invertebrates are likely to be conservative. Also there may be some additional heat produced from the inefficient utilisation of assimilated energy for productive purposes. This has been shown for homeotherms to be a function of the level of energy assimilated and also its source (Agricultural Research Council, 1965).

Assimilation efficiency is a function of both the animal and its diet. The free-living animal may exercise a considerable degree of dietary selection. Preferences for particular species, parts of plants and for living versus dead herbage have been shown for domestic herbivores (Arnold, 1964; Hamilton *et al.*, 1973). Comparable preferences have been shown for some invertebrates, e.g. earthworms (Barley, 1959), grasshoppers (Bailey & Riegert, 1971). The scope for dietary selection is restricted in laboratory feeding experiments and, hence, the values given in Table 24.4 for the ratios A/I may be underestimates for some invertebrate groups in the field.

The mean assimilation efficiency for the saprophage group was 32 %, compared with 44 % for the phytophages. Assimilation by the saprophages might be expected to be lower, since the detritus represents the less tractable residues from the higher food chain. However, this is not always so, since

Table 24.4. *Energy conversion parameters from published data; mean values are given where more than one authority is cited*

Group	Assimilation efficiency A/I		Production efficiency P/A		Ecological efficiency P/I
Phytophages					
Coleoptera (adult)	0.49	1, 2	0.21	2	0.10
Coleoptera (larva)	0.20	3	0.43	17	0.09
Scarabaeidae (larva)	0.31	4	0.43	17	0.13
Formicidae	0.57	5	0.07	18, 19	0.04
Lepidoptera (adult)	0.57	5	0.43	17	0.25
Lepidoptera (larva)	0.37	6	0.43	17	0.16
Orthoptera	0.33	7, 9, 10	0.27	7, 9, 10, 20, 21	0.09
Diptera (adult)	0.57	5	0.43	17	0.25
Diptera (larva)	0.57	5	0.37	22	0.21
Hempitera	0.40	8, 11	0.42	8, 11, 21, 23	0.17
Saprophages					
Diplopoda	0.35	12	0.43	17	0.15
Oligochaeta (large)	0.15	5	0.27	17	0.04
Enchytraeidae	0.15	5	0.43	17	0.07
Nematoda	0.60	15	0.38	15	0.23
Acarina	0.47	14	0.18	24	0.08
Collembola	0.40	13	0.46	13	0.18
Dermaptera	0.15	5	0.43	17	0.07
Predators					
Araneae	0.93	7, 16	0.29	7	0.27
Chilopoda	0.66	5	0.27	17	0.18
Sheep (wethers 2 to 5 yr)	0.55		0.06		0.03

1. Klekowski *et al.* (1967), 2. Van Hook & Dodson (1974), 3. Kitazawa (1967), 4. Nakamura (1965), 5. Reichle (1969), 6. Smith (1972), 7. Van Hook (1971), 8. Wiegert (1964), 9. Smalley (1960), 10. Wiegert (1965), 11. McNeill (1971), 12. Striganova & Rachmanov (1972), 13. Healey (1967), 14. Webb & Elmes (1972), 15. Marchant & Nicholas (1974), 16. Moulder & Reichle (1972), 17. McNeill & Lawton (1970), 18. Petal (1967), 19. Rogers *et al.* (1972), 20. Odum *et al.* (1962), 21. Menhinick (1967), 22. Maclean (1973), 23. Hinton (1971), 24. Englemann (1961).

the detritus is more subject to attack by micro-organisms, which may produce a local concentration of more soluble carbohydrate material or may themselves form a substantial part of the diet, particularly of nematodes (Nielsen, 1949), mites (Luxton, 1972) and Collembola (McMillan & Healey, 1971). Phytophages generally have a high assimilation efficiency, particularly where the diet is in liquid form and does not contain large amounts of indigestible cell-wall material (e.g., Hemiptera (McNeill, 1971)). Predators have a high assimilation efficiency due to the nature of their food sources (Lawton, 1970).

High values of production relative to respiration are commonly reported in laboratory feeding studies where temperature and moisture conditions are near the optima and where nutrition is not limiting (e.g., Klekowski, Prus & Zyromska-Rudzka, 1967); short-term growth studies also give high values

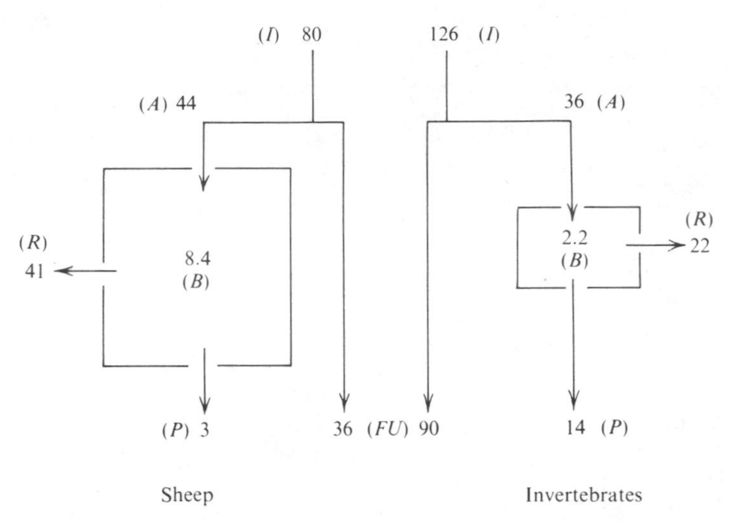

Fig. 24.2. Mean annual energy budgets (GigaJ/ha) for consumer groups (phytophages and saprophages) of sown grasslands at Armidale grazed year-long with 10 sheep/hectare. Symbols as in Fig. 24.1.

of P/R. The production efficiency parameters compiled in Table 24.4 have been drawn mainly from field studies that include the full cycle of production. McNeill & Lawton (1970) have drawn attention to the effect of life history on production efficiency (P/A). Short-lived poikilotherms have low respiratory resting stages, hence, their life cycle production efficiency is high. Long-lived poikilotherms (e.g. Chilopoda) have relatively high respiratory costs and low production efficiency. Social insects, e.g. ants, have specialised non-reproducing castes and a low production efficiency overall (Englemann, 1966). For domestic herbivores there are comparable effects of lifespan on production efficiency. Under management practices, where growth is delayed, extended maintenance respiration results in low production efficiency; non-reproducing adults, e.g. wether sheep, have a lower production efficiency than breeding and young, rapidly-growing animals.

Ecological efficiency (P/I) is the product of assimilation and production efficiencies. The ecological efficiencies derived in Table 24.4 range from 0.03 to 0.27; this is consistent with the limits proposed by Engelmann (1961) and Slobodkin (1962) and the derived values would appear to be realistic.

The mean annual energy budgets for the consumer groups of the Armidale sites are given in Table 24.3 and summarised for sheep *versus* invertebrate herbivores in Fig. 24.2. The energy value of the mean biomass of the sheep flock was about four times the biomass energy for the invertebrate herbivores. This was determined mainly by a higher energy content of fresh biomass for sheep (16.4 kJ/g) *versus* fresh invertebrates (4.8 kJ/g). The respiratory energy costs of the sheep flock were almost twice as high as the invertebrates despite the relatively small body size of invertebrates. There were substantial reductions for the invertebrate respiration due to low temperatures. The amounts of energy assimilated were comparable, but the lower assimilation efficiency of invertebrates resulted in higher excretal return and a higher level of ingestion. At the stocking level reported, the invertebrate herbivores were more important than sheep as consumers; this emphasises the need for production studies at the community level.

25. Micro-organisms

R. L. DAVIDSON

The organic matter of primary production provides the first link in the chain of food for herbivores, carnivores, parasites, predators, detritivores and decomposers that inevitably ends in soil organic matter. In pastures the domestic animals do not constitute the major part of the energy or nutrients in the system. The animal products harvested (meat, milk and wool) are only a small fraction of the nutrients and energy, the bulk being returned to the soil in excreta.

The edaphic environment may be visualised as the substrate on which crops and pastures are grown. But the concept of the soil system should include a sensitive organic complex, with the driving forces of micro-climate and a variable input of organic detritus regulating the rate of mineralisation of nutrients (Richards, 1974). Antibiosis and competition for spaces free of toxic residues should be considered (Forbes, 1974), and also the effects of the soil micro-fauna (Satchell, 1974).

In sown pastures (arable grasslands) where the grass and clover species have been chosen for high yield, these species are usually responsive to high available nutrient status, and nutrients are frequently limiting to growth in these pastures. Rapid cycling of nutrients is desirable, and rapid breakdown of excreta, litter and moribund roots should be facilitated by management practices. Microbial activity controls the gate through which all minerals must pass before they can be recycled to plants. An understanding of the driving forces controlling microbial activity is essential in the modelling of pastoral ecosystems, especially sown grasslands in soil disturbed by cultivation.

Populations

The paucity of taxonomic records of micro-flora from sown grasslands makes it impossible to give a meaningful summary of characteristic species. Ubiquitous micro-organisms can colonise a wide variety of herbaceous litter (Bell, 1974). Specialism in response to particular environments has been reported, but no clear distinction can be drawn between tropical, temperate or other grasslands. Representatives of all microbial groups such as aerobes, anaerobes, nitrifiers, denitrifiers, ammonifiers and thermophiles are present in all soils and litter. The frequency of microbes of any group varies widely over the seasons in each region, and the frequency varies also in response to topographic differences and micro-sites.

Activity

The following discussion concerning microbial activity is based on studies at the Armidale site, as no other data of this nature were received from IBP investigations of arable grassland. Studies at Armidale are useful in defining the environmental factors contributing to fluctuations in microbial activity.

Total soil respiration from manometric determinations is a convenient measure of microbial activity (Chase & Gray, 1957; Chase, 1958). It is not a measure of microbial biomass in the field, unless such conditions as temperature and moisture tension during respiration measurements are similar to conditions in the field. Bacterial populations respond very quickly to a change in these conditions, and within a few hours reach an equilibrium. In the field day-to-day changes in micro-climate are accompanied by large changes in microbial biomass. A practical approach to the monitoring of microbes in pastoral soils is to develop a simulation model from the response curves of microbial activity to environmental factors. Interaction between species and environment would occur, and floristic studies should be incorporated in the model.

Measurements of oxygen consumption were made using constant-volume respirometers (Umbreit, Burris & Stauffer, 1964). Where the effect of temperatures below 35 °C was to be measured, and the response to organic matter, fertiliser, plant species and scarab larvae determined, the soils were kept in glasshouse pots for 3.5–5 months under the experimental conditions prior to measurements of respiration. Soil was not dried before these measurements, but moisture determinations of samples from a thoroughly mixed batch of each soil enabled oven-dry equivalents to be weighed out. Changes in pressure were measured over 40–100 h after a few hours for initial settling down.

Temperature response

Total metabolic activity (measured as oxygen consumption per unit weight of pasture soil) increased exponentially from 10° to 50 °C (Fig. 25.1) with a Q_{10} of 2. Above 50 °C there was a slight increase in activity to a peak about 55 °C, and a slow decline to about 80 °C. The thermophilic species which are dominant at high temperatures have a greater overall activity than the more varied micro-flora at lower temperatures. Below 10 °C microbial activity declined slowly, and in the presence of adequate organic matter activity remained high at low temperatures. The addition of 4 % of manure to soil increased respiration to ten times the rate without manure at low temperatures (5–15 °C) and to five times at medium temperatures (Fig. 25.1).

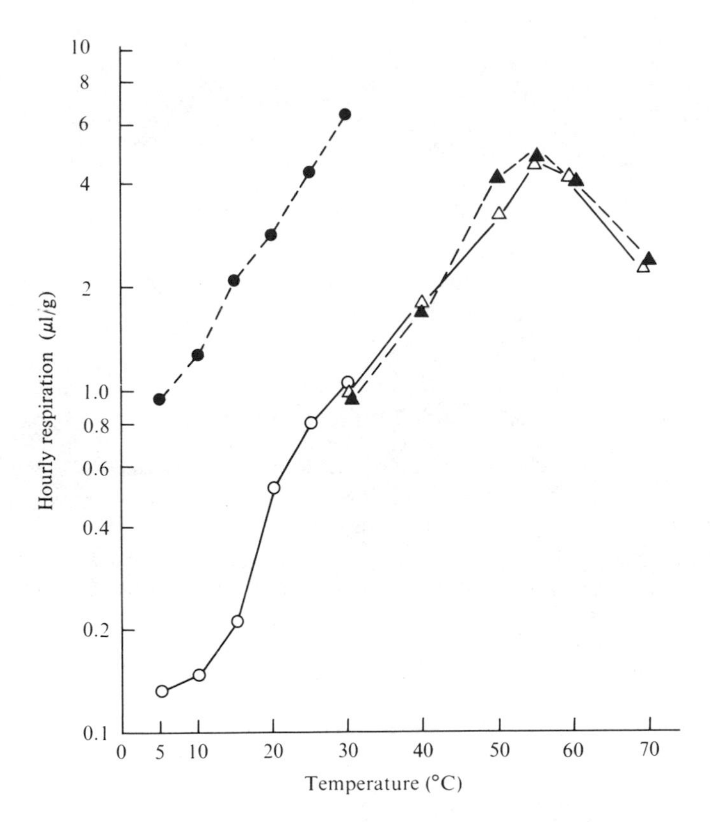

Fig. 25.1. Respiration rates (oxygen consumption) in soil from semi-natural grassland (○, △), sown *Phalaris–Trifolium* (phalaris–white clover) pasture (▲), and semi-natural grassland with 4 % manure (●). Soil moisture tensions were pF 3.0 (△), 3.3 (▲) and 3.1 (○, ●).

Response to soil moisture

Soil respiration was highest in soil when the moisture tension was approximately 1 atmosphere (pF 2.7), i.e. slightly below field capacity (Fig. 25.2). Respiration rates were relatively high over a wide range of soil moisture tensions (Fig. 25.3), and even at wilting point (pF 4.2, 16 atmospheres tension) there was little decline in oxygen uptake. At higher moisture tensions respiration declined to a very low level (Fig. 25.2). Respiration was strongly inhibited when the moisture content of soil increased above field capacity, and at pF 2.2 (28 % moisture in Armidale soil) the oxygen uptake was about half of the peak respiration.

From micro-climatic data it is estimated that microbial activity would be inhibited for a considerable part of each year by dry and wet soil conditions.

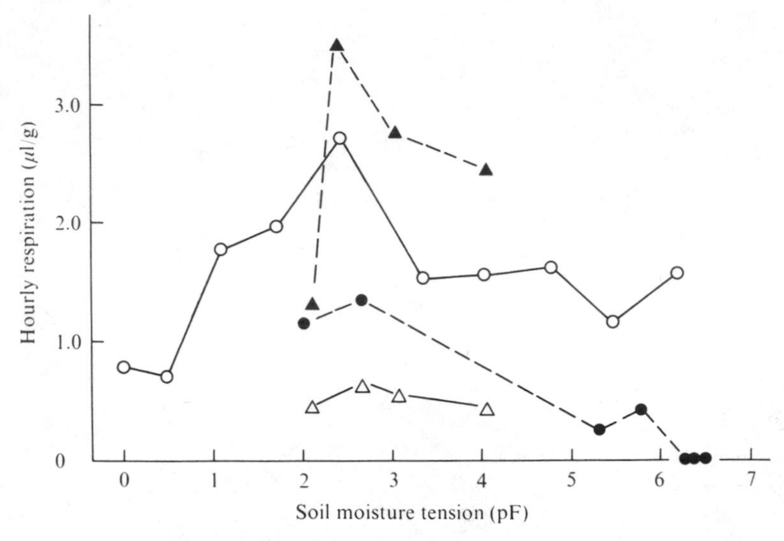

Fig. 25.2. Soil respiration at 20 °C at different moisture tensions in Armidale semi-natural grassland, without (△) and with (▲) 4 % manure, Heron Island grassland (●) forest (○).

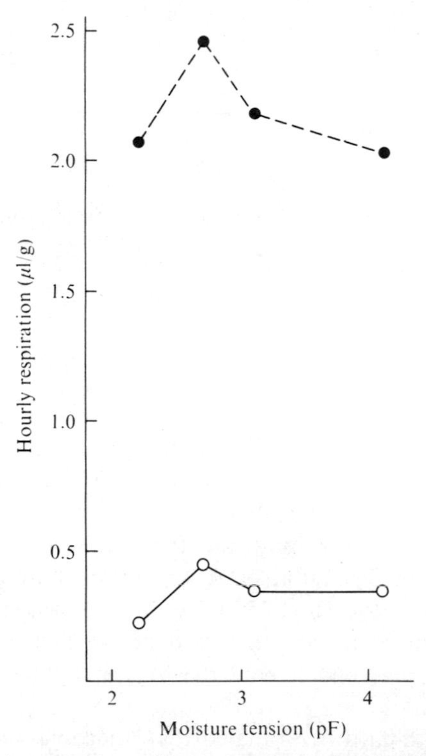

Fig. 25.3. Respiration rates in semi-natural grassland with 4 % manure (●) and without manure (○), at various soil moisture tensions, mean values of rates at 5–25 °C.

Soil organic matter, plant species, fertiliser and manure

The microbial respiration rate of soils from native grassland (ungrazed, unfertilised) and adjacent soil planted to *Phalaris aquatica* L. (phalaris) and *Trifolium repens* L. (white clover), fertilised and grazed by cattle for eight years was almost identical (Figs. 25.1, 25.4). in spite of a difference in the readily oxidisable organic matter (Walkley–Black) values (semi-natural grassland 1.7 %, sown pasture 2.6 %). However, the moisture characteristics of the two soils vary slightly, the moisture content at pF 2.4 being 20.6 and 23.2 %, respectively. At 30 °C the soil respiration was measured at 13.6

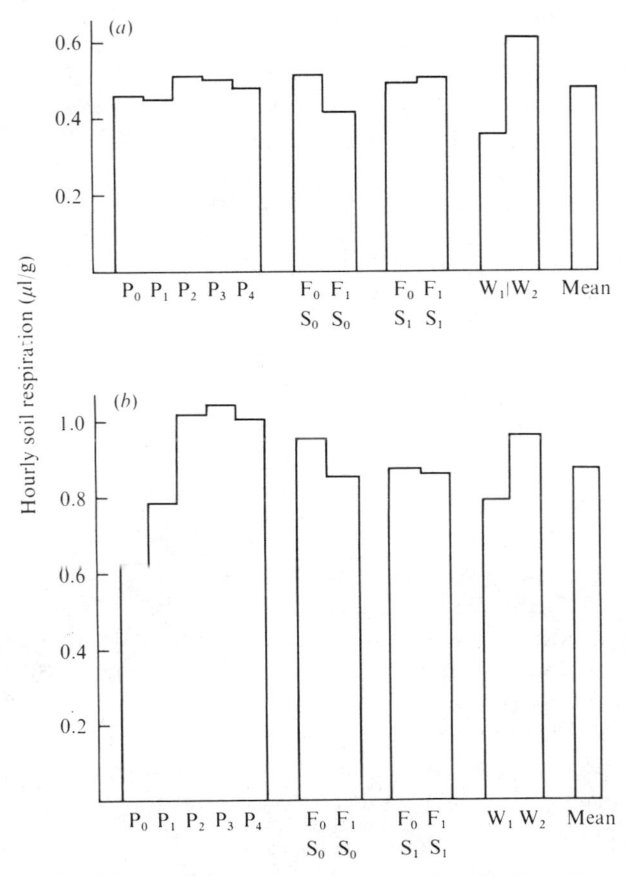

Fig. 25.4. Soil respiration (with roots removed) after five months under various treatments in pots: P_0, no plants; P_1, native grasses; P_2, *Lolium perenne* L. (perennial rye grass), P_3, *Phalaris aquatica* L. (phalaris); P_4, *Trifolium repens* L. (white clover); S_0, soil from semi-natural grassland; S_1, soil from sown pasture; F_0, no fertiliser; F_1, fertilised with NPS; W_1 10 % moisture; W_2, 20 % moisture. (*a*), seedlings; (*b*), older plants. Respiration measured at 20 °C.

and 27.5 % moisture, and if the respiration rate at pF 2.4 is interpolated the hourly rate for native grassland was 1.15 μl/g and for improved pasture was 1.27 μl/g.

These same soils were maintained in glasshouse pots with pasture scarab larvae for 5 months, with various pasture species, fertiliser, moisture and age of plant treatments. The rates of respiration of these soils (after roots were removed) are shown in Fig. 25.4. Fertiliser (N, P, S) did not have a consistent effect on soil respiration. It decreased the respiration rate of semi-natural grassland soil, while having little effect on the microbial respiration of sown pasture soil. This interaction was most marked in soil without plants, where respiration in semi-natural grassland soil was depressed 32 % by the fertiliser treatment. Soil respiration rates were higher in soils in which plants had been growing than in soil without plants. There was a positive correlation between the weight of roots in the pots and the rate of soil respiration after roots were removed, presumably a response of microbes to root exudates. In these experiments the readily oxidisable organic matter was not increased significantly by any of the treatments.

When cattle manure was added to native pasture soil (4 % by weight) the soil respiration rate increased to from five to 10 times the rate without manure, depending on the temperature (Fig. 25.1). Localised concentrations of manure from grazing animals occur in pastures, and the faeces are incorporated in the soil especially where dung beetles (*Onthophagus australis* Guérin, *O. granulatus* Boheman, and other species) are active in burying dung. Hilder & Mottershead (1963) have shown that available P, K and N in pastures is influenced by the redistribution of excreta by grazing animals.

Antibiosis

Research has not been done on antibiotics in arable grassland in relation to mineralisation, but earlier studies in climax and seral grassland in South Africa (Stiven, 1957) showed that nitrate production from climax grassland soil was only one-fifth the rate of mineralisation in soil under seral grassland on adjacent old fallows. Antibiotic substances found in the roots of climax grasses and forbs (Stiven, 1952) inhibited species of *Bacillus*, *Staphylococcus* and *Escherichia*. The products of decomposition may be regulators of mineralisation processes, and accumulation of inorganic nutrients, during periods when plant growth is slow because of unfavourable weather, may inhibit microbial activity.

Pesticides

Sown pastures in all parts of the world have populations of foliage-feeding and root-damaging insects, mites and nematodes which sporadically reach pest status. In some countries heavy and repeated use of pesticides is widespread, including (in the past) the persistent chlorinated hydrocarbons. Pesticide residues in soil have various effects on non-target organisms (Martin, 1966; Moore, 1967; Matsumura & Boush, 1971; Alexander, 1971, 1975). Another possible source of toxic substances is anthelmintics and their breakdown products excreted by domestic animals at pasture.

The toxic substances added to pasture soil appear to cause an imbalance in soil fauna and micro-flora because of the differential toxicity of even the broad spectrum pesticides. The killing of earthworms, insects and micro-arthropods appears to provide a *temporary increase* in fertility, presumably by providing an increase in the respirable substrate. In *Lolium perenne* grassland application of azinphos-ethyl, dieldrin and lindane increased foliage growth by about 30 % (R. L. Davidson *et al.*, unpublished data). In glasshouse experiments with the same soil similar increases were obtained, with smaller responses on the second application of insecticide six months later (A. Shackley, personal communication). Total soil respiration rates are increased by ethylene dibromide fumigation at temperatures of 10–30 °C, but at higher temperatures (which will kill the soil fauna by heat stress) the soil respiration rate is reduced by ethylene dibromide (R. L. Davidson, unpublished data). Increased yields of crop plants have been reported after application of insecticides to soil, associated with increased rates of nitrification (Shaw & Robinson, 1960). Bartha, Lanzilotta & Pramer (1967) have shown that insecticides can give increases in nitrification. Organophosphate insecticides can stimulate nitrifiers and nitrogen fixers (Tu, 1970).

Erroneous conclusions are likely to have been drawn about the efficacy of pesticide treatment of pastures from these anomalous increases in pasture growth.

Simulation model of activity

A simple model of soil respiration can be constructed from the regression equations derived from the data in Figs. 25.1 to 25.5. A flow diagram of the simulation of total microbial respiration (excluding the respiration of roots and macro-arthropods) is given in Fig. 25.6. There is difficulty in getting precise estimates of respiration at temperatures below 5 °C, because there appear to be interactions between the effects of temperature, moisture and organic matter. The equations applicable at temperatures 5–50 °C to derive the hourly respiration rate follow.

(i) For temperature response at pF 2.7 and organic matter *ca* 2 % in the absence of roots:

$$y = 0.94 + 0.033x \qquad (1)$$

where $y = \log_{10}$ respiration in μl/100 g

$x =$ temperature in °C (over range 5–50 °C).

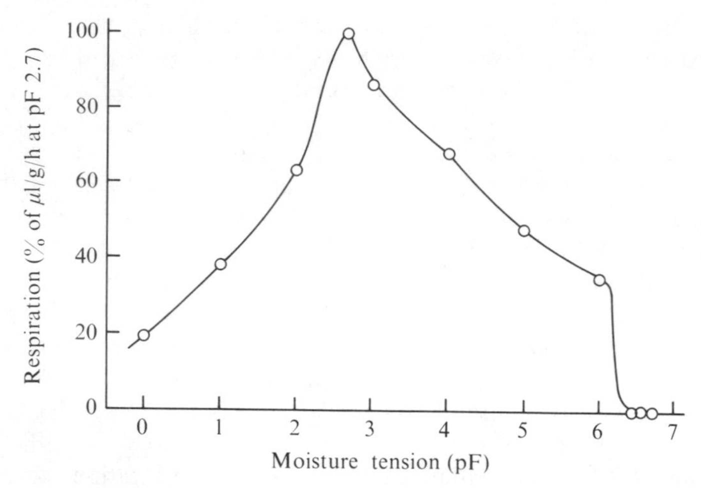

Fig. 25.5. Generalised effect of soil moisture tension on microbial respiration (percentage of peak respiration at pF 2.7), mean of four soils, including coral sand from Heron Island, 23° 27′ S, 151° 57′ E.

(ii) For response to organic matter (excluding roots):

$$y = 0.13 + 0.28x \tag{2}$$

where y = respiration in μl/g at 20 °C

x = per cent readily oxidisable organic matter (Walkley–Black).

(iii) For stimulation of microbial respiration by root mass in soil:

$$y = 0.54 + 0.044x \tag{3}$$

where y = respiration in μl/g

x = weight of roots in g/kg soil

intercept = soil respiration in absence of roots.

(iv) The effect of soil moisture tension on microbial respiration can be interpolated from Fig. 25.5 or the equation approximating this curve:

$$y = 20 - 115.9x + 174.5x^2 - 70.2x^3 + 11.14x^4 - 0.623x^5 \tag{4}$$

where y = respiration as per cent μl/g at pF 2.7 and 20 °C

x = soil moisture tension (pF).

Fluctuation of temperature has a predictable effect on total respiration, but further work is needed to define the response of different species groups. The response to high soil moisture will be better defined by studies of the carbon dioxide and oxygen tensions and diffusion rates under pasture. Seasonal patterns of root growth, exudates and senescence in relation to microbial activity need definition. Nutrient uptake by plants would affect

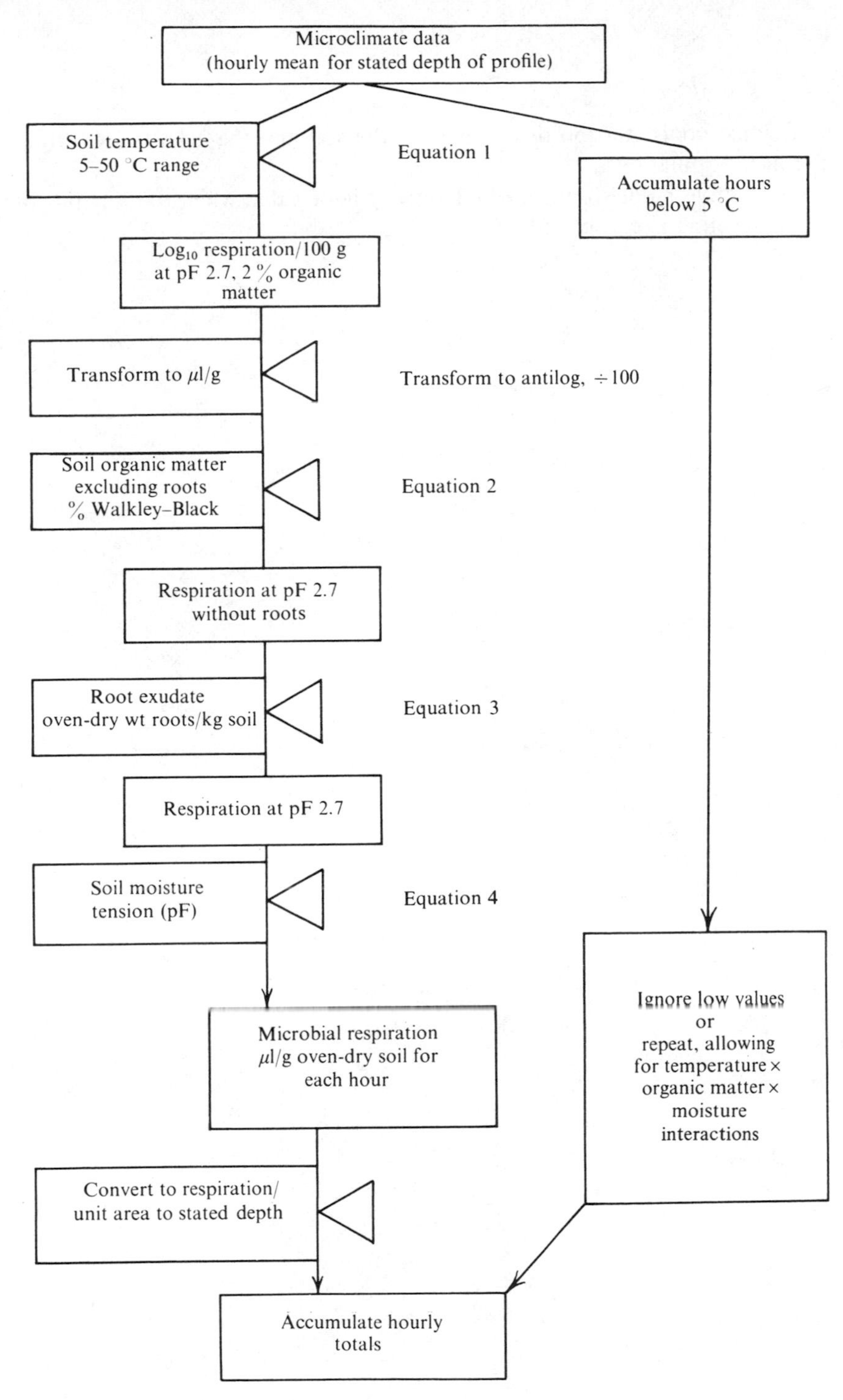

Fig. 25.6. Flow diagram for simulation of microbial respiration under pasture. See text for eqns (1) to (4).

the residual concentration of minerals in the soil and the activity of certain microbial populations.

A predictive model for microbial activity is of value when the kinetics of nutrient cycling is considered.

26. Nutrient cycling

A. R. TILL

In the grazing system the minerals in the soil essential for plant growth cycle from the soil through the plants and then return to the soil either directly or through the animals. In addition, there are also potential sources of loss and acquisition, e.g. leaching and fixation in unavailable forms, and inputs via rainfall and fertiliser application (Fig. 26.1). Some aspects of nutrient cycling and use of fertilisers in seeded grasslands have been considered by Hutchinson & King (Chapter 12 of Breymeyer & Van Dyne, 1979).

The availability to plants of essential minerals exercises considerable control over plant growth, and through this on the number of domestic animals that can be supported and their production per head. Consequently, within the environmental constraints the manipulation of nutrient availability, usually by fertiliser application, and subsequent adjustment of stocking rates offers the greatest potential for increased animal production. However fertilisers represent a large investment and it is essential that in the short term they provide good economic returns, and in the long term do not lead to degradation of the environment.

Efficient fertiliser use demands a knowledge of processes and amounts in an interacting system of various pools of nutrients in the soil, plants and animals, which are influenced by uncontrolled environmental variables. In general, nutrient utilisation has been studied in two ways. In one, the complete grazing situation is studied, usually in terms of fertiliser input and product output type relationships; in the other, detailed studies are made of separated parts of the system under controlled conditions.

In the short term, whole system studies provide the only reliable information for immediate use in the region of the experimental area, but the results may not apply to other areas and give very little information on the processes and changes taking place. Detailed studies of parts of the system, together with simulation studies, will eventually lead to more widely applicable results, but a consideration of the dynamics of even a greatly simplified nutrient cycle (Fig. 26.1) immediately stresses the importance of the interactions between components. For example, the domestic animal introduces two additional recycling pathways which are shown by the heavy arrows in Fig. 26.1.

There have been many studies of parts of the system and models proposed for various nutrient cycles (e.g., Henzell & Ross, 1973; Reuss, Cole & Innis, 1973; May, Till & Cumming, 1972; Wilkinson & Lowrey, 1973), but interactions with the grazing animal frequently have been ignored, largely because of lack of suitable techniques.

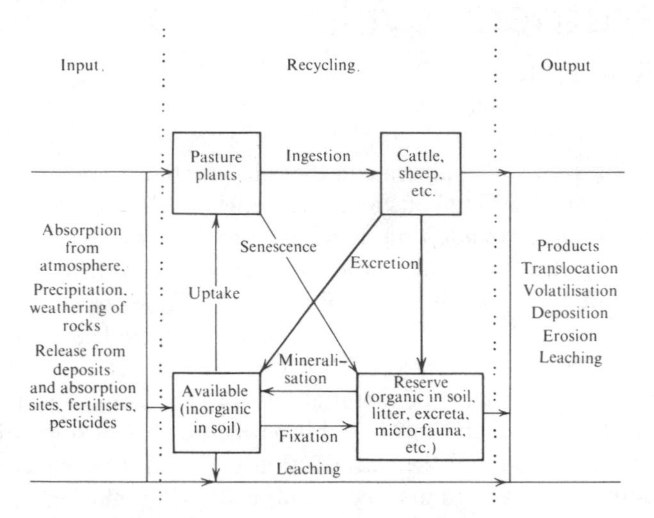

Fig. 26.1. Simplified representation of nutrient flows in grazed pastures. The rectangles represent pools of nutrient-containing materials and arrows show the flows by various processes. The heavy lines emphasise the additional pathways introduced by the grazing animal.

In their natural state most Australian soils are deficient in phosphorus, but in northern New South Wales and south-eastern Queensland there are large areas where the major response is to applied sulphur. As the Armidale site is situated in a region of sulphur deficiency the major effort in minerals research has been devoted to studies of sulphur utilisation. Radio tracer techniques have been developed which, when used together with other established techniques, allow direct measurements to be made of the location, form and recycling of sulphur in pastures grazed by sheep while the interactions of other components proceed uninterrupted.

Distribution of sulphur in the ecosystem

Radio tracer studies of nutrient recycling were carried out at Armidale on two replicates at two stocking rates (10 and 20 sheep/ha) on a gently sloping site. Each paddock (0.4 ha) was treated with sulphur-35 (100 mCi) by the strip-labelling technique as described by Till & May (1970). Samples of wool, plants and soil were collected at intervals of about 21 days and analysed for sulphur and radioactivity (Rocks, Lutton & McCabe, 1973). All results were corrected for radioactive decay to time of application which is referred to as the start of experiment (Day 0).

Soil

There were no significant differences between the stocking rates in the total sulphur (to a depth of 10 cm) or the available sulphur in the uppermost 7.5 cm of soil. There were considerable variations in the total soil sulphur and specific radioactivity SR during the year but no overall upward or downward trend (Fig. 26.2), and it is assumed that the variability reflects the movement of sulphur between the soil and other component pools. The available sulphur in the top 7.5 cm fluctuated widely (36–6 μg/g) during the year, being high immediately after fertiliser application and declining rapidly as the available sulphate was used by plants and micro-organisms. These results emphasise that the size of the available sulphur pool does not provide a satisfactory basis on which to assess the soil nutrient status and fertiliser requirements of grazed pastures. The yearly mean concentrations of total

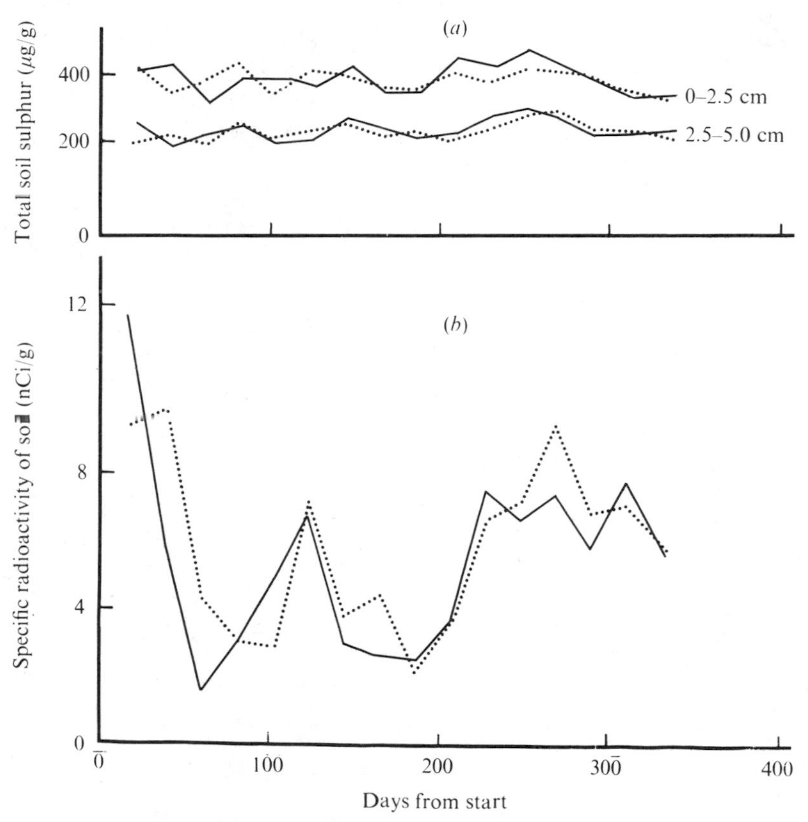

Fig. 26.2. Total soil sulphur (*a*) and specific radioactivity (*b*), (——) 10 sheep/ha, (·········) 20 sheep/ha.

sulphur in the soil declined from 397 ± 57 μg/g in the uppermost 2.5 cm of soil, to 235 ± 31 μg/g, to 204 ± 33 μg/g, and 168 ± 33 μg/g in successively deeper 2.5 cm layers. The available sulphur concentration was about 10 μg/g (0–7.5 cm). There did not appear to be any appreciable penetration of applied sulphur into the soil, as in the 7.5–10 cm layer the specific radioactivity of sulphur was low and it contained an overall average of only 1.8 ± 1.5 % of the applied radioactivity.

Plant

The varying seasonal conditions caused large variations in the sulphur concentrations and total amounts in the plant populations (Fig. 26.3), but only in the case of the litter pool size was there a significant difference between stocking rates. This is apparently due to the measurements showing only what the animals had not eaten rather than the production rate. The yearly averages for the pool size are shown in Fig. 26.7. Immediately after the application of fertiliser sulphur, the specific radioactivity of sulphur in plant shoots was high, showing the rapid uptake of fertiliser sulphur. However, there was also rapid incorporation into other organic materials in the soil and after about 200 days the fertiliser sulphur had very largely become mixed

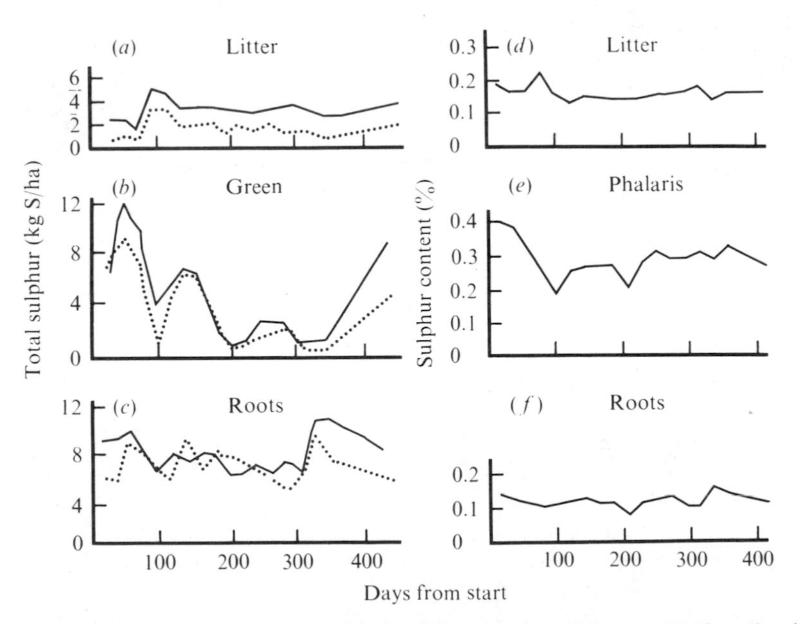

Fig. 26.3. Sulphur in plant materials. Pool sizes (*a*), (*b*), (*c*) (——) 10 sheep/ha, (······) 20 sheep/ha. The stocking rate had no significant effect on the sulphur concentrations (*d*, *e*, *f*) so the mean values are given.

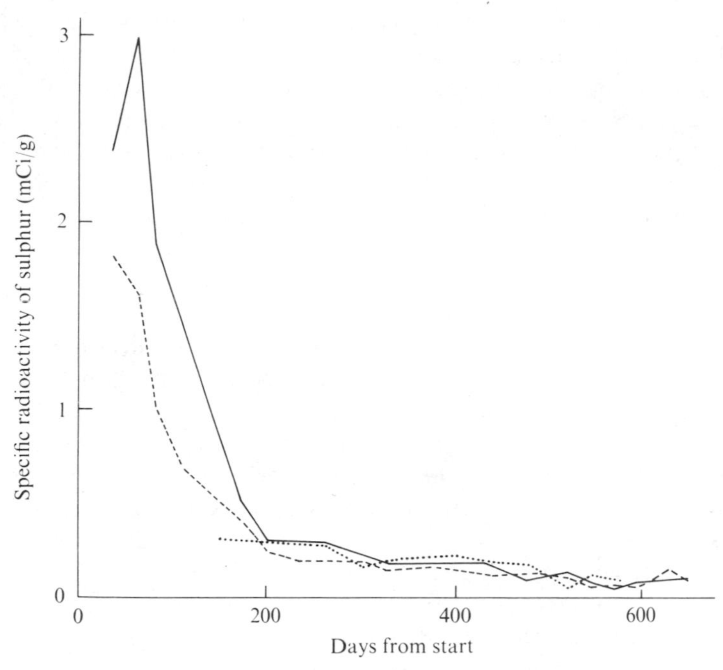

Fig. 26.4. Specific radioactivity of sulphur in clover (——), phalaris (- - - - -) and the available soil sulphur pools (······).

Table 26.1. *Mean pool sizes, sulphur concentrations and specific radioactivity (SR) of some system components*

	mg S/m²	% S	SR (μCi/g S)
Scarabaeidae	13.4±8.5	0.41±0.11	50±27
Diplopoda	9.3±4.4	0.40±0.08	34±14
Oligochaeta (large)	—	0.42±0.19	44±26
Sheep[a]	—	3.39±0.25	62±13
Plants[b]	—	0.28±0.05	40±13

[a] Wool SR over plateau (Fig. 26.6), corrected for proportion of area treated with labelled fertiliser, but not for recycling of translocated S.
[b] Plateau values on initial labelled area.

into the system, and the specific radioactivity of plant sulphur declined to similar values to that in the other components (Fig. 26.4, Table 26.1).

Shortly after fertiliser application the specific radioactivity of root sulphur was lower than that in the corresponding tops, presumably due to the dilution of new root materials by old roots already present. The ratio of the specific radioactivity of sulphur in roots and tops gradually converged until after about 120 days they had reached similar values.

The changes in the specific radioactivity ratios are due to many factors, such as the changes in the specific radioactivity in the soil-available sulphate pool and the rate of root replacement; but it appears that between the time of initial rapid change and before root replacement had become significant the ratio could give an indication of the relative amounts of active and old roots.

Sheep

At the higher stocking rate the cumulative wool sulphur production per head was only about 10 % lower than at the lower stocking rate. Consequently, total wool production per hectare at the higher stocking rate was almost double that at the lower stocking rate. As the efficiency of utilisation of sulphur would be about the same at both stocking rates studied, total sulphur intakes and recycling were much greater at the higher stocking rate (Fig. 26.7). Following the initial rapid pasture growth and during the time the litter pool was large (Fig. 26.3), the sheep on the higher stocking rate had greater wool growth rates (Fig. 26.5). This is probably a direct result of under-utilisation of pasture at the low stocking rate, leading to poorer quality material on offer. The mean specific radioactivity of wool sulphur (Fig. 26.6) showed an initial rapid rise corresponding to the high specific radioactivity of plant material, followed by a decline to a plateau level as the sulphur became mixed in other pools.

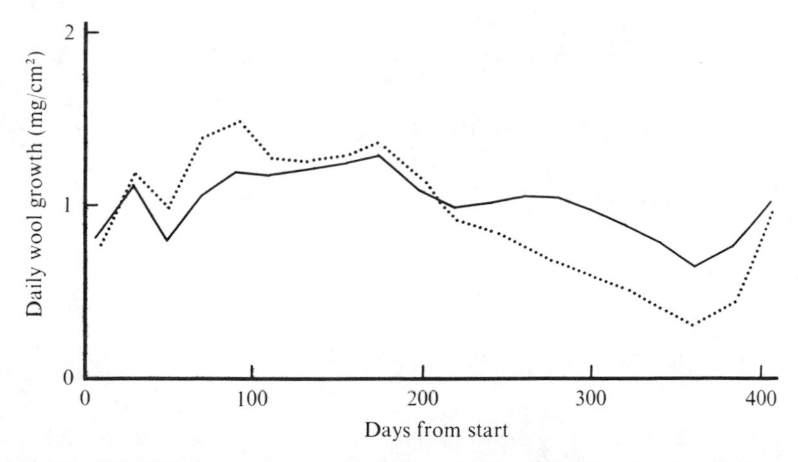

Fig. 26.5. Wool growth rates in relation to area of skin (——), 10 sheep/ha; (·····), 20 sheep/ha.

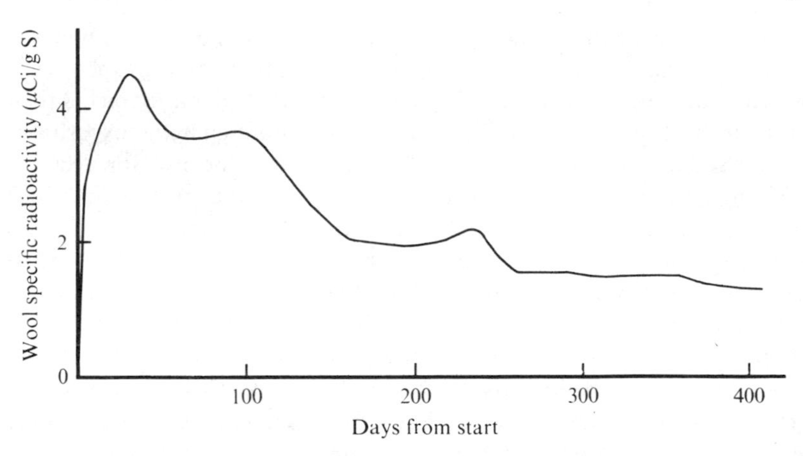

Fig. 26.6. Variation in specific radioactivity of wool after application of labelled fertiliser.

Other consumers

Estimates of the size and specific radioactivity of sulphur pools in various other consumers were very 'noisy'. The yearly averages for Scarabaeidae, Diplopoda and Olegochaeta are presented in Table 26.1. The 'noise' is the result of several interacting factors, such as the aggregation of some species, localised grazing domains, irregularities of fertiliser application, and the small numbers in some samples.

System process and recycling

The use of radioactive fertiliser allows the utilisation of applied nutrient to be traced in a system which contains all the normally interacting components. Fertiliser sulphur will supply some of the plant's needs, but the rest must come from the mineralisation processes in the cycle. The incorporation of radioactivity into plant sulphur shows that the plant sulphur that was obtained from the fertiliser was about 40 % at the peak (immediately after fertiliser application) and 10% when the fertiliser and soil sulphur became mixed, and averaged about 15 % over the year. When these proportions are taken together with the total plant production, the overall efficiency of utilisation of fertiliser sulphur by plants can be estimated to be about 5 %. Similar studies in other soils and under different climatic conditions (Till, 1976) have shown that sulphur application rate (x kg/ha) is related to the proportion of the plant sulphur that is derived from applied fertiliser (y %) by $y = 26.5\,(1-e^{-0.044x})$, $r = 0.82$. This is very different from many crop systems where strategic placement of fertiliser and well defined growing periods with high demands for nutrients lead to more efficient use. In the grazing system the rest of the applied sulphur must either have been wasted (e.g. leached) or used to build up a sulphur bank (mainly organic material) in the soil. Other grazing experiments on the Armidale site (Till & May, 1971) have shown that plants extracted their sulphur from the top 7.5 cm of soil (Fig. 26.4) and, although in this experiment the specific radioactivity of the available sulphur was too low for accurate measurements, the results indicated that this was again the region from which the plants obtained their sulphur. As only a small amount (about 2 %) of the applied sulphur-35 could be detected in the 7.5–10 cm soil horizon and its specific radioactivity was low, it appears that there was no significant leaching of the applied fertiliser to depths beyond those explored by plant roots. In some samples very small amounts of radioactivity were detected 10 cm from the lower side of the labelled areas. This suggests that there was probably some movement of sulphur in surface flows of water, but the amounts were so small that no reliable estimates could be made of total movement.

Diplopoda and large Oligochaeta rapidly incorporated sulphur-35 and over the corresponding periods had similar specific radioactivity of sulphur to those in plant shoots and in the wool of the grazing sheep (Table 26.1). The Scarabaeidae, which are root feeders, also had similar specific radio-

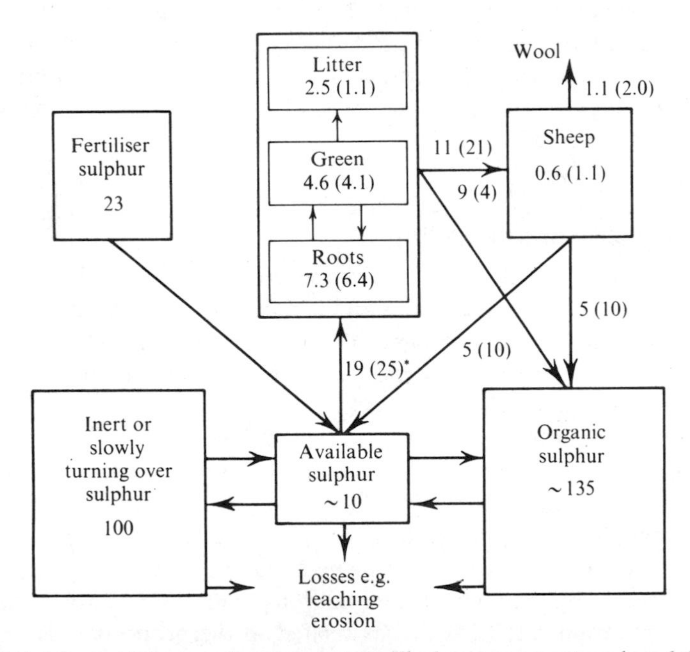

Fig. 26.7. Sulphur cycle in grazed pasture system. The boxes represent pools and the arrows flows of sulphur. Flows are in kg/ha per year and pool sizes in kg/ha at 10 sheep/ha. Values in parentheses show where there are measurable differences for 20 sheep/ha.

* These are minimum values only as no allowance has been made for consumers other than sheep.

activity of sulphur to that in plant shoots, even though the roots had lower overall specific radioactivities of sulphur. It appears, therefore, that the Scarabaeidae were feeding only on fresh roots which would have sulphur of a similar specific radioactivity to the fresh tops. Apparently the fertiliser mixes in the cycling pool and the plant and litter feeders are consuming materials that have similar specific radioactivity.

The sheep, and other consumers, interact strongly with other components of the ecosystem. This interaction is not only via the mechanical effects on soil and plants, and their subsequent effects on plant growth patterns and competition (Edmund & Hoveland, 1972), micro-climate and micro-flora (Freney & Spencer, 1960; Hall & Grossbard, 1972), but also by providing additional pathways for redistribution and mineralisation (Fig. 26.1). The yearly average pool size and annual sulphur flows are shown in Fig. 26.7. These values show the enhanced recycling of sulphur at the higher stocking rate and suggest that the observed increases in net primary production (Vickery, 1972) at the higher stocking rate may be due (in part) to more efficient use of nutrients.

In most studies the individual effects of the different interactions have not been evaluated, so they cannot be incorporated into models of grazing systems. The Armidale studies using tracer techniques allow transient responses in different pools to be followed in systems in which the interactive processes have been proceeding normally. These transients have been used, together with the other system measurements, to study the dynamics of the processes taking place in grazed pastures (May *et al.*, 1972, 1973).

27. Use, management and conservation

W. M. WILLOUGHBY & R. L. DAVIDSON

Grassland is a major contributor to world animal production, and such systems are potentially low cost using minimum inputs of support energy. Because of the extent of grassland and its location in catchments, its usage can have a large influence on the quantity and quality of the world's water supplies. Therefore, grassland should be evaluated in terms of soil and water conservation as well as animal production.

In the extensive natural grasslands of the world, pasture management aims to preserve climax grass species, because the stability of the climax vegetation is considered to be a sure way of conserving soil and water resources. Successional retrogression caused by heavy grazing is equated with pasture degradation of natural grassland (Dyksterhuis, 1949). However, in all parts of the world where domestic stock are grazed the species composition of natural grassland has been changed, as discussed in Parts II–IV of this volume, dealing with natural and semi-natural grasslands, by grazing, trampling, burning and fertilisation. The greatest impact has come from cultivation and re-seeding, usually with grasses and clover and invariably with repeated applications of fertiliser. These grasslands on arable land are ecosystems established by sowing species selected for high yield in response to nutrients released from cultivated soil and applied to redress minor as well as major element deficiencies. The aim of management is to sustain the sown species as relatively stable disclimaxes which will support greater numbers of animals and yet conserve soil and water resources. Because of the high input of economic resources (fertiliser, labour and livestock), the animal production expected is also high. The effects of management inputs on structure and function of sown grasslands have been discussed by Hutchinson & King (Chapter 12, Breymeyer & Van Dyne, 1979).

Establishment of arable grassland

Compared with the long-standing meadows and grazing lands of Europe, many of which are sown on self-seeded cultivated land originally under forest, seeded pastures in Australia and New Zealand are relatively new historically. Their success derives in part from this earlier European experience, in part from the development of techniques for clearing woodland and scrub, but in the main from research into leguminous pasture plants, into their nitrogen-fixing bacteria, and into the nutrients necessary for the symbiotic growth of both. From this research there has been developed, for each of an extremely wide range of soil and climatic environments with mean rainfall in excess of 500 mm, species and strains of pasture legumes, strains of rhizobium and information on required levels of application of phosphate or phosphate–minor element mixtures. These together have provided the means for overcoming the paucity of legumes in the grasslands and the widespread deficiencies of nitrogen, phosphorus and sulphur, and frequently of molybdenum, copper, zinc and boron, in the soils.

In the 20 years from 1945 the area of legume-based grasslands in the temperate areas of Australia, most of which have a fairly reliable winter rainfall, increased from 4000000 to 16000000 hectares. By 1963 13000000 hectares were being topdressed annually with 1800000 tonnes of superphosphate. Carter (1965) estimated that, quite apart from the additional animal numbers and production per head gained (see below), the input of nitrogen by the legumes had a value at that time of US $450000000 per year, or five times the cost of the seed and superphosphate.

In eastern temperate Australia sown grasslands have replaced *Eucalyptus* spp. woodland and dry sclerophyll forest – or the semi-natural grassland that followed their clearing – over wide areas between latitude 27° S and 39° S, and parts of Tasmania between latitude 41° S and 42° S. The native grasses associated with the woodland are *Themeda australis* Stapf. (kangaroo grass), the genera *Poa*, *Danthonia*, *Stipa*, *Bothriochloa* and many others (Moore, 1970). The native grassland is far less productive in winter compared with sown pasture, because many of the native species are subtropical in temperature response (Davidson, 1969) and are therefore winter dormant. However, these native grasses still provide summer grazing where native species have not been completely replaced by exotics. Pasture seeding has been preceded by timber clearing, but some of the land is not cultivated prior to seeding. Fertiliser is commonly applied from aircraft and seeding is often done in this way also. The species sown are *Trifolium repens* L. (white clover) or *T. subterraneum* L. (subterranean clover), alone or together with *Lolium perenne* L. (perennial ryegrass), *Phalaris aquatica* L. (phalaris), *Festuca arundinacea* Schreb. (tall fescue), or *Dactylis glomerata* L. (cocksfoot), although the latter three species are satisfactorily established only on fully prepared seedbeds.

The Armidale site is representative of the temperate woodland zone. Data from the Bureau of Agricultural Economics Report (Cook & Malecky, 1974) illustrate the increase in stock-carrying capacity associated with improvement of pasture by seeding and fertilising (Fig. 27.1). The temperate woodland zone is stocked mainly with Merino sheep for wool production, but meat production, particularly from cattle, is increasing.

There are sown pastures in Australia in the regions with rainfall exceeding 890 mm per annum, between latitudes 32° S and 43.5° S, where the climax

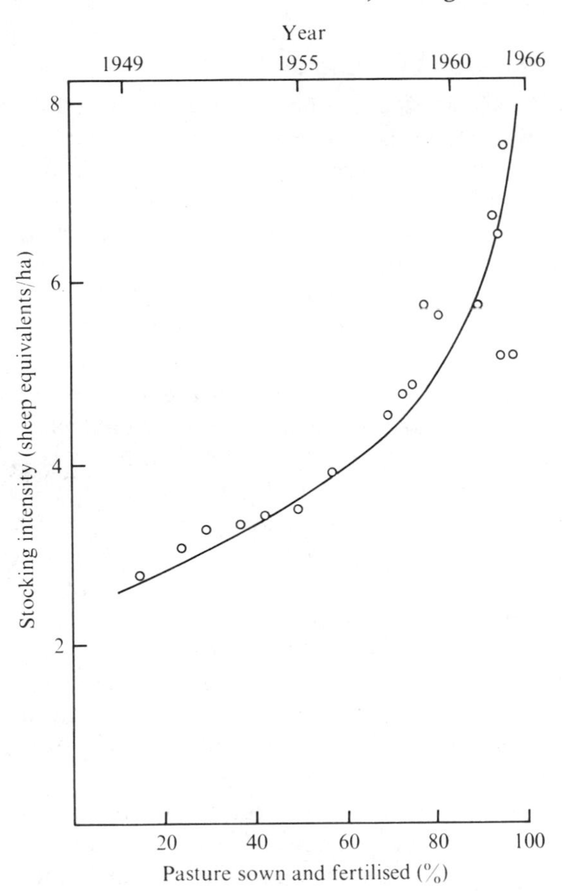

Fig. 27.1. Relationships between stock carried (sheep and cattle as dry sheep equivalents) and percentage of pasture sown and fertilised on 11 survey properties between 1949 and 1966, drawn from data in Cook & Malecky (1974).

vegetation was temperate rain forest or wet sclerophyll. These areas were sown to crops after the forest was cleared, but most of the area is now sown to pasture, mainly *Trifolium repens* L. (white clover), *Lolium perenne* L. (perennial rye grass), *Dactylis glomerata* L. (cocksfoot) and occasionally *Phalaris aquatica* L. (phalaris). Two warm-season grasses, *Paspalum dilatatum* Poir. (paspalum) and *Pennisetum clandestinum* Hochst. (kikuyu grass) are naturalised and widespread in parts of the region. Dairying is the main livestock industry in the wet temperate forest zone, now carrying 80 % of the dairy cows in Tasmania and Victoria and 30 % in New South Wales (Paton & Hosking, 1970).

Cropland regeneration by grassland

Grassland sown on arable land is a feature of agriculture in many parts of the world. It is unfortunate that there were few IBP study sites with grass leys or temporary fallows, because they are so common in both intensive agriculture and shifting cultivation (Davidson, 1964*a*).

Legumes are important in grassland regeneration of cropland, not only because the greater the proportion of legume in the diet of animals the greater the production of meat, milk or wool per hectare (Reed, 1972), but also for the nitrogen they contribute to the soil for any subsequent cropping phase. A 36-year experiment, 120 km from and in a similar environment to Armidale, compared maize–oats rotations with and without intervening phases of *Trifolium pratense* L. (red clover). Inclusion of the legume phase reduced the number of maize and oat crops from 18 to 12 each, but over the period the total maize yield was increased by 3950 kg/ha, the oats gave the same total yield of hay (milk stage), and 168 months of grazing by 7 sheep/ha were gained (Mead & Blunt, 1964).

At the Migda site in Israel, with variable rainfall averaging 260 mm per annum, abandoned cropland colonised by annual grasses and legumes provides year-round grazing for mutton Merino sheep at stocking intensities of 0.6 to 1.0/ha (Tadmor *et al.*, 1974). Experiments are in progress with croplands seeded to legumes, saltbush and cereals for pasture. In ungrazed plots the native annuals respond to fertiliser by 100 to 300 % increases in plant yield, and similar responses are shown to sheep excreta.

Where legumes are not available for inclusion in grasslands, fertilisers may be more important in relation to animal production than species or strains of grasses. In eight years of grazing experiments in South Africa on abandoned cropland in areas of 750 mm rainfall, the application of nitrogen and phosphate fertiliser increased the stocking capacity of the volunteer grassland, mainly *Eragrostis* spp. (love grasses) and *Cynodon dactylon* Pers. (Bermuda grass), but made production per head at the higher stocking rate more sensitive to good and poor seasons. However, at the same level of fertiliser, no further increases in animal productivity per hectare were obtained by replacing the volunteer grassland with sown grassland, even strains of *Eragrostis curvula* (Schrad.) Nees (love grass) selected for their plant productivity (Davidson, 1965).

Probable future usage

The most likely alternative use of arable grassland is as cropland. As the nitrogen brought in by the legumes over the years accumulates in a grassland system, total plant productivity may increase, but inevitably the proportion of non-legumes increases. The reduction in the propoıtion of legumes inhibits increases in animal production parallel with total plant productivity and reduces the further input of nitrogen. But the well-developed soil structure and fertility can now provide an excellent medium for a few years of cropping. Crop yield is enhanced by the nitrogen and other nutrients extracted and exported from the system as cash crops. When the area is resown to pasture, the competition from non-legumes is reduced by lowered fertility; the legumes are stimulated in growth, fix nitrogen which accumulates in the system, and are themselves beneficial to animal production. The alternation of arable grassland with cropping (the traditional ley farming concept), therefore, has many advantages over full-time conversion to croplands with its accompanying necessity to fertilise with nitrogen and protect the land from erosion.

Management

The simplest and least costly form of grassland management use for animal production is to allow a flock or herd of livestock full grazing access to all the area and for this to maintain them for all or as much of the year as possible. Additional practices that are employed include fertiliser, sowing or resowing of pasture, subdivision fencing, restriction of grazing to portions of the area from time to time (pasture management), weed control, conserving forage, growing fodder crops for grazing or conservation, seasonally adjusting livestock numbers to pasture growth by sale, purchase or agistment elsewhere, purchase of feed, acquiring improved breeds of livestock, techniques of animal husbandry, and control of animal parasites. These practices all require additional inputs. To be of value they need to increase output, i.e. the number of productive livestock multiplied by their productivity per head, to a degree greater than the cost of the additional input.

Few accepted or recommended practices have been tested for their ability to do this. For most practices recommended as permitting higher stocking intensity, there is little or no evidence as to how much of the increased production per hectare could have been achieved by merely increasing the stocking intensity without adopting the practice. For many of the practices or animal types which give greater production per head, there is little or no evidence as to whether or by how much their use reduces the potential stocking intensity.

Most practices are based on experience or experiments concerned with ungrazed pastures, non-grazing livestock or with grazing situations different from what really occurs in the year-long association between a particular area of vegetation and the particular flock or herd it supports. Almost totally overlooked is the substantial and continuing effect that the number, type, size and reproductive state of the animals have on the pasture, and on its composition and growth, the controlling effect that the pasture has on the size, productivity and often reproductivity of each animal, and the dominant effect that the number of animals has on production per head and per hectare.

The only means of obtaining biologically meaningful data, of avoiding confounding of treatments, and of providing for the measurement of all the functions and mechanisms contributing to the productivity achieved is to design field experiments providing statistically valid long-term comparisons between whole grazing systems, with and without the practice under examination and each at a range of stocking intensities. The productivity achieved needs to be measured in terms of the number of livestock that can be supported and their productivity per head, the stability of the system's productivity over the years and the effects on the environment, e.g. quantity and quality of stream flow.

Such biological information is an essential prerequisite to understanding how the whole system functions, and how and by how much man can improve its productivity within the environmental limits available. This in turn will enable calculation of the economic worth of a practice in various climatic and cost/price circumstances and the stocking intensity required to maximise benefit/cost ratios.

Stocking intensity of continuously grazed grassland

Productivity of a heavily fertilised sown pasture at Armidale over a wide range of stocking intensities was examined over five years (Langlands & Bennett, 1973*b*). The estimated stocking intensitiy at maximum live-weight gain per hectare varied between years from 3.4 to 19.9 sheep/ha. Wool production per hectare increased to the maximum intensity attained (37.1 sheep/ha), but deterioration of pasture and inability to sustain flocks without excessive deaths at the high intensity make these levels unrealistic. In meat production the live-weight gain per animal is an important consideration, and stocking intensities providing the best economic returns are generally well below those giving maximum production per unit area. The relationships between animal gain and stocking intensity are reviewed by Jones & Sandland (1974). A summary of the effect of stocking intensity on wool productivity of semi-natural and sown pastures and on weight change and survival on sown pastures at Armidale is given in Figs. 27.2 and 27.3, respectively.

Other experiments at the Armidale site provide data on the stocking intensity attainable without deterioration of sown pasture (discussed below under Resource Conservation). The variability of rainfall within and between years causes large differences in potential pasture productivity, but it is impractical to increase stock numbers to take advantage of peak production and to reduce stock numbers in dry periods. The farmer has to strike a balance. From the data in Figs. 27.1 and 27.2 this appears to be about eight

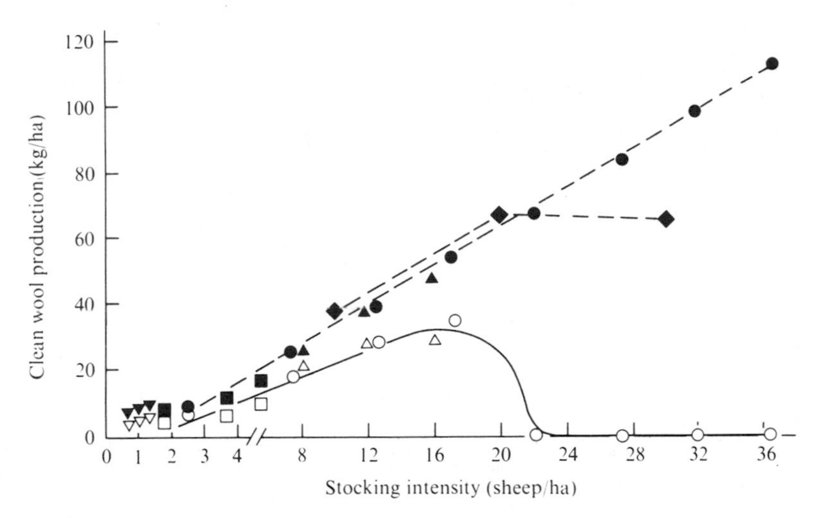

Fig. 27.2. Wool production (clean wool, kg/ha) in best years (solid symbols) and worst years (open symbols) at various stocking intensities in continuous grazing experiments. ▽, ▼, Merino wethers on unfertilised, semi-natural pasture 1949–1952 (Roe *et al.*, 1959); △, ▲, Merino ewes and lambs on sown, fertilised pasture 1964–73 (J. M. George, unpublished data); ○, ●, unmated Merino ewes on sown, heavily fertilised pasture 1963–8 (Langlands & Bennett, 1973*b*); □, ■, unmated Merino ewes on unfertilised, semi-natural pasture 1964–9 (Langlands & Bowles, 1974); ◆, Merino wethers on sown, fertilised pasture 1964–73 (K. J. Hutchinson and K. King, unpublished).

sheep equivalents per hectare on fully improved properties, where the annual application of superphosphate is 100 to 200 kg/hectare.

Fertilisers

Application of fertiliser to sown pastures in Australia is essential because of the common soil deficiencies of phosphorus, sulphur and nitrogen, and sometimes potassium and/or minor nutrients. While nutrient requirements of the plants sown in arable grassland have been well researched in agronomic experiments, the nutrient requirements of grazed systems can be determined only from field grazing experiments. It has been shown in Chapters 24 and 26 (this volume) that the greater part of the nutrients in pastoral systems is not in the domestic animals, but in other consumers and decomposers. Rapid cycling of nutrients is obviously desirable, and management procedures should be sought to achieve this. In fertilised, sown pastures there is a progressive increase in soil organic matter, since only a small proportion of primary production is incorporated in saleable animal products. In deciding on the kind and amount of fertiliser to use, a distinction should be drawn between the *initial build-up* of nutrients in circulation, and the *long-term maintenance* of nutrient levels providing maximum recovery of fertiliser

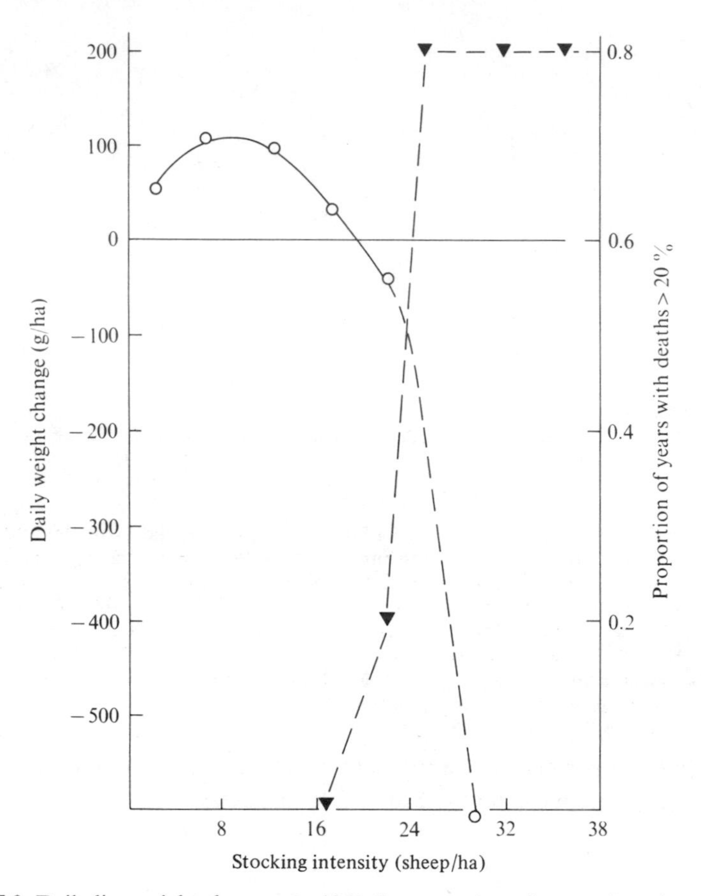

Fig. 27.3. Daily live-weight change, ○, 1963–8; proportion of years when sheep could not survive, ▼, at various stocking intensities. The sheep were unmated Merino ewes on sown, fertilised pasture (Langlands & Bowles, 1974).

(Davidson, 1964*b*). Only broad-based ecological studies will provide a sound basis for these policy decisions.

Conservation

Hay, fodder crops and grain are commonly conserved and used to supplement animals at pasture during the seasonal periods of insufficiency, but experiments comparing whole systems with and without these high cost practices have been very few. Where the supplement is grown within the farm it has had little or no effect on production, except during infrequent drought conditions. In an experimental evaluation at Armidale of fodder conservation in wool production from four temperate sown grass pastures (Hamilton,

Hutchinson & Swain, 1972), there were significant increases in wool production and sheep survival in a year of very low rainfall following a satisfactory hay crop in spring, but in normal years there was no evidence of gains from fodder conservation. In *Phalaris–Trifolium* (phalaris–white clover) pasture at Armidale, Hutchinson (1971) obtained increased wool production from fodder conservation only in a drought year. The increase was appreciable at a moderate stocking intensity, but less at high intensity. Despite the high potential net energy value of the hay crop, fodder conservation had little effect on annual net energy flow through the sheep, except in a low rainfall year at the higher stocking intensity where the response was associated with the prevention of sheep deaths from undernutrition.

Annual fodder crops

Cereals, crucifers and legumes (for grazing or conservation) may be grown on portions of the area available for animal production; for a review of Australian research see Dann (1972). In the high grassland plateau of southern Africa, where winter rainfall is low and sown pasture growth is inadequate for the maintenance of stock, hay is conserved and supplemented with silage, urea and autumn-sown fodder crops (Davidson, 1965). Such practices greatly increase the cost of animal production. At Armidale, Wheeler (1965) compared, at three stocking intensities of breeding ewes, year-round grazing of *Lolium–Trifolium* (perennial rye grass–white clover) grassland with systems in which part of the grassland was replaced each autumn with an annual crop for grazing in the winter period of pasture insufficiency. This arable grassland proved just as good – or bad – as the crop, depending on the season. The system with the crop gave no increase in ewe or lamb productivity, the ewes merely regaining on the crop the losses incurred through having less grassland during growing of the crop and re-establishment of the grassland.

Other management procedures

A number of other practices have been or are being studied at Armidale for their effectiveness or otherwise on the functioning and productivity of year-round grazing systems. These include alternate species of grass and types of pasture management (Wheeler, 1965), subdivision of pastures (Southcott *et al.*, 1972), windbreaks (Lynch & Marshall, 1969), type of sheep and time of lambing (J. M. George and W. M. Willoughby, personal communication), and age of weaning (Corbett & Furnival, 1976). In these and other experiments the chief objective has been to gain a better understanding of the component mechanisms within the system (most of which are commonly studied in isolated parts of the system), the interactions between mechanisms and, thereby, the reasons why the practice does or does not improve the stability and/or productivity of the system.

Resource conservation

Perennial pasture provides not only a productive crop, but also a feature of the landscape which contributes to soil stability and regulation of run-off rainfall. At Armidale, Langlands & Bennett (1973*a*) have shown that bulk density of soil increased and pore space decreased with increasing stocking intensity. Water infiltration time increased with increasing stocking intensity, the percolation time being three times as long at 17 sheep/ha as at 2.5 sheep/ha and 16 times as long as 32 sheep/ha (Fig. 27.4).

Plant cover declined with increasing stocking intensity. Basal cover provides a measure of species changes, and in *Phalaris–Trifolium* (phalaris–white clover) pasture over four years at stocking intensities of 10, 20 and 30 sheep/ha (Hutchinson, 1970) basal cover of grass showed a significant effect of stocking intensity, with a relatively large decline in years of low rainfall at high stocking intensities. The legume had the highest basal cover at the intermediate stocking intensity, but in all experiments it tended to die out during dry periods each year, and regenerated from self-sown seed. Annual grasses increased in all grazed pastures, and after four years represented about 40 % of the total grass cover.

Invasion of sown pastures by *Hordeum leporinum* Link (barley grass) and *Cirsium vulgare* Ten. (spear thistle) frequently occurs in sown pastures. *C. vulgare* invaded monospecific swards of *Lolium perenne* L. (perennial rye grass), *Dactylis glomerata* L. (cocksfoot) and *Festuca arundinacea* Schreb. (tall fescue) pastures in inverse proportion to basal cover, which declined with increasing stocking intensity (George, Hutchinson & Mottershead, 1970). *Festuca* (fescue) was not invaded at low stocking intensity, and *Phalaris* (phalaris) was least affected.

No measurements of run-off or erosion are available from grazing experiments on sown pastures at the Armidale site. Soil under sown pastures tends to be drier than under native pasture (Begg, 1959), and farm dams fill less frequently when the watershed has been sown to introduced species (Donald, 1970). The effects of grassland management on the quantity and quality of stream water should not be neglected in deciding on future land usage in the arable grassland and cropping zones.

Little information is available on the interrelations between grasslands and other nearby ecosystems, e.g. forests, marshes and croplands. However, from the few ecological studies of agricultural systems it is clear that the biota other than domestic stock are essential components of the agricultural production process. Management practices should not override ecosystem processes. Failure to realise the importance of consumers other than domestic stock, the role of decomposers, and the control of injurious pests by birds and other predators has led to biologically and economically wasteful practices in pastoral production.

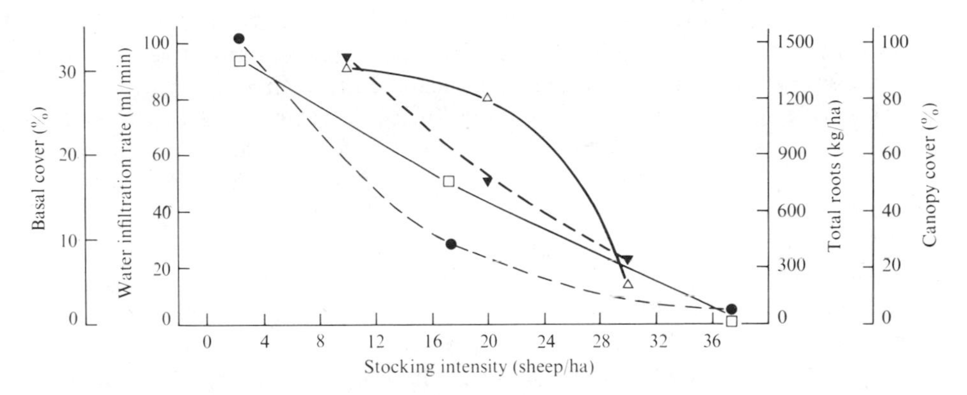

Fig. 27.4. Water infiltration rate (●), total canopy cover □, basal cover of sown species △ total root weight ▼ at various stocking intensities with Merino sheep. (Data from references cited in text, for the period 1963/4–1967/8, except basal cover and roots after treatments maintained 1964–73).

More serious than the wastefulness of some management practices is the disruption of ecosystem processes that contribute to productivity. Overgrazing results in a decline in root growth and a reduction in foliage and causes a fall in productivity, in addition to the long-term effect on soil fertility. The side effects of insecticides on pastures are far reaching. The pollution of water supplies by run-off and leaching of mineral fertiliser and excreta are potential dangers.

The long-term trends are slow and tedious to study experimentally, yet the dynamics of all processes in pastoral systems should be part of the research on animal productivity (Willoughby, 1970). The magnitude of the impact of grazing animals on environmental parameters measured after only a few years at various stocking intensities (Fig. 27.4) shows the need for critical experiments of much longer duration, if the results are to be meaningful. Short-term research which is fragmented into discipline-oriented groups does not provide a sound basis for the economic use of arable grassland consistent with conservation of associated natural resources.

Aesthetic considerations

Sown grasslands, generally, have longer growth periods than natural grasslands, that is they are green and more pleasant in aspect for longer periods of the year. Often, too, a greater proportion is edible, so at times there is less unused and untidy looking material left standing. Sown grasslands generally consist of a few species, so they frequently give a relatively uniform, and to some eyes, pleasing surface. However, in many situations or times sown grasslands are more prone than natural grasslands to invasion by weeds. The replacement of native vegetation with grasslands foreign to the environment also displaces birds and other fauna associated with the natural vegetation. These are often the most pleasing component of the environment.

The widespread clearing of trees in the Australian temperate zones, ranging from temperate rain forest to eucalypt woodland, has had a far-reaching impact on the ecosystems concerned. Diversity of habitat has been reduced, and the viability of relict natural communities is not assured. Regeneration of vegetation should be studied, and the trends in faunal populations given urgent consideration. There may be conflicts of interest between commercial production and recreational needs. Most farmers are concerned about the aesthetics of their environment, and support for ecological studies of the stability of natural and man-made ecosystems comes from all sections of the population. The greatest need is for understanding of the very complex inter-relationships in the highly sensitive, dynamic systems.

Part VI. Croplands

Subeditor: L. RYSZKOWSKI

28. Introduction

L. RYSZKOWSKI

Although considerable effort was made during the IBP to study natural and semi-natural terrestrial situations on a total ecosystem basis, such investigations of agro-ecosystems were rare. Until recently agricultural research was directed to examination of the components of ecosystems that were considered to be limiting factors in production, especially problems of nutrient supply and pest control. The objective in Chapters 28 to 32 of this volume is to review the present state of knowledge concerning structure and function in croplands, particularly in relation to the flow of energy and cycling of nutrients through them.

The first major orientation towards a truly ecological perspective in agricultural research was that of Azzi (1956). This was followed by the work of Tischler (1965) and others. These works formed the basis for an ecological consideration of crop adaptations, pest control, distribution of organisms in cultivated fields and land husbandry. However, these studies did not imply an ecosystem approach to analysis of complete patterns of energy flow and cycling of matter in agricultural systems. Except for the reports resulting from the Polish study considered here, no other publications are known to the authors dealing with total characteristics of agro-ecosystems, with consideration not only to parameters directly determining crop production, but taking into account the interrelations between physical, chemical and biological processes.

The evaluation of structure and function of agro-ecosystems discussed here is based mainly on results obtained in the Polish study of energy flow and mineral cycling in cropland. This investigation was conducted by 14 scientific organisations coordinated by the Polish Academy of Sciences through its Department of Agrobiology at Poznań. The location of the fields studied is the Agroecological Station in Turew (Plate 28.1), near Koscian (about 40 km from Poznań), within the geographical coordinates: 16°45′–16°50′ E Longitude and 52°00′–52°05′ N Latitude. Where possible, the results of this intensive study are supplemented by comparative data in the literature concerning portions of the ecosystem.

The climate of Poznań region (Wielkopolska), where Turew is situated, is one of the warmest in Poland, with a mean annual temperature of 8 °C and the highest summer temperatures (the mean for July being 18.5 °C). The climate is also one of the driest in Poland. The mean annual precipitation is 533 mm, of which 375 mm falls in the vegetative period, and 197 mm during the summer (June to August). A marked predominance of summer over winter precipitation and the preponderance of autumn over spring rainfall indicate a certain continentality of the climate (Table 28.1).

Plate 28.1. Rye and potato fields that were studied as total ecosystems at Turew, Poland. (Photograph by J. Karg.)

The Turew area lies within a plain formed by ground moraine with elevation of 80–90 m above sea level. Light boulder clays, covered with a 50–90 cm layer of light loamy sands, constitute the parent material of the soils. Cultivated soils are pseudo-podzolic with a very low content of organic carbon (0.6 %) and almost without calcium carbonate. The pH of the plough layer depends on the agronomic measures that maintain it. The water economics of the soil of the investigated fields is mainly ombrophilous. The moisture supply is not influenced by capillary rise of ground water, the level of which lies below 3 m. In dry years precipitation from April to the end of August does not make up for the loss by evapotranspiration, so a water deficit is observed. The uppermost layer of soil shows the greatest deficit. Water is thus the main factor limiting agricultural production.

In the neighbourhood of Turew, the potential natural vegetation (that is, the vegetation that would develop if human interference ceased) would be represented by poor forms of deciduous forest composed of oak and ironwood (Querco–Carpinetum medioeuropaeum R.Tx.) and mixed forest of pine and oak (Pino–Quercetum; Kozł.). However, in cultivated fields weedy associations occur in cereals – *Papaveretum agremones* (Krusem and Vlieger) and in root crops – Echinochloo–Setarietum (Kruseum and Vlieger). The Poznań region is a land of old and superior traditions in agriculture. Today, cultivated fields cover 56 % of the area, orchards 1 %, meadows and pastures 11 %, forest 23 %, water surface 2 % and 7 % of the land is used for other miscellaneous purposes.

Table 28.1. *Selected meterological data by month at Turew (after Radomski* et al., *1974); temperature data are for 1925–65 and precipitation data for 1891–1930*

	Jan	Feb.	Mar.	Apr.	May	June	July	Aug.	Sept.	Oct.	Nov.	Dec.	Year
Air temperature (°C)													
Mean	−2.3	−2.0	1.8	7.8	13.1	16.7	18.5	17.6	13.9	8.5	3.1	−0.5	8.0
Mean extreme maximum[a]	11.2	14.2	20.6	28.5	33.3	34.7	38.2	37.0	32.5	25.5	18.0	15.0	38.2
Mean extreme minimum[a]	−26.4	−31.0	−26.6	−9.5	−3.0	−0.5	3.8	4.4	−1.6	−6.4	−17.2	−23.9	−31.0
Precipitation (mm)													
Mean	32.5	29.5	28.4	35.6	52.4	62.4	67.4	67.3	45.9	43.9	35.9	32.0	533.2
Extreme daily maximum[b]	36.3	20.9	19.4	21.8	41.7	44.6	58.0	111.9	60.4	62.0	29.8	116.6	—
Number of days with precipitation	14	12	11	12	13	12	13	12	10	11	12	13	145

[a] These are mean extreme values for each month and year. [b] These are extreme values for entire period of observation.

29. Producers

Z. WÓJCIK

The investigation at Turew was conducted in fields of rye (*Secale cereale* L.) that were sown in the fall and of potato (*Solanum tuberosum* L.) planted in the spring. These crops represent typical agricultural conditions in Wielkopolska, where they are cultivated on a large scale and yield large crops. In this discussion published reports of the study have been drawn upon. These include those of Kukielska (1972, 1973*a*, *b*) and of Wójcik (1973).

The fields under study were given the usual management treatments. Fertilising was differentiated between crops. Applications of manure (used only for potatoes) ranged from 25000 to 38500 kg/ha (fresh weight) and of NPK (used for both crops) from 162.5 to 623 kg/ha. Insecticides were used to control Colorado potato beetle (*Leptinotarsa decemlineata* Say) and herbicides were used to control weeds. Three varieties of rye and two of potatoes were included in the fields studied. Crop rotation was simplified to two crops: rye and potatoes. In some fields potato was planted as a sole crop, in others after a fore-crop* of fall rye for green forage (Ryszkowski, 1974).

Net primary production was estimated by growth analysis and included all plant growth, both of weeds and crop, that took place during the whole year. Production of the principal crops was investigated by a sampling of shoots and their attached under-ground parts on three occasions during development; for rye, growth in the fall was included. Growth curves were constructed from the data, and these were the basis of estimating production. For weeds, mosses, after-crops and volunteer crop plants the maximum standing crop, of above-ground and under-ground parts, was used as the estimate of production. The autumn production of after-crop was included as part of the analysis of rye fields, while the spring production of fore-crop was included in potato fields. Production of soil algae was not considered. The investigation lasted for five years and included a total of 10 estimates in crops of rye and 10 in crops of potato. Some fields were studied in two different years, but most of the estimates were made in different fields each year.

Net primary production in rye fields ranged approximately from 1000 to 1500 g/m^2, with a mean of 1258 g/m^2. An average of 95 % of production was accounted for by the main crop, with 26 % being in the form of grain; weeds contributed 3 % of production (Table 29.1).

* Sometimes a fall-seeded crop of rye is inserted in the rotation between the main crop of rye and the crop of potato. This is harvested for feed in the spring or is ploughed under as a 'green manure' treatment. This is referred to here as an 'after-crop' in relation to rye and a 'fore-crop' in relation to potato.

Table 29.1. *Mean values for annual net primary production in cropland at Turew (after Kukielska, 1973*a, b*; Wójcik, 1973*)

	Rye field		Potato field	
Elements of production	g/m²	kJ/m²	g/m²	kJ/m²
Crop				
Above ground				
Grain	327	5587	—	—
Stems and foliage	646	11034	283	4704
Below ground				
Tubers	—	—	469	7779
Roots	232	3957	74	1225
Total crop	1206	20578	826	13707
Weeds	33	567	231	3833
Mosses	4	56	—	—
Volunteer crop	15	253	—	—
Total for crop season	1258	21453	1057	17540
After-crop (autumn)	71	1214	—	—
Fore-crop (spring)	—	—	640	10921
Total for year	1329	22667	1697	28462

The range of net primary production in the potato fields approximated 600 to 1600 g/m², with an average of 1057 g/m². Of this the crop contributed 78 % and tubers alone 44 %. Weeds provided a mean of 22 % of net production (Table 29.1).

The coefficient of variation ($\sigma/\bar{x}$) of net production between fields (and years) was 15 % for rye and 26 % for potato, while that of the harvestable portion of crops was 20 % for rye (grain) and 32 % for potato (tubers). Thus, net primary production was more uniform than crop yield. The variability of net production and its component parts was distinctly less in rye fields than in potato fields.

The major climatic factor influencing the rate of primary production is precipitation, while temperature is of secondary importance. During the period of rye growth (March–July) the amount of precipitation directly affected both net production and grain yields of rye. Deficient precipitation in June is responsible for poor crops of rye (kernels not being filled). Warm weather in late fall and winter (November–February) contributes towards an increase in total net production in rye fields; high temperatures from June onwards are harmful to net production in the field, as well as to the yield of grain. Deficient precipitation in June is responsible for poor growth of potatoes, while a warm September with low precipitation results in early drying of shoots and decrease in biomass of tubers.

Fertilising with NPK alone had only a slight effect on the rate of primary production in fields of rye, under all weather conditions. However, the effect of combined manure and mineral treatments was marked when it coincided with favourable weather, particularly when above-average precipitation was distributed to coincide with the periodic needs of the crop.

Introducing a fore-crop of rye for green forage before planting potatoes results in an increase in total net primary production of potato fields of about 640 g/m²; production rises on the average from 1057 to 1697 g/m². In this way the vegetation of the field can produce organic matter during the whole spring. Potatoes are planted in May and the first shoots appear at the beginning of June. If there is no fore-crop, the field is devoid of plant growth during this period. Thus, total annual net production in potato fields with a fore-crop of rye is greater than that of rye fields, in which after-crop production is low (Table 29.1).

The relative proportion contributed by each component of primary production was changed in potato fields by altering the density of potato plants. When the density of potato plants was reduced below average, the proportion of production contributed by the crop declined, while that by weeds increased. Depending on the density of potatoes, the yearly infestation with weeds ranged from 4 to 59 % of total net production of the fields.

Of the organic matter produced by herbaceous plants, man takes a considerable portion in addition to the grain or tubers that he harvests. A mean value of 461 g/m² (range 388–596 g/m²) of organic matter was left in rye fields, representing only 35 % of total net primary production of the fields. This compares with 75 % of production that never reached the grain. The proportion not removed from rye fields was greatest when net production was lowest. In potato fields, 64 % of production was left after harvesting tubers (which themselves only contained 27 %); when surface debris was removed by raking, only 33 % of production was left in the fields. When a fore-crop was used, 68 % of its biomass was left after removal of feed.

The values presented here for net production of cropland at Turew are similar to those obtained by other authors. In Poland, for example, Herbich (1969) estimated the net primary production of two rye fields to be 1964 and 1013 g/m² and that of potato fields to be 988, 1027 and 984 g/m². Pasternak (1974) estimated net production of a wheat field to be 1537 g/m². Similarly, in West Germany (Köhnlein & Vetter, in Lieth (1962)) net primary production of a rye field was found to be 1002 g/m², while in the Kursk region (USSR) it amounted to 1100 g/m² in a barley field. The magnitude of primary production in croplands of the USA is also similar (Bray, Lawrence & Pearson, 1959; Bray, 1963).

A comparative study of net productivity of cropland under different management regimes has been made in Japan (Iwaki, 1974). The values obtained were generally higher than those obtained in Poland. They were as follows (g/m²): wheat fields, 1681; barley fields, 1379, 1448; oat fields, 2004; rice fields, 1449–2041; maize fields, 1765; sugar beet fields, 1712. Apparently, however, there is great variation in productivity of Japanese cropland, since the average net production of wheat fields throughout the country is estimated to be approximately 750 g/m^2.

Utilisation of photosynthetically active solar energy by the vegetation of rye and potato fields at Turew ranges from 0.7 to 1.5 %. These values are almost identical with those (0.7–1.4 %) given by Iwaki (1974) for Japan's cultivated fields.

It seems appropriate to emphasise that the vast body of data available concerning the croplands of the world does not provide a basis for estimating efficiency of the system in fixing radiant energy. In most instances harvestable yield is the only measure of output that has been made. Information concerning the amount of energy fixed by plants that is available to other biotic components of the ecosystem is generally lacking. Similarly, little is known concerning the quantity of organic debris that is left after harvest and its role in maintaining the physical well-being of the soil. The IBP studies reported here are a beginning towards a more thorough understanding of ecosystems of cropland.

30. Consumers

L. RYSZKOWSKI

Although there is abundant information concerning pests of cropland, including the life histories of individual species, few studies have been concerned with the total structure and role played by fauna in agro-ecosystems. The first synthesis of the scattered data on life histories, habitat distributions, trophic relationships and other characteristics of the animals of cultivated fields was that of Tischler (1965). Even under IBP the analysis of the total consumer trophic level in agro-ecosystems for the most part was neglected. Consequently, the evaluation presented here of structure and function of the animal component of cropland is based mainly on the results of the study at Turew, which was conducted by the Department of Agrobiology of the Polish Academy of Sciences. A comparison of populations of small herbivores in cultivated and uncultivated soils has been presented by Andrzejewska & Gyllenberg (Chapter 3, Breymeyer & Van Dyne, 1979).

Invertebrates

The number of species of invertebrates is lower in cropland sown annually than in adjacent perennial stands of grass or in wooded areas (Table 30.1). From the studies conducted at Turew the number of species in cultivated land is more greatly reduced in relatively immobile groups (Enchytraeidae, Lumbricidae, Collembola, Elateridae and Acarina), including those insects that are relatively immobile during only part of their life history (Aphididae and Orthoptera), than in the mobile groups (Diptera, Heteroptera, Hymenoptera, Carabidae, Staphylinidae, Araneae). In terms of number of individuals, Diptera forms the most important above-ground group of insects. In cropland at Turew 55–80 % of insects emerging from the soil surface belong to this order (Table 30.2). Diptera also contributes the largest proportion of flying insects. In comparing the relative abundance of Diptera in croplands and meadows, Dabrowska-Prot, Karg & Ryszkowski (1974) concluded that with increasing interference by man (i.e. increasing intensity of agro-technology) the proportion contributed by Diptera to the total insects increases. In addition, they found that a simultaneous decrease occurs in the total number of individuals and of species of Diptera. However, an increase in the proportion of herbivorous forms in the trophic structure of Diptera was observed.

Generally the decrease in abundance of herbivorous forms was less in agro-ecosystems than in adjacent uncultivated areas (meadows, forests, shelterbelts) as compared to predatory and parasitic organisms. In the

Table 30.1. *Comparison of numbers of species of various groups of invertebrates observed in cropland and in a tree shelterbelt* (after Aleynikova & Utrobina, 1969)

Systematic group	Fallow (bare soil)	Corn (maize)	Wheat	Perennial arable grassland	Shelter-belt	Area as a whole
Carabidae	23	38	29	42	57	70
Staphylinidae	17	21	29	21	31	61
Curculionidae	7	10	15	15	46	55
Elateridae	4	5	5	7	10	13
Histeridae	6	6	6	12	11	20
Scarabaeidae	2	2	2	2	4	6
Collembola	4	11	19	16	—	29
Araneae	8	13	11	12	20	35
Oribatei	7	18	14	13	—	31
Gamasoidea	7	—	7	17	—	19
Lumbricidae	1	4	3	4	7	7

Table 30.2. *Number of individuals (per square metre) of the principal orders of insects that emerged from the soil surface during the whole vegetative period*[a] *in different years under various crops. The data were collected under cages that were moved every few days* (J. Karg, personal communication)

	Rye			Potato				Barley	Sugar beets	Corn (maize)
Order	1972	1973	1974	1972	1973	1974*a*	1974*b*	1974	1974	1974
Diptera	237.1	534.6	200.9	181.9	239.1	148.6	220.8	414.0	340.8	191.6
Coleoptera	105.6	100.4	38.0	69.2	43.2	34.0	66.0	52.0	65.2	88.0
Heteroptera	27.2	17.3	12.4	10.8	34.2	5.2	27.6	11.2	11.2	2.4
Hymenoptera	11.9	0.6	1.2	10.1	1.2	1.6	2.8	10.0	7.6	2.0
Other	42.2	8.8	11.2	6.1	7.8	2.0	7.2	67.6	3.6	3.2
Total	418.6	661.7	263.7	278.1	325.5	191.4	324.4	554.8	428.4	287.2

[a] Mid-April to mid-October.

cultivated fields studied at Turew, predatory forms of Nematoda were lacking, while those of Acarina and Diptera (especially Syrphidae), Homoptera Parasitica, and Araneae were of comparatively low abundance. Less decline in abundance, compared to adjacent meadows and forests, was recorded in cultivated fields for predaceous forms of Staphylinidae, Carabidae and Coccinellidae, while Nabidae (predatory Heteroptera) and Neuroptera were often of greater abundance in cropland than in uncultivated areas. Ants, snails, millipedes and centipedes were found only sporadically in cultivated fields, while they were quite numerous in adjacent forests and meadows.

Beneath the soil surface large organisms are relatively much less abundant than smaller ones. For example, the mean densities of Lumbricidae (earthworms) at Turew ranged from 6.7 individuals/m² in rye fields to 8.7/m² in potato fields; Enchytraeidae ranged from 10900/m² in rye fields to 16900/m² in potato fields; Collembola from 27200/m² in potato fields to 40600/m² in rye fields; and Nematoda from 5497000/m² in potato fields to 8618000/m² in rye fields. The averages mentioned above are intraseasonal means and for Enchytraeidae, Collembola and Nematoda are high in comparison with those previously published in respect to cropland (e.g. Tischler, 1965).

At Turew, detailed studies were made of the trophic structure of important systematic groups in cropland. These revealed that herbivores comprised the most important component of Diptera (Table 30.3), while omnivorous

Table 30.3. *Populations, standing crop and production of Diptera in various trophic groups in cropland at Turew* (after Dabrowska-Prot and Karg, 1975)

	Emerging[a] from soil surface				Standing crop[b]			
	Rye field		Potato field		Rye field		Potato field	
Feeding category	No./m²	mg/m²	No./m²	mg/m²	No./m²	mg/m²	No./m²	mg/m²
Herbivores	98	242	118	381	3.1	4.3	1.8	2.0
Saprophages	98	81	47	51	0.9	0.5	0.2	0.6
Predators and parasites	36	51	16	32	0.5	0.9	0.3	0.4
Total	232	374	181	464	4.5	5.7	2.3	3.0

[a] Emergence data are accumulations for the vegetative season (mid-April–mid-October) of individuals that emerged from the soil surface and were trapped in cages. Cages were moved every 5 to 6 days. These estimates approximate production, excluding only larvae that died before emerging.

[b] Standing crop data were obtained by quick traps and are the means of several collections during the vegetative season.

Table 30.4. *Mean populations and energy values of Heteroptera in cropland at Turew, classified as to trophic level* (after Karg, 1975)

	Rye field		Potato field	
Feeding category	No./m²	kJ/m²	No./m²	kJ/m²
Omnivores	12.8	0.469	5.5	0.205
Herbivores	7.1	0.272	1.6	0.092
Predators	4.8	0.163	3.0	0.100
Total	24.7	0.904	10.1	0.397

Data were obtained by averaging several collections by quick traps made throughout the vegetative season (180 days).

Table 30.5. *Mean populations and standing crop*[a] *of Nematoda in cropland soil at Turew, classified according to trophic level* (after Wasilewska, 1974)

	Rye field		Potato field	
Feeding category[b]	No./m² × 10³	mg/m²	No./m² × 10³	mg/m²
Microbivores	4375	169	2969	131
Fungivores	2178	40	460	6
Herbivores (parasites of higher plants)	1963	57	1843	42
Omnivores	102	53	225	22
Total	8618	319	5497	201

[a] Dry weights have been calculated as 30 % of live weight.
[b] Predators were not present.

forms (of the genus *Lygus*) made up more than 50 % of all Heteroptera (Table 30.4). In considering subterranean groups, among Nematoda the microbivorous forms were dominant over other groups (Table 30.5), due partly, at least, to the greater abundance of bacteria rather than fungi in the soil of the agro-ecosystems studied.

The data obtained in the Turew study provide a basis for estimating the total invertebrate biomass in cropland (Table 30.6). Although numerous errors are undoubtedly introduced into these estimates (due to varying levels of accuracy of sampling methods used for different groups and because of some short-cuts that were made in the calculations), the estimates probably give a representative picture of the invertebrate structure in croplands of the area. The mean standing crop was approximately 5.7 g/m². The bulk (98 %) of the invertebrate biomass was in the soil, with Protozoa and Lumbricidae comprising about 87 % of the total standing crop biomass of invertebrates. Among insects, the most important groups are saprophagous forms (such as Collembola) and herbivores (such as aphids and the Colorado potato beetle). The latter two groups of herbivores are important pests of crops and their fluctuating populations cause considerable change from year to year in abundance and biomass of the above-surface fauna of cropland.

Vertebrates

To evaluate the trophic structure of vertebrate populations in a meaningful way, it is necessary to consider an area large enough to include their daily range of movement. Thus, in the Turew study the mean vertebrate density and biomass was estimated in an area of 3100 ha where croplands totalled 79.5 % of the area and shelterbelts, small forests and land used for miscellaneous other purposes made up the remainder.

Approximately 140 species of vertebrates occurred within this area. These

Table 30.6. *Mean standing crop estimates in mg/m² of invertebrates during the vegetative season*[a] *in cropland at Turew*

Taxon	Depth of sampling (cm)	Rye field	Potato field	Mean	Source[b] of data
In soil					
Protozoa	25	—	—	~ 3000	Very rough estimate by J. Eismond-Karabin
Nematoda	25	319	201	260	Wasilewska (1974)
Enchytraeidae	30	138	260	199	B. Iwanowska
Lumbricidae	30	2100	1900	2000	Jopkiewicz (1972)
Acarina	—	7	5	6	Wasylik (1975)
Collembola	25	81	54	67	A. Czarnecki
Carabidae[c] larvae + imago	—	36	31	34	Kabacik-Wasylik (1975)
Other insects mainly larvae	30	73	58	66	J. Karg
Subtotal				5632	
On soil surface and in canopy					
Aphidiidae	—	40	19	29	B. Gałecka
Diptera	—	6	3	4	Dąbrowska-Prot & Karg (1975)
Leptinotarsa decemlineata	—	—	168	84	J. Karg
Other insects	—	7	4	5	J. Karg and (for Thysanoptera) J. Kot
Araneae	—	9	6	7	Łucrak (1975)
Subtotal				129	
Total				5761	

[a] Mid-April to mid-October.
[b] Personal communication, except where year is given.
[c] Sampled in cages when emerged from soil.

included breeding birds (100 species), mammals (30 species), amphibians (10 species) and one lizard. Approximately 65 % of vertebrate species were predators. These included all the amphibians, 65 % of the bird species and 30 % of the mammalian species both of which feed mainly on insects, and 15 % of the mammalian species and 5 % of the bird species which prey mainly on vertebrates. Omnivores comprised 23 % of all vertebrate species, including granivorous birds which seasonally changed their feeding habits to feed on insects. Herbivorous species made up 12 % of the total.

Several species attain sufficiently high populations to cause damage to crops. The common vole (*Microtus arvalis* Pallas), the populations of which cycle, is particularly damaging at certain times, particularly in meadows and perennial crops (such as alfalfa (*Medicago*)) and to a lesser degree in cereal crops and rapeseed (*Brassica*). It is practically absent from fields of potato, sugar beet and maize (*Zea*). Other abundant vertebrate species of cropland include: fox (*Vulpes vulpes* L.), roe deer (*Capreolus capreolus* L.), wild boar (*Sus scrofa* L.) and hare (*Lepus europaeus* Pallas) among the mammals, while abundant birds are rook (*Corvus frugilegus* L.), sparrow (*Passer domesticus* L. and *P. montanus* L.), starling (*Sturnus vulgaris* L.) and lapwing (*Vanellus vanellus* L.). The long-term trend (10–20 yr) was towards a decrease in numbers in some predatory birds (*Accipiter gentilis* L. and *Falco tinnunculus* L.), while others (*Buteo buteo* L.) increased.

Shelterbelts provide breeding places for many species, but many of these range into cropland. Among 12 species of rodents that are present, only the common vole is restricted in its activity to cropland. Of 44 species of birds that breed in small shelterbelts, four also nest in fields. Only three species of birds nest only in fields.

In order to compare the relative importance of vertebrates and invertebrates in cropland, a very rough estimate of the standing crop of vertebrates was made for the arable portion of the area under study of Turew. Only those species of vertebrates that obtained a substantial portion of their diet in cultivated fields were considered. Densities from several years of observations in various crops located throughout the study area were used in calculations. The estimate arrived at was a standing crop of 200 mg/m^2. Three animals made up three-quarters of this standing crop. These were: (i) hare (66 mg/m^2), an important game animal; (ii) common vole (46 mg/m^2), a pest in crops; and (iii) roe deer (38 mg/m^2), another important game animal. The mean biomass of all vertebrate predators (insectivorous birds and mammals, amphibians, and such animals as foxes, weasels, martens, owls and buzzards that prey on other vertebrates) was estimated to be approximately 38 mg/m^2. These estimates suggest that the ratio of invertebrates to vertebrate biomass in this cropland is of the order of 28:1. However, within the surface and above-ground fauna, vertebrates contribute about twice as much biomass as do invertebrates.

Role of animals

In order to evaluate the role of animals in agro-ecosystems it is necessary to consider both the structure of the faunal components and their functional relationships. Generally the activity of any subunit of an ecosystem in performing a function is proportional to its biomass, its turnover rate and the amount of energy available for the activity. In considering the structure of an ecosystem it is necessary to take into account both the internal structure of the recognized subunits and the organisation of the subunits into the whole ecosystem. For example, the difference in the internal structure (body–size distribution) of the two largest components of the soil fauna (Protozoa and Lumbricidae) in the Turew cropland (Table 30.6) determine their bioenergetic

Table 30.7. *Mean residence time of adults (in days) of various Diptera groups in cropland at Turew* (E. Dabrowska-Prot, personal communication)

Crop	Year	Antomy idae	Muscidae	Lycoridae	Chloropidae	Empididae	Drosphilidae	Syrphidae	Bibionidae
Rye	1972	2.2	3.2	0.7	8.8	0.7	8.0	5.4	3.9
	1973	1.4	4.3	1.1	3.2	0.8	2.5	11.3	8.2
Potato	1972	0.7	3.1	0.2	14.6	0.7	11.2	2.0	1.9
	1973	0.9	1.1	1.0	6.3	0.1	12.8	0.6	0.9

Table 30.8. *Energy values* (*in kJ/m²*) *of the total insect emergence from soil in cultivated land supporting two different crops at Turew* (J. Karg, personal communication)

Year	Rye	Potato
1970	22.5	10.5
1971	35.1	22.9
1972	28.9	18.2
1973	21.4	12.0
1974	8.9	6.3
Mean	23.3	14.0

Insects emerged from mid-April to mid-October.

parameters, while the subunits comprised of animals of small size influence the rate of organic matter mineralisation.

Only meagre information is available concerning the rates of turnover of animals in ecosystems, including agro-ecosystems. Detailed field studies are necessary to determine such characteristics as survival rates, reproduction rates and the degree of movement across ecosystem boundaries, all of which are important in estimating turnover rates. Each of these characteristics is dependent on the interplay of many factors and varies greatly from time to time in the same system. For example, the mean time of residency of some families belonging to the order Diptera was considerably longer in 1972 than in 1973, while the reverse was true for several other families. Nor was there uniformity between crops in either the mean residence time or the relative length of these times in 1972 and 1973 (Table 30.7). Generally, variability with time in density, rate of turnover and production is much greater in small, clearly defined subunits than within more complex systematic groups. Thus, during a five-year period of study in cropland at Turew the energy content of the maximum total annual insect fauna emerging from soil was four times that of the minimum year (Table 30.8), while the maximum obtained for individual species was more than 10 times that of minimum values. This phenomenon implies the existence of compensation processes between animal components in agro-ecosystems.

The direct influence of herbivores on plants through consumption is usually small and does not reach 10 %, except in outbreak situations. However, outbreaks of some herbivores that do cause infestations of crops are often widespread; the damage that occurs depends on the standard of agricultural practices. Pest control is one of man's important management roles in cropland. The role of animals in cropland is best known in respect to such pests. The other well-known role of animals in cropland is to bring about the pollination of plants. Problems have arisen in Poland because chemicals used to control pests have interfered with pollinating organisms, particularly those of rapeseed (which is an important vegetable-oil crop).

The functions of animals in cropland can be grouped into those concerned with: (i) direct impact on primary producers; (ii) movement of materials; (iii) alteration of the environment; and (iv) impact on other heterotrophs. All of these functions influence (directly or indirectly) productivity processes in agro-ecosystems. However, because of meagre information, it is at present only possible to evaluate the importance of animals in respect to the first function (i).

The function of animals in the breakdown of plant debris has been recognised for a long time. Animals have been shown in many studies to speed up the rate of decay of organic matter. However, several studies have failed to demonstrate any effect of animals on the rate of mineralisation of organic matter in ecosystems intensively managed by man. The effect of animals on the rate of primary production and on the transference and dislodging of plant materials into and within the soil profile is poorly understood. Animal activities result in redistribution of nutrients in the ecosystem and often cause horizontal concentrations to local areas (Table 30.9).

Table 30.9. *Effect of insects on concentration of mineral elements. Data are in terms of mg/100 g of dry excrement or soil* (after Kozlovskaya, 1965)

Material	NH_4-N	P_2O_5	K_2O
Excreta of Tipulidae larvae	56.8	9.4	109.0
Soil	41.0	4.9	78.6
Excreta of Sciaridae larvae	44.9	17.4	127.8
Soil	8.8	12.9	144.6
Excreta of Elateridae	61.6	5.6	141.6
Soil	23.8	9.0	84.4
Soil of ant hill	30.1	0.4	116.1
Soil	18.1	0.3	212.9

The redistribution of plant debris within the soil profile that results from the activities of certain soil fauna modifies environmental conditions. Soil-water storage capacity and soil temperature are particularly affected. Burrowing activity transfers nutrients more efficiently than exposure to the metabolic processes that follow ingestion.

In considering the total role of animals in agro-ecosystems it is necessary to include their activities as vectors of diseases and in predator–prey relationships. Many of the heterotrophs of agro-ecosystems are vectors of diseases and affect other organisms in the system in this way. The activities of animals in affecting biological control mechanisms must be considered in relation to their roles in predator–prey relationships and in competition (Pielowski *et al.*, 1974).

The contribution of animals to the rate of energy flow through agro-ecosystems will be evaluated in Chapter 32 of this volume.

31. Micro-organisms

J. GOŁĘBIOWSKA

Considerable research has been undertaken on the role of micro-organisms in the productivity of various natural ecosystems. An understanding of the functions of micro-organisms and of the degree of their activity in various biochemical processes in soil is also of importance in developing management regimes in agro-ecosystems.

The biochemical processes that concern us here can be represented by the flows indicated in the following model.

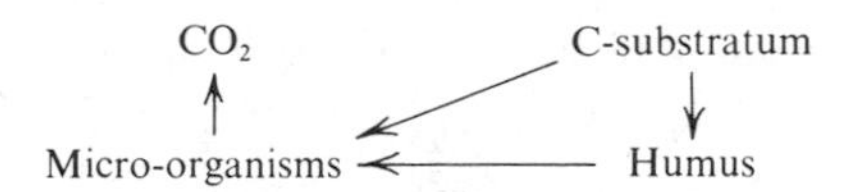

For micro-organisms, it is important to know the biomass and rate of secondary production and for humus, the standing crop and rate of turnover. The biochemical processes proceed at various rates. In order to determine these rates it is necessary to measure them all in the same ecosystem. In Poland, the preliminary test was conducted in cropland at Turew.

Determination of the annual rate of input of carbon into the soil is important for the quantitative presentation of our model. It is important to distinguish between the nature of the input of plant debris in natural ecosystems and that in cropland ecosystems. Natural ecosystems are much more uniform from year to year in the amount that is added and in its qualitative characteristics. This is because in natural ecosystems organic matter is produced by the same plants each year (except for slow successional changes in plant populations) and in some ecosystems the transfer of plant debris to soils is largely from a layer of raw humus on the surface and is not dependent on annual fluctuations in plant production. As a result, the microbial populations in natural communities are in a more-or-less balanced state. On the other hand, the amount and quality of the annual addition of plant debris to cultivated soil varies greatly because of crop rotation and the use of organic fertilisers. Quantities of carbon (calculated as 50 % of plant mass) introduced into the Turew soil were 190–300 g/m^2 in rye fields and 170–450 g/m^2 in potato fields. When fore-crops were harvested before potatoes, an additional amount of about 220 g/m^2 was added to the soil. Carbon introduced with manure amounted to 200 g/m^2.

The carbon content in soil is comparatively stable, that in the Turew arable soil averaging 0.504 % on a dry matter basis. This value ranged from 0.41 to 0.61 %, partly due to spatial variation, but probably also to differences in intensity of fertilisation and temporal differences in the rate of transfer of plant debris to the soil (W. Loginow, personal communication). Uncertainty exists concerning the proportion of carbon from plant debris that is transferred to the humus fraction of the soil. The general opinion, supported by experiments with radioactive carbon, is that 0.5–1 % of humus compounds are reproduced each year (Paul, Biederbeck & Rosha, 1970), indicating that probably 10 % of the carbon introduced into the soil develops into humus and the rest is lost during respiration processes of soil organisms. It is expected that our knowledge of these transfers will soon be greatly increased because of the development of new methods to determine the susceptibility of humus fractions to oxidation.

It is not clear, however, whether the carbon used in respiration processes always comes directly from decomposition of fresh organic matter. Some more complicated processes may occur, such as biological sorbtion, autolysis and resynthesis of micro-organisms, as well as the action of enzymes (according to the Kleinhempel, Freytag & Steinbrenner (1971) hypothesis). These processes may proceed at various speeds, depending on the environmental conditions that exist at different seasons of the year.

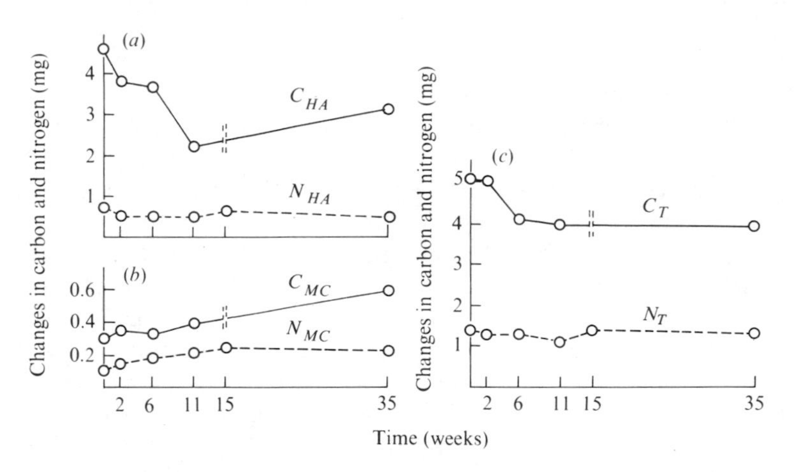

Fig. 31.1. Changes in carbon (——) and nitrogen (– – –) contents of humic acid (*a*) and micro-organisms (*b*) in culture with humic acid. (*c*), total changes. (K. Dębosz, personal communication.)

It has been ascertained that carbon and nitrogen appear in the humus a relatively short time after addition of plant material to soil. Also, it has been possible to show that synthesis of humus substances occurs through the formation of microbial biomass and biochemical processes coupled with autolysis (Martin & Haider, 1971; Freytag, 1971; McGill *et al.*, 1973; Shields, Paul & Lowe, 1974). The results of preliminary experiments with inoculum from Turew soil (Dębosz, 1975) agreed with these findings (using

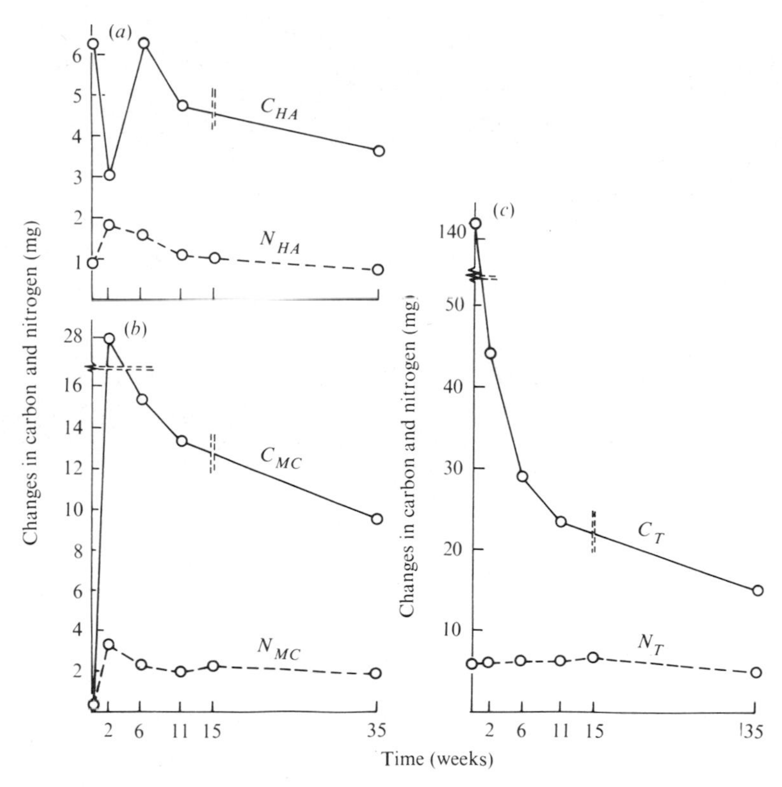

Fig. 31.2. Changes in carbon (——) and nitrogen (– – –) contents of humic acid (*a*) and micro-organisms (*b*) in culture with glucose and amonium sulphate. (*c*), total changes. (K. Dębosz, personal communication.)

a culture in a basic medium with the addition of humic acids only (0.06 %) or glucose (1 %) and ammonium sulphate (0.05 %). A decrease in carbon and nitrogen in humic acids and their accumulation in the cells of micro-organisms occurred during the first 10–15 weeks of the experiment when incubated with humic acid only (Fig. 31.1). In the presence of glucose and ammonium sulphate (Fig. 31.2), carbon and nitrogen accumulated in considerable quantities in microbial biomass, and the rates of decomposition and resynthesis were greater than with humic acid alone.

Populations and biomass

Determination of biomass of soil micro-organisms is based on numbers. Estimates of populations have been made by both the plate method and the direct count method, with the latter giving higher estimates. The merit of the direct method relates to the opportunity of counting all cells occurring in the soil, while its weakness lies in the difficulty in distinguishing living cells from dead ones (Bucherer, 1966) and in distinguishing microbial cells from colloidal particles (Casida, 1971). With this reservation, many authors (e.g. Babiuk & Paul, 1970) are of the opinion that data obtained with the plate method relate more closely to the number of metabolising cells. In soil of the Turew cropland the plate method suggested a population of 2 to 18×10^6/g of dry soil, while the direct method estimated 1.2 to 5.9×10^9 cells. After an appropriate conversion, these populations correspond to a biomass of about 0.2–1.8 g/m^2 from the plate count and 90–206 g/m^2 (with a mean of 160 g/m^2) from direct counts.

Table 31.1. *Content of ATP in microbial cells based on population estimates by the direct method* (after Kaszubiak *et al.*, 1975)

		1×10^{-10} μg ATP/cell		
		Average for vegetative season	Range of variation	
Crop	Field		from	to
Potato	A	0.99	0.47	1.73
	B	0.88	0.25	1.54
Rye	C	0.87	0.43	1.63
	D	1.02	0.51	1.65

Such a wide discrepancy in estimates of biomass imposes the necessity of verifying these values by other methods. Accordingly, we determined the ATP content of soil. Lee *et al.* (1971*b*) have shown that a mean of 1 to 4×10^{-10} μg of ATP occurs in each living microbial cell. This has been confirmed by investigations in aquatic environments, sediments and other materials (Holm-Hansen & Bootch, 1966; Holm-Hansen, 1970; Lee *et al.*, 1971*a*; Greaves *et al.*, 1973). Kaczmarek, Kaszubiak & Pędziwilk (1975) have found similar quantities of ATP in microbial cells in Turew soil (Table 31.1) if the number of micro-organisms in the soil is assumed to be that obtained by the direct method. However, if we assume a population of micro-organisms corresponding with that obtained by the plate method, the ATP content per cell would be almost a thousand times as great as that reported by other workers. It seems reasonable, therefore, to accept the higher population estimate.

The studies at Turew indicate the necessity for further research on the energy metabolism of micro-organisms in the soil environment. In particular, we must learn more about the proportion of the microbial community which is in a state of anabiosis and the conditions which determine that state. Likewise, we need to know more about the part which is metabolically active under field conditions.

Table 31.2. *Estimates of mean monthly microbial production during the vegetative period in g/m² based on direct counts converted on the basis that dry weight is 10 % of live weight* (H. Kaszubiak, personal communication)

Sample	Method of estimation (after) Mihaylova & Nikitina (1972)	Aristovskaya (1972)
1	740	1100
2	430	520
3	360	340
4	600	630
5	820	880
6	560	710
7	600	300
8	740	620
Mean	606	637

Aristovskaya (1972) assumes that a standing crop of micro-organisms turns over several times annually (or even monthly). In their resynthesis, products of metabolism and partially-decayed organic matter may be reutilised. In other words, organic compounds introduced into the soil with plant residues recycle within populations before being respired in the form of carbon dioxide. Mihaylova & Nikitina (1972) have used the concept of generation time to calculate productivity of micro-organisms in the soil, while Aristovskaya estimates microbial production on a daily fluctuating rhythm from differences in standing crops of micro-organisms measured daily. Both methods were used in the Turew investigations (Table 31.2, Fig. 31.3). By these techniques, estimates of monthly production are about 3 to 4 times the mean standing crop of micro-organisms. These values seem high, but are not beyond the range of probability. They indicate, however, that carbon recycles in biochemical processes in the soil several times each year. In laboratory conditions resynthesis of dead microbial cells (killed by heat) requires only a few days and the addition of glucose stimulates this process (Kaszubiak, Kaczmarek & Durska, 1975). However, it is not known to what extent under field conditions resynthesis depends on the flow of fresh organic matter. Nor do we know the significance of root exudates that maintain a high level of metabolising cells in the rhizosphere.

Activities

The composition of microbial communities acting as decomposers in cultivated soils differs from that of natural ecosystems. Although few investigations have been made to demonstrate these differences, indirect inferences suggest that soil fungi play a more important role in the transformation of

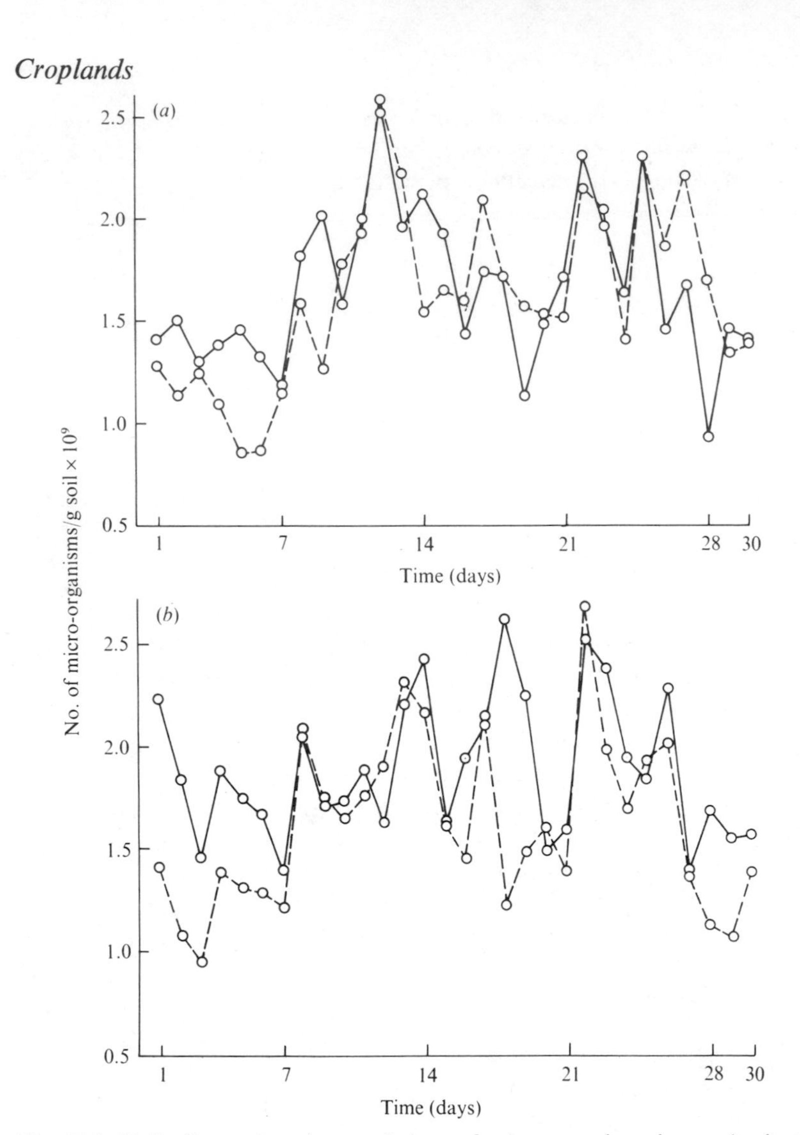

Fig. 31.3. Daily fluctuations in populations of micro-organisms in cropland at Turew. (*a*), potato field; (*b*), rye field. Each line represents one series of observations. (H. Kaszubiak, personal communication.)

organic matter in forest than in cultivated soils. For example, the number of fungi in the soil of the experimental field at Turew ranged from a few to over a hundred thousand per gram of dry soil; in soils of forests and grasslands, however, they are much more numerous. If one takes into account the greater length of hyphae in uncultivated soil, then still greater differences are obtained in the estimates of biomass (Parkinson, Gray & Williams, 1971).

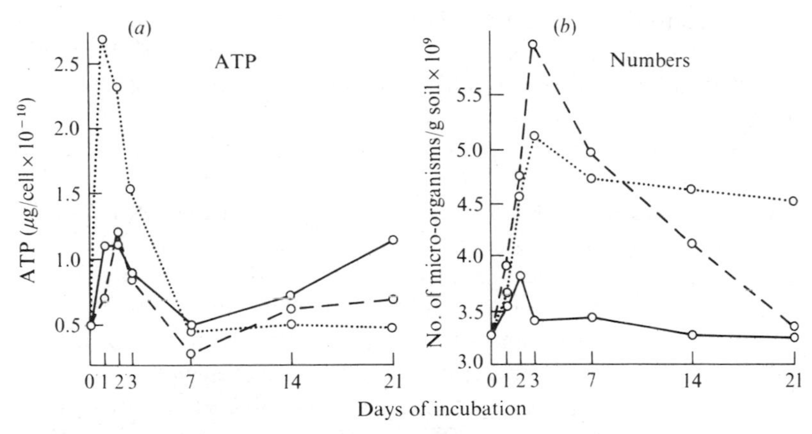

Fig. 31.4. Influence of organic matter on (*a*) ATP content and (*b*) number of micro-organisms in Turew cropland soil. Control, no additions, (——), sugar-beet leaves added (– – –), caseine added (····). After Kaszubiak *et al.*, 1975.

The rate of evolution of carbon dioxide is an index of metabolic activity and is apparently related to the level of ATP. Kaczmarek *et al.* (1975) compared the number of micro-organisms, ATP level and rate of carbon dioxide evolution, but could not demonstrate a correlation among these factors. This may be explained partly by the observation that, when soil was inoculated with an organic substance, ATP levels increased faster than did the number of micro-organisms (Fig. 31.4). Also, the rate of evolution of carbon dioxide may reach its maximum at times other than that of the level of ATP and of the number of micro-organisms. Measurements of this rate of carbon dioxide evolution from soil revealed values from five to seven times as great as estimates of heterotroph respiration obtained by carbon balance of the soil system (Gołębiowska, Margowski & Ryszkowski, 1974). This result is in accord with findings of Macfadyen (1971) and of Minderman & Vulto (1973), but the reason for this apparent discrepancy is unknown. The factor that is causing concern in respect to crop production is the apparent decline in availability of nutrients in the soil. In many instances there has been a decrease in the effectiveness of fertilisation with mineral nitrogen so that the crop yield resulting from a given annual level of application of nitrogen has not been maintained.

Laboratory experiments have shown that nitrogen added to the soil without fresh organic matter does not change the number of micro-organisms (Muszynska, 1974; Muszynska & Andrzejewski, 1974). Addition of mineral nitrogen together with easily-assimilated forms of carbon leads to differences in the numbers of proteolytic and nitrate-assimilating bacteria. The content of mineral nitrogen in soil after inoculation with easily-assimilated forms of carbon exceeds, in some instances, the amount of this form of nitrogen which was added to the soil. Perhaps, in these conditions, nitrogen is released from humus. The addition of plant residues with large C:N ratios into soil inhibited nitrification, whereas residues with high protein content favoured the formation of nitrates.

These results suggest that the same micro-organisms within the soil may assimilate different forms of nitrogen. Accordingly, nitrogen nutrition of several strains of bacteria isolated from cropland soil and the rhizosphere of rye plants was studied at Turew (Gołębiowska, Sawicka & Skalski, 1972). Most of the bacteria were able to assimilate various compounds of organic, ammonium or nitrate nitrogen, depending on the concentration of these various forms, the source of energy and the C:N ratio of the medium. This suggests that micro-organisms control the transformation of nitrogen in soil by changes in their community structure or by changes in their biochemical processes.

The issues discussed above are only some of those relating to the activity of micro-organisms in agro-ecosystems. The particular aim was to draw attention to the dynamic character of biological changes in the soil in which micro-organisms participate and to the necessity of presenting these dynamic processes in quantitative terms. In addition, the composition of populations of micro-organisms in various ecological conditions, as well as the character of interrelationships of microbial growth and the significance of rhizospheric populations, should become matters for further study in agro-ecosystems.

32. Ecosystem synthesis

L. RYSZKOWSKI

Cultivated fields are ecosystems maintained by man at an early stage of succession. They exhibit simple structure, few possibilities for modifying the effect of climatic factors, low energy costs in production of biomass, and have open cycles of mineral circulation. In these ecosystems the cost of maintaining stability is borne by man. Man influences practically all ecosystem processes.

One characteristic of a stable ecosystem is that inputs must balance outputs, so that the greatest proportion of material exchanges are internal. However, agro-ecosystems are managed to provide a substantial export of crop products. Thus, there is a need to provide additional inputs to the system to maintain productivity. However, many of these inputs exceed the capacity of the system and they become outputs or wastes added to other systems.

Recently a more holistic view of regulation and cycling in agricultural ecosystems has developed because of three new areas of concern:

(i) a recognition that the agricultural landscape serves as an unintended sink for many industrial contaminants which have a potential impact on agricultural productivity;

(ii) evidence that heavy applications of chemical fertilisers have led to upsets in the agricultural landscape, such as eutrophication of aquatic ecosystems and the development of hazardous levels of nitrate in water supplies; and

(iii) indications that a serious shortage in world food supplies is developing that will result in an effort to increase agricultural production to much higher levels.

The relative importance of these factors varies from region to region, but there appears to be agreement that much more fundamental research will be needed in relation to structure and function of agro-ecosystems on a total system basis. The limited information available permits only a few general comments on functional characteristics of these systems.

Energy flow

Different types of terrestrial ecosystems have different patterns of heat and water balance. The study of Rauner (1972) in the USSR (Table 32.1) is of particular interest in the present context because it compares the energy and water economy in a barley field with that in oak forest and steppe, all under similar conditions of climate and soils (chernozem). The forest intercepted the highest amount of radiant energy, while the cultivated field intercepted

Table 32.1. *Photosynthetic and water economy characteristics (on an annual basis) of three different ecosystems in the USSR* (after Rauner, 1972)

Parameter	Oak forest	Steppe	Barley field
Photosynthetically active radiation (PhAR) (kJ/m²)	1883000	1883000	1883000
Intercepted PhAR (kJ/m²)	1046000	753000	460000
$PhAR_{intercepted}:PhAR_{radiant}$	0.56	0.40	0.24
Precipitation (mm)	750	680	680
Transpiration (mm)	500	300	170
Transpiration:precipitation	0.67	0.44	0.25
Net production (kJ/m²)	26778	21757	18410
Net production: $PhAR_{intercepted}$ (× 100)	2.6	2.9	4.0
Net production: transpiration × L (× 100)	2.2	2.9	4.4

L, latent heat of water vaporisation (about 600 cal/g).

least. The amount of water transpired decreased in the same direction. However, efficiency of photosynthesis (ratio of net production to energy intercepted) was greatest in cropland and least in forest, as was the ratio of net production to the energetic cost of respiration. If these results can be confirmed in other situations, we may characterise cropland ecosystems as ones able to utilise intercepted solar energy and water more efficiently in production processes than uncultivated ecosystems.

The studies carried on at Turew concerning primary production of potato and rye fields have shown that the variability of net primary production is less than the variability of yield. This seems to indicate that agricultural technology influences yield more than production in the absence of deficits

Table 32.2. *Annual energy flow (in kJ/m²) in cropland at Turew* (after Ryszkowski, 1975)

Photosynthetically active radiation (PhAR)	1891000
Net primary production (Pp)	14088–29535
Pp:PhAR	0.7–1.6
Input of plant debris into soil	6046–19648
Respiration of herbivores	8–59
Total respiration of decomposers	8117–9121
Protozoa	?
Nematoda	125–167
Lumbricidae	20–25
Enchytraeidae	109–125
Other invertebrates	33
Micro-organisms	7824–8703

in water and plant nutrients. This is an important ecological principle that needs verification elsewhere.

The basic heterotrophic components of the Turew cropland are comprised of standing crops of 160 g/m^2 of micro-organisms, 5.7 g/m^2 of invertebrates and 0.2 g/m^2 of vertebrates. Thus, the ratio of microbial biomass to animal biomass approximates 27. If respiration activity of heterotrophs is estimated on carbon balance, the ratio of microbial respiration to animal respiration is about 20 (Table 32.2). Although direct respiration measurements (carbon dioxide evolution) were about five to seven times as high for microbial respiration, if that estimate is used in evaluation of total energy flow then microbial respiration is much higher than inputs of plant debris to the soil. This should be reflected in a rapid change in the content of organic matter into soil. Since the errors in the methods used for estimates of biomass and respiration are not known, it is impossible at this preliminary stage to rely much on the values obtained. Further research in this field is badly needed. Nevertheless, the results obtained raise several important questions, such as the need for more exact methods of estimating heterotrophic metabolic activity under field conditions and of estimating rates of turnover.

Nitrogen cycling

The studies at Turew have included measurements of the rate of leaching of nitrogen into the ground water. Changes in the amount of assimilable nitrogen (determined every 10 days using the Brauner–Keeney method) and of nitrate nitrogen in the arable layer do not correspond with the changes in concentration of these compounds in ground water (W. Loginow, personal communication). In soil the greatest concentrations of assimilable nitrogen are found in spring (0.63–0.83 mg/100 g soil). In summer (June–August) they are lower (0.54–0.58 mg), and in September they increase again (to about 0.65 mg). However, in ground water the highest concentrations (about 1 mg/l) are observed in summer. Nitrate nitrogen in the ground water decreases in concentration from over 12 mg/l in spring to about 7 to 8 mg in autumn. In the soil the amounts of nitrate nitrogen are more stable, with a maximum in June and a minimum in August and September. These divergences may be partly explained by intensive utilisation of nitrogen from the humus horizon of plants. The other explanation is that leaching nitrogen out of the humus horizon into ground water takes several months. Thus, the analysis of water extracts from soil have shown waves of nitrate nitrogen concentration within the soil profile (Fig. 32.1). The results obtained under field conditions can be supported by the observations of Vömel (1974) who has shown in lysimeters a delay in nitrate leaching amounting to six months, but the length of time depending on the soil. Z. Margowski and A. Barto-

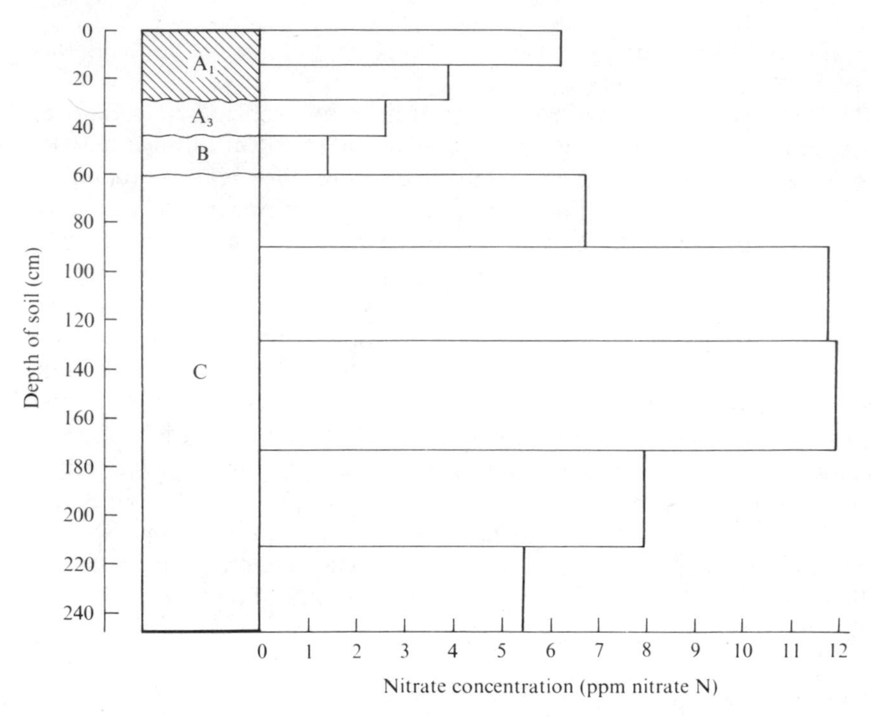

Fig. 32.1. Concentration of nitrates in water extracts of soil from various depths in Turew cropland. (Z. Margowski & A. Bartoszewicz, personal communication.)

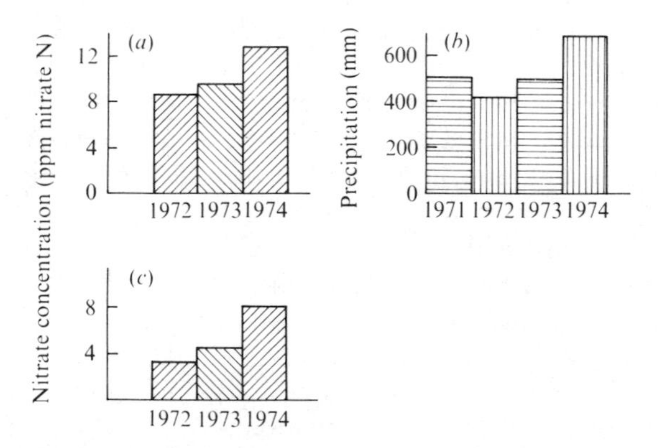

Fig. 32.2. Concentration of nitrates in ground water and in water of drainage canal in Turew cropland. (*a*), ground water; (*b*), drainage-canal water; (*c*), precipitation. (Z. Margowski & Bartoszewics, personal communication.)

Table 32.3. *A comparison of nitrogen concentration (in ppm) in ground water and drainage canals in cultivated fields at Turew and a forest 50 km from Turew on similar soil* (after Z. Margowski and A. Bartoszewicz, unpublished)

	Cultivated fields		Pine forest	
Forms of nitrogen	Ground water	Drainage canal	Ground water	Drainage canal
Ammonium	1.6	0.9	2.7	1.3
Nitrate	10.9	6.5	0.3	< 0.1
Organic	0.9	1.3	1.5	2.4
Total	13.4	8.7	4.5	3.7
Mineral (% of total)	93	85	66	35

szewicz (personal communication) have found a correlation between rainfall and concentration of nitrate nitrogen in ground water, as well as a correlation of rainfall with the amount of nitrates in the water of a drainage canal (Fig. 32.2). The concentration of nitrates in the water of a drainage canal was only about half that observed in the ground water in the field. The concentration of total nitrogen in both ground water and a drainage canal was higher in arable fields than in pine forest situated in soils similar to the Turew cropland (Table 32.3). In cropland, nitrate was the most abundant form of mineral nitrogen present, while in forest the ammonium form predominated.

Thus, the inputs of nitrogen fertilisers in cultivated fields result in leaching of mineral forms of nitrogen, indicating that inputs exceed the storage capacity of the system. These and other results indicate that agro-ecosystems have more open cycles of mineral circulation than do other terrestrial ecosystems.

Part VII. Conclusion

33. Conclusion

R. T. COUPLAND

The objective here is to make some generalisations regarding structure and production in various types of grasslands. It must be remembered that, because of differences in methodology and intensity of study, identical parameters were not measured in all sites. Furthermore, the degree of completeness of reporting varied considerably from project to project. However, the data summarised here are so voluminous and are from such a large number of grassland types that they should typify a large share of the world's grasslands. Since only gross comparisons are possible, equal weight is given to values from one year of study and those which are means obtained in investigations which extended over several years. The data summaries in this chapter include only the least disturbed treatment in each study. These stands were usually ungrazed or unmown (at least in the year of study prior to measurement) and not fertilised or irrigated. As elsewhere in this volume, all biomass values are in terms of dry matter unless otherwise indicated.

Producers

Above-ground biomass

Maximum green- (live-)shoot biomass attained in the various grasslands reported in this volume ranges from 76 to 2088 g/m^2 (Table 33.1). The lowest values are typical of semiarid temperate to tropical climates where the growth restriction imposed by low precipitation is accentuated by delay in spring growth (at high latitudes and elevations) and by early termination of the growing season by the onset of xeric conditions (at all latitudes). Greatest live-shoot biomass is generally associated with subhumid tropical climate, with maritime climate, or with temperate habitats receiving natural irrigation from run-off, under-ground water or surface inundation. Intermediate values for maximum green-shoot biomass are typical of semi-natural and natural temperate grasslands that do not receive water from sources other than precipitation. In the absence of natural irrigation maximum herbage biomass in arid, semiarid and dry subhumid climates increases with increasing precipitation (Fig. 33.1). However, there is a poor relationship between precipitation and maximum green-shoot biomass in more humid situations. Apparently maximum development of green canopy biomass can be expected in the tropics under less than maximum moisture conditions. The additional moisture provided in humid tropical climates (e.g. Ambikapur, India) is less favourable to herbaceous growth than is subhumid climate. Indeed two of the dry subhumid Indian sites (Ujjain and Kurukshetra) supported more

Table 33.1. *Canopy biomass (in g/m²) in various grasslands, listed in approximate increasing order of biomass of live shoots*

Location	Dominant taxa	Annual precipitation (mm)	Live-shoot biomass		Dead-shoot biomass			Maximum canopy biomass (live plus dead)	Ratio of mean-dead to mean-live biomass
			Maximum	Mean	At live maximum	Maximum	Mean		
Pilani, India	*Cenchrus biflorus*	391	76	55	10	27	10	86	0.18
Pawnee, USA	*Bouteloua gracilis, Buchloe dactyloides*	311	101	70	—	—	65	166[a]	0.93
ALE, USA	*Agropyron spicatum*	254	108	52	—	—	107	215[a]	2.06
Jornada, USA	*Bouteloua eriopoda*	228	125	49	—	—	33	158[a]	0.67
Matador, Canada	*Agropyron dasystachyum*	388	131	90	—	—	411	542[a]	4.57
Jodhpur, India	*Lasiurus indicus*	311	—	—	—	—	—	164	—
Baraba, Karachi, USSR	*Puccinellia tenuifolia, P. distans*	438	170	—	184	329	—	354[a]	—
Csévharaszt, Hungary	*Festuca vaginata, Koeleria glauca*	515	172	—	185	234	—	357[a]	—
Pantex, USA	*Bouteloua gracilis, Buchloe dactyloides*	533	176	70	—	—	129	305[a]	1.84
Baraba, Karachi, USSR	*Festuca pseudovina, Artemisia pontica*	438	180	—	216	290	—	396[a]	—
Cottonwood, USA	*Agropyron smithii*	384	188	93	—	—	141	329[a]	1.52
Bison, USA	*Festuca scabrella*	330	220	120	—	—	84	304[a]	0.70
Baraba, Karachi, USSR	*Phleum phleoides, Koeleria gracilis*	438	220	—	254	426	—	474[a]	—
Bridger, USA	*Phleum alpinum*	980	222	181	—	—	21	243[a]	0.12
Hays, USA	*Andropogon gerardi*	582	225	79	—	—	117	342[a]	1.48
Ambikapur, India	*Bothriochloa*	1379	—	—	—	—	—	423	—
Rajkot, India	*Cenchrus ciliaris*	242	228	95	95	430	195	432	2.05

Baraba, Karachi, USSR	*Calamagrostis epigeios, Galatella biflora*	438	240	—	349	490	—	589[a]	—
Lanžhot, Czechoslovakia	*Festuca sulcata*	550	252	—	112	164	—	364[a]	—
Dickinson, USA	*Stipa comata*	398	270	178	—	—	369	639[a]	2.07
Skowronno, Poland	*Brachypodium pinnatum*	730	282	—	244	378	—	526[a]	—
Nanashigure, Japan	*Zoysia*	—	284	—	84	110	—	368[a]	—
Osage, USA	*Andropogon gerardi*	930	287	152	—	—	446	733[a]	2.93
Baraba, Karachi, USSR	*Calamagrostis neglecta, Carex gracilis*	438	290	—	534	650	—	824[a]	—
Terschelling, Netherlands	*Juncus gerardii*	760	302	—	123	190	—	425[a]	—
Solling, West Germany	*Festuca rubra*	1100	316	—	8	148	—	324[a]	—
Ojców, Poland	*Alopecurus pratensis*	654	332	—	208	398	—	540[a]	—
Sagar, India	*Dichanthium annulatum*	1410	337	—	—	—	—	—	—
Armidale, Australia	*Phalaris aquatica, Trifolium repens*	859	346	—	—	—	260	645	—
Ratlam, India	*Sehima nervosum*	1257	363	92	105	316	211	468	2.29
Ispina, Poland	*Arrhenatherum elatius*	708	383	—	—	—	—	—	—
Ispina, Poland	*Carex vesicaria*	708	389	—	155	203	—	544[a]	—
Ruwenzori Park, Uganda	*Hyparrhenia filipendula–Themeda triandra*	900	405	—	—	364	—	—	—
Terschelling, Netherlands	*Puccinellia maritima*	760	407	—	88	206	—	495[a]	—
Lanžhot, Czechoslovakia	*Alopecurus pratensis*	550	413	—	93	124	—	506[a]	—
Ojców, Poland	*Brachypodium pinnatum*	654	420	—	512	700	—	932[a]	—
Delhi, India	*Heteropogon contortus*	800	—	—	—	—	—	771	—
Kazuń, Poland	*Festuca pratensis, Holcus lanatus*	613	426	—	62	—	—	488	—
Kazuń, Poland	*Dactylis glomerata, Festuca pratensis*	613	435	—	91	586	—	586[a]	—
Terschelling, Netherlands	*Plantago maritima*	760	438	—	167	314	—	605	—
San Joaquin, USA	*Bromus mollis*	559	440	—	—	—	—	440	—
Ujjain, India	*Dichanthium annulatum*	1030	457	159	127	422	164	584	1.03

Table 33.1 (*cont.*)

Location	Dominant taxa	Annual precipitation (mm)	Live-shoot biomass		Dead-shoot biomass			Maximum canopy biomass (live plus dead)	Ratio of mean-dead to mean-live biomass
			Maximum	Mean	At live maximum	Maximum	Mean		
Lanžhot, Czechoslovakia	*Phalaris arundinacea*	550	469	—	90	186	—	559[a]	—
Kampinos, Poland	*Carex fusca*, *Deschampsia caespitosa*, moss	613	476	—	241	394	—	717	—
Ispina, Poland	*Equisetum limosum*	708	562	—	244	416	—	806[a]	—
Sagar, India	*Heteropogon contortus*	1410	572	203	399	518	338	1006	1.67
Jhansi, India	*Sehima nervosum*, *Heteropogon contortus*	936	—	—	—	—	—	1408	—
Migda, Israel	*Brachypodium*, *Elymus*, *Hordeum*	—	—	—	—	—	—	629	—
Kawatabi, Japan	*Miscanthus sinensis*	2335	679	—	49[a]	100[a]	—	728[a]	—
Lamto, Ivory Coast	Palm savanna (*Loudetia simplex*)	1158	690	—	—	—	—	—	—
Tonomine, Japan	*Miscanthus*, *Pleioblastus*	—	699	—	—	—	—	—	—
Welgevonden, South Africa	*Panicum*, *Digitaria*, *Eragrostis*	436	753	—	—	421	—	—	—
Ujszentmargita, Hungary	*Artemisia maritima*, *Festuca pseudovina*	527	787	—	14	121	—	801[a]	—
Lanžhot, Czechoslovakia	*Glyceria maxima*	550	948	—	183	639	—	1131[a]	—
Aso-Kuju, Japan	*Miscanthus*, *Pleioblastus*	—	1306	—	—	—	—	—	—
Kurukshetra, India	*Panicum miliare*	790	1974	827	142	1268	506	2216	0.61
Panama, Panama	*Hyparrhenia rufa*	—	2088	—	762	2905	—	3406	—
Varanasi, India	*Desmostachya bipinnata*	1134	—	—	—	—	—	2360	—
Varanasi, India	*Eragrostis nutans*	1134	—	—	—	—	—	3296	—

[a] Estimated or transposed values.

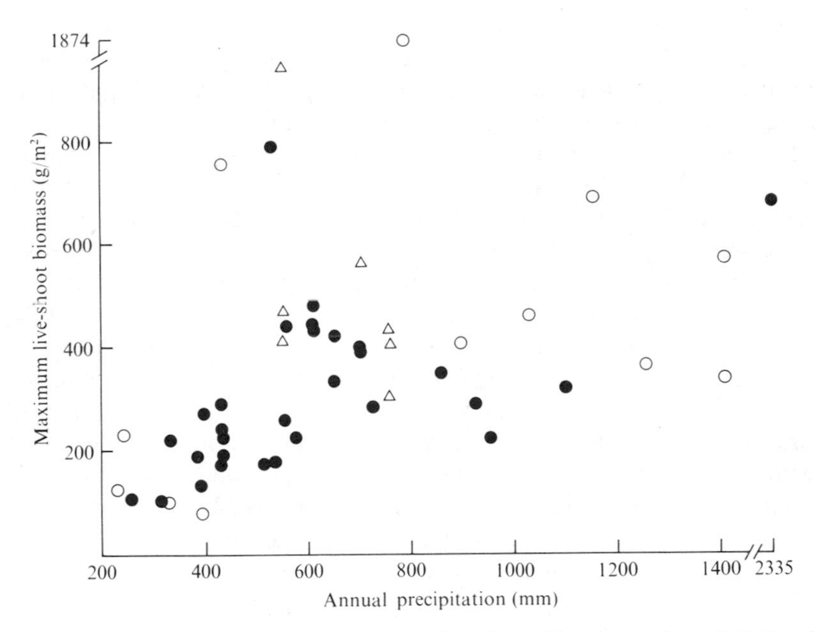

Fig. 33.1. Relationship between maximum live-shoot biomass and precipitation in the grasslands listed in Table 33.1. Symbols indicate the type of grassland (○, tropical; ●, temperate) and moisture supply above normal for the region because of inundation, surface flooding or from under-ground water (△).

green biomass than several moist subhumid sites. Shrubs and trees apparently reduce herbage production of the grass layer in more humid areas. In some sites within regions of favourable climate, growth is restricted by the physical (e.g. texture) or chemical (saline) condition of soils.

The information available in this volume is not sufficient to estimate how these sites would rank under the same degree of protection from human influence. Undoubtedly past heavy grazing by livestock is an important depressant to maximum canopy development in many sites. Also native herbivores are apparently not active in foliage removal in proportion to vegetation stature and density. Some herbivores (e.g. grasshoppers) are apparently more abundant in situations where canopy biomass is reduced by grazing.

Mean green-shoot biomass in the growing season ranged from about 50 g/m² in the most arid grasslands to a high of 827 g/m² in the dry subhumid climate of Kurukshetra, India, where a green component exists in the canopy throughout the year.

The maximum values for dead-shoot biomass presented in Table 33.1 are probably much less comparable among sites than maximum values for green-shoot biomass, since clipping schedules were planned to embrace the latter. Dead-shoot biomass values at time of maximum green standing crop are useful mainly in estimating maximum canopy biomass (where this has not been recorded). The most useful value for comparing dead-shoot biomass among sites is the time-weighted mean, but it must be emphasised that values are averaged only for the growing season. Reported mean values range from 10 to 506 g/m², with no apparent trend being evident in relation to habitat. Consideration should be given in future studies to estimating the mean standing crop of dead shoots throughout the year, as a basis for comparison of this canopy compartment among sites.

A comparison, among sites, of maximum canopy biomass (green plus dead shoots) is useful because this parameter can be used in estimating turnover rates. This value, not reported for most sites, has been estimated here (Table 33.1) either as the sum of maximum green biomass and dead biomass at the time of maximum green biomass or as the sum of the maximum green biomass and mean dead biomass. These values range from 86 to 3406 g/m². Maximum total canopy biomass ranges widely in tropical grasslands, averaging about 130 g/m² in semiarid regions, between 400 and 450 g/m² in humid regions, and about 1300 g/m² in subhumid climate. In temperate regions the average maximum total canopy biomass is in the 350 to 400 g/m² range in natural grasslands and 550 to 600 g/m² in semi-natural types.

The relative proportions of green shoots and dead shoots in the canopy of various grasslands can be compared by the ratio of mean dead biomass to mean green biomass. This ratio ranges from as low as 0.12 to a high of 4.57, with a median of 1.5 (Table 33.1). No trends are apparent. Important factors, presumably, that affect the relative amount of dead material to green material in the canopy are current weather conditions, the intensity of grazing by livestock prior to protection for study, and the rate of turnover of dead shoots, as well as the length of the past period of protection from mowing and fire.

From the reports in this volume it appears that the materials from the canopy that fall to the litter layer on the soil surface were not studied extensively except in the North American grasslands. There the mean biomass of litter in various sites ranged from about 70 to 900 g/m², and on average was slightly greater than total canopy biomass. The amount of litter present is affected by the same factors as determine the amount of dead canopy material, but the differing rate of decomposition in various habitats probably has more effect in this layer. Rate of decomposition of litter is influenced by the same factors that affect plant growth.

Under-ground biomass

Mean under-ground biomass values for the various grasslands reported in this volume range from 45 to 4707 g/m² (Table 33.2). Mean under-ground biomass in most tropical grasslands is less than 1000 g/m² (median of about

Table 33.2. *Under-ground plant biomass (in g/m²) in various grasslands, listed in approximate order of increasing biomass; approximate ratios of under-ground plant biomass to maximum canopy biomass are also given*

Location	Dominant plant taxa	Under-ground plant biomass[a]	Ratio of under-ground to canopy biomass
Pilani, India	*Cenchrus biflorus*	45	0.5
Jornada, USA	*Bouteloua eriopoda*	187	1.2
Jhansi, India	*Sehima nervosum, Heteropogon contortus*	227	0.2
Welgevonden, South Africa	*Panicum, Digitaria, Eragrostis*	297[b]	—
Jodhpur, India	*Lasiurus indicus*	428	2.6
Varanasi, India	*Desmostachya*	493	0.2
Tvärminne, Finland	*Helictotrichon pubescens*	558	—
San Joaquin, USA	*Bromus mollis*	637	1.4
Ratlam, India	*Sehima nervosum*	698	1.5
Ujjain, India	*Dichanthium annulatum*	717	1.2
Nanashigure, Japan	*Zoysia*	759	—
Kurukshetra, India	*Panicum miliare*	800	0.4
Varanasi, India	*Eragrostis nutans*	806	0.2
Kampinos, Poland	*Carex fusca, Deschampsia caespitosa,* moss	823	1.1
Armidale, Australia	*Phalaris aquatica, Trifolium repens*	836	1.3
Csévharaszt, Hungary	*Festuca vaginata, Koeleria glauca*	863	2.4
Solling, West Germany	*Festuca rubra*	946	2.9
Sagar, India	*Heteropogon contortus*	1004	1.0
Pantex, USA	*Bouteloua gracilis, Buchloe dactyloides*	1033	3.4
Kawatabi, Japan	*Miscanthus sinensis*	1196	1.6
Lanžhot, Czechoslovakia	*Alopecurus pratensis*	1221	2.4
Ruwenzori Park, Uganda	*Hyparrhenia filipendula, Themeda*	1260[b]	—
Hays, USA	*Andropogon gerardi*	1319	3.9
Osage, USA	*Andropogon gerardi*	1435	2.0
Baraba, Karachi, USSR	*Calamagrostis neglecta, Carex gracilis*	1436	1.7
Tonomine, Japan	*Miscanthus*	1510	—
Lanžhot, Czechoslovakia	*Festuca sulcata*	1556	4.3
Baraba, Karachi, USSR	*Puccinellia tenuifolia, P. distans*	1580	4.5
Ujszentmargita, Hungary	*Artemisia maritima, Festuca pseudovina*	1598	2.0
Moscow, USSR	*Phleum phleoides, Festuca rubra*	1610	3.6
Ispina, Poland	*Arrhenatherum*	1655	—
Ojców, Poland	*Alopecurus pratensis*	1664	3.1
Lanžhot, Czechoslovakia	*Phalaris arundinacea*	1664	3.0
Dickinson, USA	*Stipa comata*	1715	2.7
Pawnee, USA	*Bouteloua gracilis, Buchloe dactyloides*	1716	10.3
Terschelling, Netherlands	*Juncus gerardii*	1813	4.3
Leningrad, USSR	*Agrostis tenuis*	1820	3.3
Bridger, USA	*Phleum alpinum*	1928	7.9
Cottonwood, USA	*Agropyron smithii*	1963	6.0

Table 33.2 (*cont.*)

Location	Dominant plant taxa	Under-ground plant biomass[a]	Ratio of under-ground to canopy biomass
Lamto, Ivory Coast	Palm savanna (*Loudetia simplex*)	2100[b]	—
Aso-Kuju, Japan	*Miscanthus, Pleioblastus*	2195	—
Tambovskaya, USSR	*Bromus riparia, Poa angustifolia*	2200	6.7
Baraba, Karachi, USSR	*Phleum phleoides, Koeleria gracilis*	2350	5.0
Terschelling, Netherlands	*Plantago maritima*	2522	4.2
Baraba, Karachi, USSR	*Calamagrostis epigeios, Galatella biflora*	2610	4.4
Lanžhot, Czechoslovakia	*Glyceria maxima*	2640	2.3
Baraba, Karachi, USSR	*Festuca pseudovina, Artemisia pontica*	2740	6.9
Matador, Canada	*Agropyron dasystachyum*	2763	5.1
Ojców, Poland	*Brachypodium pinnatum*	3212	3.4
Ispina, Poland	*Equisetum limosum*	4707	5.8

[a] In most instances these are time-weighted mean values.
[b] Estimated as mean of maximum and minimum values mentioned in text.

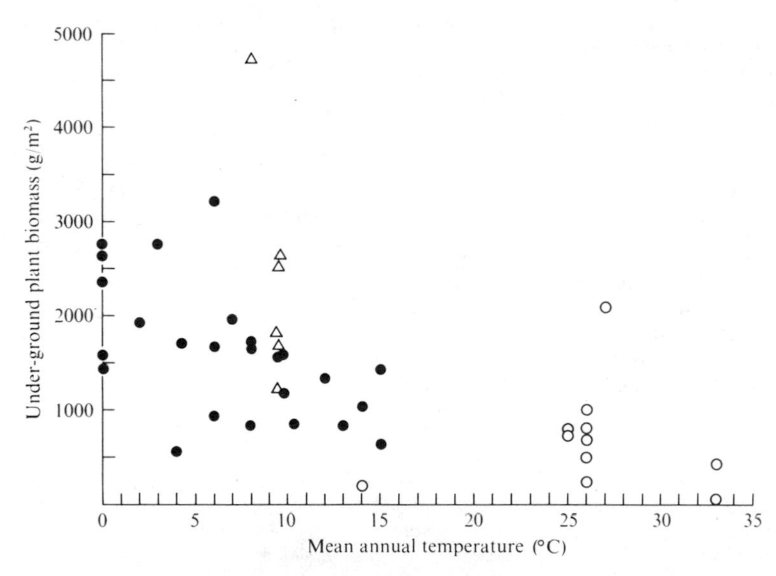

Fig. 33.2. Relationship between under-ground plant biomass and mean annual temperature in the grasslands listed in Table 33.2. Symbols indicate the type of grassland (○, tropical; ●, temperate) and moisture supply above normal for the region because of inundation, surface flooding or from under-ground water (△).

700 g/m^2). The lowest tropical values are from semiarid regions, while maximum values occur in subhumid climate. Only a few of the temperate grasslands are below 1000 g/m^2, with median biomass in the 1600 to 1700 g/m^2 range in both natural temperate and semi-natural temperate sites. The higher values in temperate than in tropical situations are probably due in part to a slower rate of turnover that is associated with lower temperature. Under-ground biomass is inversely related to temperature (Fig. 33.2).

Comparability of under-ground plant biomass among sites depends on either constancy in depth of sampling or constancy of sampling to the maximum depth of rooting. Neither of these criteria is met by the data in Table 33.2. In several reports in this volume authors have failed to note either the absolute depth of sampling or its relation to depth of rooting. In other instances estimates have been made of the proportion of roots removed, but the data in Table 33.2 do not take the unsampled layers into account. There is general agreement that the time of season when maximum biomass is reached is variable between years and is influenced by past grazing pressure.

Ratio of mean under-ground plant biomass to maximum canopy biomass ranges among sites from 0.2 to 10.3 (Table 33.2). This ratio is much smaller in the tropical grasslands (0.2–2.6; mean 0.8) than in both the semi-natural temperate sites (1.1–6.9; mean 3.3) and the natural temperate grasslands (1.4–10.3; mean 4.4). Although the lowest values in the tropics are from subhumid areas and the highest from arid climate, there the data available give no constant relationship to moisture supply; the highest values occur under moderate grazing. In the temperate zone the wet meadows (at Lanžhot and Baraba, Karachi) have lower ratios than the driest sites in the same study areas. In the above analysis total canopy biomass is used because it seems to be a more appropriate basis of comparison with under-ground plant parts, which also include both dead and live structures. However, when ratios were calculated using green-shoot biomass in the denominator, similar relationships appeared.

Productivity

Estimates of total annual net primary production of biomass range from 239 g/m^2 in semiarid tropical grassland to 4557 g/m^2 in subhumid tropical areas (Table 33.3). The estimates in the natural and semi-natural temperate grasslands are in a narrower range, from 702 to 3470 g/m^2. The values from temperate cropland (Poland) are also in this range, being 1329 g/m^2 for rye and 1697 g/m^2 for potato, while arable grassland at Armidale (Australia) produced 2164 g/m^2.

According to reports in this volume, annual above-ground biomass production in the tropics ranges from 82 to 3396 g/m^2, while in natural and semi-

Table 33.3. *Estimates of net annual primary production of biomass (in g/m²) in various grasslands and croplands, listed in approximate order of increasing production*

		Production			Percentage	
Location	Dominant plant taxa	Shoots	Underground	Total	Aboveground	Underground
Fété Olé, Senegal	*Aristida*	82	—	—	—	—
Pilani, India	*Cenchrus biflorus*	84	155	239	35	65
ALE, USA	*Agropyron spicatum*	98	—	—	—	—
Jornada, USA	*Bouteloua eriopoda*	148	147	295	50	50
Jodhpur, India	*Lasiurus indicus*	164	—	—	—	—
Ujszentmargita, Hungary	*Festuca pseudovina*	253	449	702	36	64
Bridger, USA	*Phleum alpinum*	249	471	720	35	65
Pawnee, USA	*Bouteloua gracilis, Buchloe dactyloides*	172	568	740	23	77
Udaipur, India	*Dichanthium annulatum*	184	—	—	—	—
Rajkot, India	*Cenchrus ciliaris*	244	—	—	—	—
Cottonwood, USA	*Agropyron smithii*	249	533	782	32	68
Fété Olé, Senegal	*Panicum, Chloris*	256	—	—	—	—
Bison, USA	*Festuca scabrella*	272	—	—	—	—
Novosibirsk, USSR	*Calamagrostis epigeios, Poa pratensis*	530	270	800	66	34
Solling, West Germany	*Festuca rubra*	316	496	812	39	61
Ratlam, India	*Sehima nervosum*	433	399	832	52	48
Osage, USA	*Andropogon gerardi*	346	542	888	39	61
Pantex, USA	*Bouteloua gracilis, Buchloe dactyloides*	257	634	891	29	71
San Joaquin, USA	*Bromus mollis*	441	464	905	49	51
Terschelling, Netherlands	*Juncus gerardii*	406	578	984	41	59
Ujjain, India	*Dichanthium annulatum*	520	464	984	53	47
Ambikapur, India	*Bothriochloa pertusa*	436	563	999	44	56
Fété Olé, Senegal	*Panicum, Zornia*	476	—	—	—	—
Lanžhot, Czechoslovakia	*Festuca sulcata*	706	389	1095	64	36
Matador, Canada	*Agropyron dasystachyum*	495	661	1156	43	57
Lamto, Ivory Coast	Palm savanna (*Loudetia simplex*)	498	—	—	—	—

Terschelling, Netherlands	*Puccinellia maritima*	507	760	1267	40	60
Dickinson, USA	*Stipa comata*	351	932	1283	27	73
Kazuń, Poland		574	—	—	—	—
Tambovskaya, USSR	*Bromus riparia, Poa angustifolia*	590	730	1320	45	55
Migda, Israel	*Brachypodium, Elymus, Hordeum*	629	—	—	—	—
Welgevonden, South Africa	*Panicum, Digitaria, Eragrostis*	710	—	—	—	—
Delhi, India	*Heteropogon contortus*	798	—	—	—	—
Turew, Poland	*Secale cereale*	1072	257	1329	81	19
Moscow, USSR	*Phleum phleoides, Festuca rubra*	540	800	1340	40	60
Novosibirsk, USSR	*Puccinellia distans*	330	1080	1410	23	77
Hays, USA	*Andropogon gerardi*	363	1062	1425	25	75
Tambovskaya, USSR	*Festuca sulcata, Artemisia monogyna*	360	1100	1460	25	75
Lanžhot, Czechoslovakia	*Alopecurus pratensis*	1070	407	1477	72	28
Jhansi, India	*Sehima nervosum, Heteropogon contortus*	1019	497	1516	67	33
Leningrad, USSR	*Agrostis tenuis*	630	960	1590	40	60
Terschelling, Netherlands	*Plantago maritima*	559	1092	1651	34	66
Turew, Poland	*Solanum tuberosum*	985	712	1697	58	42
Kawatabi, Japan	*Miscanthus sinensis*	—	—	1778	—	—
Novosibirsk, USSR	*Festuca pseudovina, Poa angustifolia*	380	1400	1780	21	79
Sagar, India	*Heteropogon contortus*	914	937	1851	49	51
Lanžhot, Czechoslovakia	*Phalaris arundinacea*	1597	554	2151	74	26
Armidale, Australia	*Phalaris aquatica, Trifolium repens*	940	1224	2164	43	57
Ruwenzori Park, Uganda	*Hyparrhenia filipendula, Themeda triandra*	701	1495	2196	32	68
Varanasi, India	*Dichanthium annulatum*	—	—	2200	—	—
Novosibirsk, USSR	*Calamagrostis epigeios, Poa angustifolia*	420	2680	3100	14	86
Lanžhot, Czechoslovakia	*Glyceria maxima*	2430	880	3310	73	27
Novosibirsk, USSR	*Festuca pseudovina, Artemisia pontica*	270	3200	3470	8	92
Kurukshetra, India	*Panicum miliare*	2407	1131	3538	68	32
Varanasi, India	*Desmostachya bipinnata*	2218	1377	3595	62	38
Varanasi, India	*Heteropogon contortus*	—	—	4200	—	—
Varanasi, India	*Eragrostis nutans*	3396	1161	4557	75	25

natural temperate grasslands it ranges between 98 and 2430 g/m^2. Annual Mediterranean-type grasslands in California and Israel produced 441 and 629 g/m^2 of shoots, respectively; the arable grassland at Armidale (Australia) produced 940 g/m^2 of forage and the Polish cropland 985 (potato) to 1072 (rye) g/m^2 of above-ground parts. The estimated proportion of production that accumulates above ground exceeds 50 % only in the tropical grasslands (particularly the highly productive ones), in moist areas, and in cropland.

Most estimates of primary production reported in this volume are based on very conservative means of analysis. Most of them are the summation of increments, usually on a different replicate of plots for each time period. Above-ground, these increments yield values greater than the maximum standing crop of green shoots for one or two reasons: (i) two or more peaks and troughs in green biomass may occur within the year due to seasonality of climate or different phasing of growth in different dominant species; and (ii) the increments in some instances were determined for different species or groups of species with different seasonality of development. Under-ground biomass values fluctuate considerably, so that the sum of the combined increments in each layer results in a still higher estimate of under-ground production. The estimates obtained from incremental analysis increase with frequency of sampling, which varied greatly in the sites reported in Table 33.3.

Most of the estimates of production fail to account for losses in biomass that occur concurrently with growth. Such losses occur due to the activities of heterotrophs, both above ground and under ground, but as well losses from shoot biomass occur as leaves die and fall to the ground. Estimates of shoot production are probably more reliable than those of under-ground production because of greater variability of biomass under ground and usually lower frequency of sampling.

Hence, the various estimates presented in Table 33.3 represent in varying degree the actual level of annual net primary production. We do not know to what extent each study has achieved the IBP goal of estimating biological productivity. An indication of the degree of adequacy of the estimates of shoot production may be suggested in Table 33.4, where they are compared with maximum standing crop of green shoot biomass. Annual shoot production estimates range from 0.7 to 3.8 times maximum live-shoot biomass. The values below unity presumably reflect a component of woody tissue of shrubs in the 'live' shoot biomass. The vegetation of the site at unity (San Joaquin) is dominated by annual grasses. Estimated production exceeds twice maximum green-shoot biomass only in two sites. The highest ratio (3.8 times maximum green-shoot biomass) was obtained where an attempt was made to account for all losses from the canopy during growth, as indicated in Table 33.5.

In a balanced system, the rate of turnover is determined by the relationship

Table 33.4. *Comparison between maximum standing crop of live shoots and net annual above-ground primary production of biomass in various sites (in g/m²)*

Location	Maximum live-shoot biomass (MSC)	Net annual shoot production (P)	Ratio of P to MSC
Lamto, Ivory Coast	690	498	0.72
ALE, USA	108	98	0.91
Welgevonden, South Africa	753	710	0.94
San Joaquin, USA	440	441	1.00
Rajkot, India	228	244	1.07
Pilani, India	76	84	1.11
Bridger, USA	222	249	1.12
Ujjain, India	457	520	1.14
Leningrad, USSR	550	630	1.14
Jornada, USA	125	148	1.18
Ratlam, India	363	433	1.19
Moscow, USSR	450	540	1.20
Osage, USA	287	346	1.21
Kurukshetra, India	1974	2407	1.22
Bison, USA	220	272	1.24
Terschelling, Netherlands (Puccinellia)	406	507	1.25
Terschelling, Netherlands (Plantago)	438	559	1.28
Dickinson, USA	270	351	1.30
Cottonwood, USA	188	249	1.32
Terschelling, Netherlands (Juncus)	302	406	1.34
Solling, West Germany	224	316	1.41
Pantex, USA	176	257	1.46
Novosibirsk, USSR (Festuca–Artemisia)	180	270	1.50
Novosibirsk, USSR (Puccinellia)	220	330	1.50
Sagar, India	572	914	1.60
Novosibirsk, USSR (Calamagrostis)	330	530	1.61
Hays, USA	225	363	1.61
Novosibirsk, USSR (Festuca)	230	380	1.65
Kawatabi, Japan	546	925	1.69
Pawnee, USA	101	172	1.70
Ruwenzori Park, Uganda	405	701	1.73
Novosibirsk, USSR (Calamagrostis)	240	420	1.75
Tambovskaya, USSR (Bromus–Poa)	330	590	1.79
Tambovskaya, USSR (Festuca–Artemisia)	200	360	1.80
Lanžhot, Czechoslovakia (Glyceria)	948	2430	2.56
Lanžhot, Czechoslovakia (Alopecurus)	413	1070	2.59
Armidale, Australia	346	940	2.72
Lanžhot, Czechoslovakia (Festuca)	252	706	2.80
Lanžhot, Czechoslovakia (Phalaris)	469	1597	3.41
Matador, Canada	131	495	3.78

between standing crop and rate of production. Thus the mean turnover time of above-ground parts in each site is the reciprocal of the ratio of P:MSC given in Table 33.4. This turnover time ranges from 0.26 year at Matador

Table 33.5. *Estimates of annual shoot production (in g/m²) obtained at Matador, Canada, by applying various interpretative techniques*

Method of estimation	Production	% increase over MSC
(1) Maximum green-shoot biomass	131	—
(2) Summation of individual peak biomass values for:		
Graminoids and two forb groups	135	3
Four to five compartments of graminoids and two forb groups	141	8
(3) Summation of significant increases in total green biomass	131	0
(4) As in (3) plus significant increases in biomass that could not be accounted for by losses from green	257	96
(5) As in (4) plus increases in litter biomass that could not be accounted for by losses from the canopy	294	124
(6) As in (5) plus losses in weight of leaves on death	469	258
(7) As in (6) plus estimated consumption by herbivores and wastage in grasshopper grazing (green to litter)	495	278

MSC, maximum green-shoot biomass.

to 1.4 years at Lamto. No environmental trends are easily discernible, probably because of variability in degree of use by livestock and in the methods of estimating shoot production. For under-ground parts, calculated values for turnover time are given in Table 33.6. They range from 0.36 to 4.18 yr, with the values being smaller in tropical and subtropical than in temperate regions.

Autotrophic processes and productivity in higher plants are treated in detail by Singh *et al.* (Chapter 2, in Breymeyer & Van Dyne, 1979).

Consumers

Integration of information concerning consumers in the various kinds of grasslands is very much more difficult than with producers for several reasons. Studies of consumers have been made in a relatively few sites. Most studies have not considered all of the consumer groups that are present; most frequently invertebrate populations only have been studied. Few investigations have involved more than an estimate of population density. Accordingly, few generalisations concerning consumers in grasslands can be expected to emerge in this summary beyond those that have appeared elsewhere in this volume.

It should be emphasised that contributions of IBP studies towards a quantification of consumer activities in grassland ecosystems have been much more significant than this comparative summary indicates. Many very important data resulting from individual studies have not been discussed by the authors of the various consumer chapters because of lack of comparative

Table 33.6. *Comparison between mean standing crop and annual production of under-ground plant parts (in g/m² of biomass) in various grasslands*

Location	Standing crop	Annual production	Turnover time (yr)
Varanasi, India (Desmostachya)	493	1377	0.36
Jhansi, India	227	497	0.46
Armidale, Australia	836	1224	0.68
Varanasi, India (Eragrostis)	806	1161	0.69
Kurukshetra, India	800	1131	0.71
Pilani, India	45	61	0.74
Novosibirsk, USSR (Festuca–Agropyron)	2760	3200	0.86
Novosibirsk, USSR (Calamagrostis)	2680	2300	0.87
Sagar, India	1004	937	1.07
Hays, USA	1319	1062	1.24
Jornada, USA	187	147	1.27
Kawatabi, Japan	1188	876	1.36
San Joaquin, USA	637	464	1.37
Novosibirsk, USSR (Festuca–Poa)	2170	1400	1.55
Ujjain, India	717	464	1.55
Novosibirsk, USSR (Puccinellia)	1730	1080	1.60
Pantex, USA	1033	634	1.63
Ratlam, India	608	399	1.75
Dickinson, USA	1715	932	1.84
Leningrad, USSR	1820	960	1.90
Solling, West Germany	946	496	1.91
Moscow, USSR	1610	800	2.01
Terschelling, Netherlands (Plantago)	2522	1092	2.31
Osage, USA	1435	542	2.65
Novosibirsk, USSR (Calamagrostis–Poa)	800	270	2.98
Tambovskaya, USSR (Festuca)	3300	1100	3.00
Lanžhot, Czechoslovakia (Phalaris)	1664	554	3.00
Lanžhot, Czechoslovakia (Alopecurus)	1221	407	3.00
Lanžhot, Czechoslovakia (Glyceria)	2640	880	3.00
Tambovskaya, USSR (Bromus)	2200	730	3.01
Pawnee, USA	1716	568	3.02
Terschelling, Netherlands (Juncus)	1810	578	3.13
Cottonwood, USA	1963	533	3.68
Lanžhot, Czechoslovakia (Festuca)	1556	389	4.00
Bridger, USA	1928	471	4.09
Matador, Canada	2763	661	4.18

studies elsewhere. This summary, likewise, ignores many important contributions that are discussed concerning only one type of grassland.

In-depth considerations of structure and function of various consumer subsystems have been presented in the companion volume on grasslands (Breymeyer & Van Dyne, 1979). Small herbivores have been treated by Andrzejewska & Gyllenberg (Chapter 3), large herbivores by Van Dyne *et al.* (Chapter 4), invertebrate predators by Kajak (Chapter 5), and vertebrate predators by Harris & Bowman (Chapter 6).

Populations

Orders of invertebrates that are the major contributors to above-ground populations are Diptera, Collembola, Coleoptera, Homoptera, Heteroptera and Hymenoptera (Tables 12.1, 30.2, this volume). Total numbers have been reported in the volume only for cropland in Poland, where they range from about 200 to 650 individuals/m^2 (Table 30.2, this volume).

Censusing of under-ground invertebrate populations is difficult because of the wide range in size of individuals. Care must be taken, in comparing sites, to recognise that the sampling methods used determine the relative proportions contributed to the population by small organisms and large organisms. Where small invertebrates have been sampled adequately the most numerous invertebrate groups are Nematoda, Collembola, Acarina and Enchytraeidae (belonging to Oligochaeta) (Tables 12.3, 18.2, 24.2, this volume). These minute organisms number in the hundreds of thousands/m^2 to millions/m^2 in perennial grassland (Table 7.3, this volume); however, the numbers in cropland in Poland are reported to be much lower (Table 30.5, this volume). The larger invertebrate soil organisms number in the hundreds/m^2 (Tables 12.8, 12.9, this volume). A large variety of groups are represented, with Lumbricidae, Scarabaeidae, Elateridae and Carabidae being common. The number of species of soil invertebrates present in any particular site is probably in the hundreds (Table 30.1, this volume).

While disturbance reduces populations and numbers of species of some groups of soil invertebrates (Tables 30.1, 30.5, this volume), there is some evidence of population increase in other groups (Table 24.1, this volume).

Biomass

Biomass values for above-ground invertebrates are evidently much lower than for soil invertebrates. Above-ground biomass is probably well below 1 g/m^2 in most instances (Tables 8.1, 8.2, 30.3, 30.6, this volume). However, the reported biomass values for spiders and ants in Polish meadows suggest that total biomass of all invertebrates in the canopy and on the soil surface may approach or exceed 1 g/m^2 in some instances (Table 12.2, this volume). In terms of biomass, the larger organisms, such as species of Orthoptera, contribute much more significantly than their relatively low numbers would suggest.

Biomass values for soil invertebrates are in the 1 g/m^2 to more than 50 g/m^2 range (Tables 8.1, 8.2, 12.8, 12.9, 24.2, 30.6, this volume). The highest values are reported from sites where Lumbricidae and Scarabidae are abundant (Tables 12.9, 24.2, 30.6, this volume).

Standing crop biomass of vertebrates has not been discussed much in this volume, but is considered in much greater depth by Van Dyne *et al.* (Chapter 4 in Breymeyer & Van Dyne, 1979). In much of the natural grassland in North

America estimated sustainable grazing capacity for domesticated animals is in the 2–5 g/m^2 range, a similar level to that of the total invertebrate population in the cosystem (Chapter 8, this volume); in France grazing capacities in semi-permanent grassland are in the 75 to 90 g/m^2 range, where invertebrate biomass is probably 5 to 10 g/m^2 (Chapter 12, this volume).

Energy flow

Most of the energy flow studies reported in this volume have been directed towards estimating the efficiency of assimilation of ingested plant materials and the efficiency of conversion of ingested or assimilated energy into animal tissue. An indication of a wide range of efficiencies by different organisms has emerged. Among invertebrate herbivores 9–25 % of ingested energy has been estimated to be converted into animal tissue (Tables 12.6, 24.4; Chapter 12, this volume). For sheep and cattle on pasture, estimates range from 3 to 15 % (Table 24.4; Chapter 12, this volume). Where comparisons were made in the same site, invertebrates were found to be much more efficient tissue producers than sheep or cattle (Table 24.3, Fig. 24.2; Chapters 9, 12, this volume). Maintenance energy cost of large animals is evidently higher than small ones.

Total annual energy flow through various kinds of consumers was estimated in a few instances. In one study in North American natural temperate grassland it was about 200 kJ/m^2 for invertebrates (Table 8.4). Annual consumption by cattle in a French semi-natural pasture approximated 14000 kJ/m^2; in a seeded pasture in Australia combined consumption by invertebrates and sheep was about 21000 kJ/m^2 annually (Table 24.3). The proportion of herbage production that is removed by livestock ranges from 10 % or less in rangelands of North America to about 85 % in French and Polish pastures. The proportion consumed by invertebrate herbivores, in comparison, seems to be quite modest, being probably in the 2 to 5 % range. However, there are reports in North America of grasshopper populations being triple in grazed as compared to ungrazed grasslands.

Production of meat and milk by cattle in the French pasture approximated 2100 kJ/m^2 (Chapter 12, this volume), while in Australian pasture sheep tissue production contained 260 kJ/m^2, while that of invertebrates was about 1456 kJ/m^2 (Table 24.3, this volume).

These analyses suggest that in many grasslands our present management regimes result in the utilisation of a very small proportion of primary production by consumers and that the efficiency of conversion of consumed materials falls far short of that achieved by natural herbivores.

Micro-organisms

It is difficult to generalise among reports of populations of micro-organisms for several reasons. Some studies have distinguished counts of bacteria from those of actinomycetes, while in others actinomycetes are included in the reported bacterial populations. Most estimates were made by traditional plate-count techniques, while some involved direct counts. Populations and biomass are mostly quoted in terms of numbers per gram of soil or plant material, rather than in terms of area of soil surface, as should be conventional for all organisms in ecosystem studies. In some instances live weights only are reported: for the purpose of the following discussion these are converted on the basis that 1 g of dry soil has a volume of 1 cc and that the dry-matter content of micro-organisms is 10 %.

Combined viable (plate-count) populations of bacteria and actinomycetes in the uppermost layer of soil that are quoted in this volume range from 4×10^{10} to $7 \times 10^{13}/m^2$ to a depth of 10 cm. However, the few direct counts referred to range from 1×10^{13} to $1.8 \times 10^{15}/m^2$ to a depth of 10 cm (Tables 13.2, 19.4; Fig. 7.1, this volume). In most instances where populations were compared in the same site bacteria were very much more numerous than actinomycetes.

Fungal populations in soil by plate counts are reported to be in the range of 2×10^8 to $9.7 \times 10^{10}/m^2$ to a depth of 10 cm (Tables 19.3, 19.4, 31.2, this volume).

Microbial populations decrease with increasing soil depth below 10 to 15 cm. Viable populations fluctuate on a seasonal and daily basis to the extent that the estimate on one day is sometimes double that of the next day (Fig. 31.3, this volume).

Combined populations of bacteria and actinomycetes in plant litter reported in this volume range from 2×10^6 to 2×10^{10}/g of plant material (Table 13.1, this volume), while fungal populations are reported in the range of 1×10^4 to 2×10^8/g of litter. Micro-organisms are also abundant on living plants, both above ground (Table 13.1, this volume) and under ground (Fig. 7.1, this volume).

Estimates of microbial biomass that are quoted in this volume are based on several methods of calculation. Interpreted biomass values for bacteria (including actinomycetes) (to a depth of 10 cm in the soil) range from 1 to 8 g/m² on a plate-count basis, and from 1 to 770 g/m² on a direct-count basis (Tables 7.2, 13.3, this volume). Fungal biomass is reported to be two to seven times the combined bacteria and actinomycetes biomass in two natural temperate grassland sites (Tables 7.1, 7.2, this volume).

It is evident that considerable information was obtained concerning rates of activity of micro-organisms in grassland ecosystems. Activity was measured in some studies in terms of rate of respiration (Figs. 25.1 to 25.6, this volume), while in others rate of disappearance of organic material was evaluated (Tables 19.5, 19.6, this volume). The rate of microbial cell production was estimated in one study (Table 31.2, this volume) the results of which have suggested that carbon recycles several times each year in microbial populations. Coleman *et al.* (Chapter 7 in Breymeyer & Van Dyne, 1979) have discussed decomposition processes in grasslands in considerable depth.

Energy flow

Sufficient data have been collected in some ecosystems to estimate the routes of flow of energy between the various compartments. The static models that have resulted must be considered tentative, since they are all based partly on assumptions.

The levels of photosynthetically-active radiation (PHAR) quoted for various sites (assumed to be 45–50 % of global radiation) are sometimes for only the period to maximum green-shoot biomass and sometimes for the whole year. Consequently, it is not possible to compare net photosynthetic efficiency among sites from these data. Values of net primary production are commonly 1 % or less of PHAR when total growing-season or annual measurements are used. However, much higher values are obtained when only the period of rapid plant growth is considered. For example, from the start to the peak of the cumulative growth curve net photosynthetic efficiency was estimated to be about 1.5, 2, 3 and 7 %, respectively, in dry, moist, damp and wet meadows in Czechoslovakia (Table 14.2, this volume).

The estimated proportion of net primary production that is transferred to under-ground plant parts ranges from about 30 % in Czechoslovakian meadows (Table 14.1, this volume) to 75 % in natural temperate grasslands (Table 5.8, this volume). Where studies have been made of losses from the canopy, the proportion of energy that does not appear in litter ranges from 4 % in some Indian grasslands (Table 20.2, this volume) to 39 % in natural temperate grasslands (Fig. 5.6, this volume). This is presumably lost by consumption and decomposition of leaves and stems before they drop to the soil surface.

Except in heavily stocked semi-permanent natural pastures and in grazed arable grasslands, the proportion of energy that is routed through above-ground consumers is very low, so that most shoot energy is assumed to be released through activities of micro-organisms in litter decomposition. Studies have not progressed sufficiently to partition accurately the relative activity of micro-organisms and invertebrates in release of energy below the soil surface. In one study, however, such an estimate was made (Table 8.4, this volume). In that instance about 86 % of the total energy fixed in net primary production (above ground and under ground) was estimated to be released by micro-organisms.

Table 33.7. *Standing state of nitrogen (in g/m²) in plant material and annual flow rates between plants and soil in various types of grassland*

Type of grassland	Soil pool	Soil to plants	In plants and litter	Plants to soil
India				
Semiarid	64	3	1	2
Dry subhumid	452	26	17	16
Moist subhumid	451	15	12	10
Humid	398	21	20	15
Czechoslovakia (meadows)				
Dry	—	14	12	6
Moist	—	19	12	6
Wet	—	32	36	30
USSR, Baraba, Karachi	1400	30	36	39
USA, Pawnee	334	3	18	3
Canada, Matador	540	10	20	10

Nutrient cycling

Data have been collected in several grasslands concerning the standing state of certain nutrients in the ecosystem and its changes throughout the year. These data, together with information on productivity and rates of decomposition, provide a basis for constructing static models of nutrient flow. The nutrients that have received most attention in grasslands are nitrogen, phosphorus and sulphur. The most intensive study of nutrient cycling reported in this volume was undertaken at the Baraba, Karachi site in the USSR (Fig. 14.2, this volume). Nutrient cycling has been discussed by Clark *et al.* (Chapter 8 in Brymeyer & Van Dyne, 1979), with emphasis on nitrogen and phosphorus.

The amount of nitrogen in soil organic matter (humus) (Table 33.7) ranges from 64 g/m² in semiarid Indian grassland (Table 20.3, this volume) to 1400 g/m² in a mesohalophytic meadow in the USSR (Fig. 14.2, this volume). Other values reported for temperate and tropical regions range between 300 and 550 g/m². The annual flow from soil to plants ranges from 3 g/m² in relatively arid temperate and tropical grasslands (Tables 8.6, 20.3, this volume) to over 30 g/m² in mesic to wet temperate meadows (Table 14.4, this volume). The mean amount of nitrogen in plants ranges between 1 g/m² in tropical semiarid grassland (Table 20.3, this volume) to 36 g/m² in mesic to wet temperate meadows (Table 14.4, Fig. 14.2, this volume). The amount of nitrogen that returns to the soil each year depends on the degree of usage, being high (in relation to plant uptake) under grazing but much lower in mown meadows. Micro-organisms in natural temperate grasslands are estimated to contain one-sixth to one-half as much nitrogen as plants, but that contained by animals is only in the order of 1 % of the amount of nitrogen in plants (Tables 8.1, 8.2, this volume).

A comprehensive account of the status of mineral elements in grassland ecosystems appears in Chapter 14 (this volume), while sulphur is examined in depth in Chapter 26 (this volume).

Use and management

There is a very considerable body of information concerning the effect of grazing or mowing on the structure of most grasslands. Primary productivity has been investigated most intensively in very productive semi-natural pastures and meadows and in arable pastures, but information is scarce concerning the effect of grazing on primary production in most of the world's rangelands. The most intensive study of the effects of grazing on grasslands during IBP seems to have been made in the USA. The results are discussed in Chapter 5 (this volume). No consistent effect of moderate grazing on net primary production was found, with reductions in production in some sites and increases in others. Too few reports are available on the effects of grazing on heterotrophic populations and activities to permit generalisation.

In a few sites the role of fire in grasslands was considered. Burning plays a very important role in modifying the structure of grasslands. It is an important management tool, especially in tropical regions, in maintaining an herbaceous cover by interference with invasion by trees and shrubs (Chapter 17, this volume). However, much information needs to be gathered on the effect of fire on primary productivity. In some moist subhumid temperate situations it seems beneficial in releasing nutrients from accumulated litter, while in drier areas vigour of the herbaceous vegetation is noticeably reduced for several years after burning.

Fertilisation is a standard means of increasing productivity, both primary and secondary, in intensively used arable pastures and in certain semi-natural pastures and meadows. Under extensive rangeland conditions insufficient moisture and economic feasibility have interferred with widespread use of fertilisers. In several IBP studies fertilisers were used as a stress factor; the results of these studies have not been discussed fully in this volume.

Cultivation appears to be an effective, but exploitive, means of increasing production by accelerating the rate of release of nutrients from organic matter in soil. Investigations of primary productivity in cropland are surprisingly few. Much more attention needs to be given to the relationship between inputs and outputs of nutrients in cropland of regions of natural grassland, where the current crop is harvested at the expense of soil reserves that have been accumulated over many centuries prior to the time when tillage commenced in the nineteenth and twentieth centuries.

References

References to Foreword and Part I

Beard, J. S. (1953). The savanna vegetation of northern tropical America. *Ecological Monographs*, **23**, 149–215.

Bews, J. W. (1929). *The World's Grasses: Their Differentation, Distribution, Economics and Ecology*. New York & Toronto: Longmans, Green & Co.

Bliss, L. C., Cragg, J. B., Heal, O. W. & Moore, J. J. (eds.). *Tundra ecosystems: a comparative analysis. IBP synthesis series*. Cambridge University Press (in preparation).

Borchert, J. R. (1950). The climate of the North American grassland. *Annals of the Association of American Geographers*, **40**, 1–39.

Breymeyer, A. I. & Van Dyne, G. M. (eds.) (1978). *Grasslands, Systems Analysis and Man. IBP Synthesis Series* **19**. Cambridge University Press.

Brian, M. V. (ed.) (1977). *Production Ecology of Ants and Termites. IBP Synthesis Series* **13**. Cambridge University Press.

Clark, F. E. & Paul, E. A. (1970). The microflora of grassland. *Advances in Agronomy*, **22**, 375–435.

Clements, F. E. & Shelford, V. E. (1939). *Bio-ecology*. New York: J. Wiley & Sons, Inc.

Cloudsley-Thompson, J. L. (1969). *The Zoology of Tropical Africa*. London: Wiedenfeld & Nicolson.

Coupland, R. T. (1974). Grasslands. In *The New Encyclopaedia Britannica*, **8**, 280–94.

Coupland, R. T., Ripley, E. A. & Robbins, P. D. (1973). *Description of Site: I. Floristic Composition and Canopy Architecture of the Vegetative Cover. Canadian Committee for IBP, Matador Project Technical Report* **11**. Saskatoon, Canada: University of Saskatchewan.

Coupland, R. T. & Van Dyne, G. M. (eds.). *Grassland Ecosystems: Reviews of Research. Proceedings of the Second Meeting of the PT Grassland Working Group, IBP Saskatoon and Matador, Saskatchewan, Canada. Range Science Department Science Series 7*. Fort Collins: Colorado State University.

Coupland, R. T., Willard, J. R., Ripley, E. A. & Randell, R. L. (1975). The Matador Project. In *Energy Flow – Its Biological Dimensions: A Summary of the IBP in Canada, 1964–74*, ed. L. W. Billingsley, pp. 19–50. Ottawa: Royal Society of Canada.

Coupland, R. T., Zacharuk, R. Y. & Paul, E. A. (1969). Procedures for study of grassland ecosystems. In *The Ecosystem Concept in Natural Resource Management*, ed. G. M. Van Dyne, pp. 25–57. New York & London: Academic Press Inc.

Dahlman, R. C. & Kucera, C. L. (1965). Root productivity and turnover in native prairie. *Ecology*, **46**, 84–9.

Dix, R. L. & Beidleman, R. G. (eds.) (1969). *The Grassland Ecosystem: A Preliminary Synthesis. Range Science Department Science Series 2*. Fort Collins: Colorado State University.

Dyksterhuis, E. J. (1958). Ecological principles in range evaluation. *Botanical Review*, **24**, 252–72.

Golley, F. B. & Buechner, H. K. (eds.). (1969). *A Practical Guide to the Study of the*

Productivity of Large Herbivores. IBP Handbook 7. Oxford: Blackwell Scientific Publications.

Golley, F. B., Petrusewicz, K. & Ryszkowski, L. (eds.) (1975). *Small Mammals: their Productivity and Population Dynamics. IBP Synthesis Series 5.* Cambridge University Press.

Goodall, D.W. & Perry, R. A. (1979). *Arid Land Ecosystems; Structure, Functioning and Management. IBP Synthesis Series 16.* Cambridge University Press. In preparation.

Hartley, W. (1950). The global distribution of tribes of the Gramineae in relation to historical and environmental factors. *Australian Journal of Agricultural Research*, **1**, 355–73.

Heady, H. F. (1975). *Rangeland Management.* New York: McGraw Hill-Book Co.

IBP/PT Grassland Working Group (1973). Compte-rendu du Colloque du Programme Biologique International sur les milieux herbacés d'Afrique, décembre 29, 1971 à janvier 3, 1972, Lamto, Côte d'Ivoire. *Annals of the University of Abidjan, Series E*, **6** (2).

IBP/PT Grassland Working Group (1974). *Proceedings, Symposium and Synthesis Meetings on Tropical Grassland Biome*, January 17–22, 1974. Varanasi, India: Department of Botany, Banaras Hindu University.

IBP/PT Grassland and Tundra Working Groups (1972). *Report of the Modelling and Synthesis Workshop*, August 14–27, 1972. Natural Resource Ecology Laboratory, Colorado State University.

Keller, B. (1927). Distribution of vegetation on the plains of European Russia. *Journal of Ecology*, **15**, 189–233.

Kendeigh, S. C. & Pinowski, J. (eds.) (1977). *Granivorous Birds in Ecosystems: their Evolution, Populations, Energetics, Adaptations, Impact and Control. IBP Synthesis Series 12.* Cambridge University Press.

Kucera, C. L., Dahlman, R. C. & Koelling, M. L. (1967). Total net productivity and turnover on an energy basis for tallgrass prairie. *Ecology*, **48**, 536–41.

Milner, C. & Hughes, R. E. (1968). *Methods for the Measurement of the Primary Production of Grassland. IBP Handbook 6.* Oxford: Blackwell Scientific Publications.

Moore, R. M. (ed.) (1970). *Australian Grasslands.* Canberra: Australian National University Press.

Nutman, P. S. (ed.) (1976). *Symbiotic Nitrogen Fixation in Plants. IBP Synthesis Series 7.* Cambridge University Press.

Perry, R. A. & Goodall, D. W. (eds.). *Arid Land Ecosystems; Structure, Functioning and Management. IBP Synthesis Series 17.* Cambridge University Press. In preparation.

Petrusewicz, K. (ed.) (1967). *Secondary Productivity of Terrestrial Ecosystems: Principles and Methods.* 2 vols. Warsaw: Państwowe Wydawnictwo Naukowe.

Petrusewicz, K. & Macfadyen, A. (1970). *Productivity of Terrestrial Animals – Principles and Methods. IBP Handbook 13.* Oxford: Blackwell Scientific Publications.

Phillipson, J. (ed.). (1971). *Methods of Study in Quantitative Soil Ecology: Population, Production and Energy Flow. IBP Handbook 18.* Oxford: Blackwell Scientific Publications.

Reichle, D. E. (ed.) (1979). *Dynamic properties of woodland ecosystems. IBP Synthesis Series, 23.* Cambridge University Press (in press).

Reichle, D. E., Franklin, J. F. & Goodall, D. W. (eds.) (1975). *Productivity of World Ecosystems. Proceedings of a symposium held at the V General Assembly of SCIBP on August 31–September 1, 1972 at Seattle, Washington.* Washington, DC: National Academy of Sciences.

Roseveare, G. M. (1948). *The Grasslands of Latin America. Bulletin 36.* Aberystwyth: Imperial Bureau of Pastures and Field Crops.
Shantz, H. L. & Marbut, C. F. (1923). *The Vegetation and Soils of Africa. American Geographical Society, Research Series 13.* New York: American Geographical Society.
SCIBP (1966). *IBP News 7.*
SCIBP (1969). *Section PT: Productivity Terrestrial. IBP News 13.*
Steiger, T. L. (1930). Structure of prairie vegetation. *Ecology*, **11**, 170–217.
Stewart, W. D. P. (ed.) (1976). *Nitrogen Fixation by Free-living Micro-organisms. IBP Synthesis Series 6.* Cambridge University Press.
Weaver, J. E. (1954). *North American Prairie.* Lincoln, Nebraska: Johnsen Publishing Co.
Weaver, J. E. & Albertson, F. W. (1956). *Grasslands of the Great Plains.* Lincoln, Nebraska: Johnsen Publishing Co.
Weaver, J. E. & Clements, F. E. (1938). *Plant Ecology.* New York: McGraw-Hill Book Co.
Wiegert, R. G. & Evans, F. C. (1964). Primary production and the disappearance of dead vegetation on an old field in south-eastern Michigan. *Ecology*, **45**, 49–63.
Worthington, E. B. (ed.) (1975). *The evolution of IBP. IBP Synthesis Series 1.* Cambridge University Press.

References to Part II

Abouguendia, Z. (1973). Net primary production and turnover in a mixed grass prairie. M.S. Thesis, North Dakota State University, Fargo, North Dakota.
Aldous, A. E. (1930). Effect of different clipping treatments on the yield and the vigor of prairie grass vegetation. *Ecology*, **11**, 752–9.
Allen, D. L. (1966). The preservation of endangered habitats and vertebrates in North America. In *Future Environments of North America, ed.* F. F. Darling & J. P. Milton, pp. 22–37. Garden City, New York: The Natural History Press.
Andrews, R. M., Coleman, D. C., Ellis, J. E. & Singh, J. S. (1974). Energy flow relationships in a shortgrass prairie ecosystem. In *Proceedings of the First International Congress of Ecology*, pp. 22–8. Wageningen, Netherlands: Centre for Agricultural Publishing and Documentation.
Andrzejewska, L., Breymeyer, A. Kajak, A. & Wójcik, Z. (1967). Experimental studies on trophic relationships of terrestrial invertebrates. In *Secondary Productivity of Terrestrial Ecosystems: Principles and Methods*, 2 vols, ed. K. Petrusewicz, pp. 477–96. Warsaw: Państwowe Wydawnictwo Naukowe.
Biswell, H. H. & Weaver, J. E. (1933). Effect of frequent clipping on the development of roots and tops of grasses in prairie sod. *Ecology*, **14**, 368–90.
Black, C. C. (1971). Ecological implications of dividing plants into groups with distinct photosynthetic production capacities. *Advances in Ecological Research*, **7**, 87–114.
Bokhari, U. G. & Singh, J. S. (1974). Effects of temperature and clipping on growth, carbohydrate reserves, and root exudation of western wheatgrass in hydroponic culture. *Crop Science*, **14**, 790–4.
Borchert, J. R. (1950). The climate of the North American grassland. *Annals of the Association of American Geographers*, **40**, 1–39.
Chew, R. M. (1974). Consumers as regulators of ecosystems: an alternative to energetics. *Ohio Journal of Science*, **74**, 359–70.
Christensen, M. & Scarborough, A. M. (1969). *Soil Microfungal Investigations,*

Pawnee Site, US/IBP Grassland Biome, Technical Report 23. Fort Collins: Colorado State University.

Clark, A. H. (1956). The impact of exotic invasion on the remaining New World mid-latitude grasslands. In *Man's Role in Changing the Face of the Earth*, ed. W. L. Thomas, Jr., pp. 736–62. Chicago: University of Chicago Press.

Clark, F. E. & Paul, E. A. (1970). The microflora of grassland. *Advances in Agronomy*, **22**, 375–435.

Cody, M. L. (1974). *Competition and the Structure of Bird Communities*. Princeton, NJ: Princeton University Press.

Cole, G. W. (ed.) (1976). *ELM: Version 2.0 Range Science Department Science Series 20*. Fort Collins: Colorado State University.

Cooper, J. P. (1970). Potential production and energy conversion in temperate and tropical grasses. *Herbage Abstracts*, **40**, 1–15.

Coupland, R. T. (1973*a*). *Producers: I. Dynamics of Above-ground Standing Crop. Canadian Committee for IBP, Matador Project Technical Report 27*. Saskatoon, Canada: University of Saskatchewan.

Coupland, R. T. (1973*b*). *Producers: II. Nutrient and Energy Contents of Above-ground Parts. Canadian Committee for IBP, Matador Project Technical Report 32*. Saskatoon, Canada: University of Saskatchewan.

Coupland, R. T. (1973*c*). *Producers: III. Rates of Dry Matter Production and of Nutrient and Energy Flow Through Shoots. Canadian Committee for IBP, Matador Project Technical Report 33*. Saskatoon, Canada: University of Saskatchewan.

Coupland, R. T. (1974*a*). *Producers: IV. Under-ground Plant Parts. Canadian Committee for IBP, Matador Project Technical Report 41*. Saskatoon, Canada: University of Saskatchewan.

Coupland, R. T. (1974*b*). *Producers: VI. Summary of Studies of Primary Production by Biomass and Shoot Observation Methods. Canadian Committee for IBP, Matador Project Technical Report 62*. Saskatoon, Canada: University of Saskatchewan.

Coupland, R. T. (1974*c*). Conservation of grasslands. In *Conservation in Canada: A Conspectus*, ed. J. S. Maini & A. Carlisle. Publication 1340, pp. 93–111. Ottawa: Canada Department of the Environment.

Coupland, R. T. & Abouguendia, Z. (1974). *Producers: V. Dynamics of Shoot Development in Grasses and Sedges, Canadian Committee for IBP, Matador Project Technical Report 51*. Saskatoon, Canada: University of Saskatchewan.

Coupland, R. T., Ripley, E. A. & Robbins, P. C. (1973). *Description of Site: I. Floristic Composition and Canopy Architecture of the Vegetative Cover. Canadian Committee for IBP, Matador Project Technical Report 11*. Saskatoon, Canada: University of Saskatchewan.

Coupland, R. T., Willard, J. R., Ripley, E. A. & Randell, R. L. (1975). The Matador Project. In *Energy Flow – Its Biological Dimensions: A Summary of the IBP in Canada, 1964–74*, ed. L. W. Billingsley, pp. 19–50. Ottawa: Royal Society of Canada.

Cressler, L. (1942). The effect of different intensities and times of grazing and the degree of dusting upon the vegetation of range land in west central Kansas. *Transaction of the Kansas Academy of Science*, **45**, 75–91.

Curtis, J. T. (1956). The modification of mid-latitude grasslands and forests by man. In *Man's Role in Changing the Face of the Earth*, ed. W. L. Thomas, Jr., pp. 721–36. Chicago: University of Chicago Press.

Dahlman, R. C. & Kucera, C. L. (1965). Root productivity and turnover in native prairie. *Ecology*, **46**, 84–9.

Dahlman, R. C. & Kucera, C. L. (1969). Carbon-24 cycling in the root and soil components of a prairie ecosystem. In *Proceedings of the Second National Symposium on Radioecology*, ed. D. J. Nelson & F. C. Evans, pp. 652–60. Springfield, Virginia: Division of Technical Information, USAECTIO–4500 (Conf–670503).

Doxtader, K. G. (1969). *Microbial Biomass Measurements at the Pawnee Site: Preliminary Methodology and Results, US/IBP Grassland Biome, Technical Report 21.* Fort Collins: Colorado State University.

Driscoll, R. S. (1967). *Managing Public Rangelands: Effective Livestock Grazing Practices and Systems for National Forests and National Grasslands.* United States Department of Agriculture, AIB–315.

Dubbs, A. L. (1966). *Yield, Crude Protein, and Palatability of Dryland Grasses in Central Montana.* Bulletin 604. Montana Agricultural Experiment Station.

Duncan, D. A. (1975*a*). The San Joaquin site of the grassland biome; its relation to the annual grassland ecosystem synthesis. In *The California Grassland Ecosystem*, ed. R. M. Love, pp. 9–15. California Chapter, American Society of Agronomy. A symposium, Anaheim, California, 30 January, 1975, Institute of Ecology Publication 7. Davis, University of California: American Society of Agronomy.

Duncan, D. A. (1975*b*). *Comprehensive Network Site Description, San Joaquin. US/IBP Grassland Biome Technical Report 296.* Fort Collins: Colorado State University.

Dyer, M. I. & Bokhari, U. G. (1976). Plant–animal interactions: Studies of the effects of grasshopper grazing on blue grama grass. *Ecology*, **57**, 762–72.

Dyksterhuis, E. J. (1949). Condition and management of rangeland based on quantitative ecology. *Journal of Range Management*, **2**, 104–15.

Freeland, W. J. & Janzen, D. H. (1974). Strategies in herbivory by mammals: the role of plant secondary compounds. *American Midland Naturalist*, **108**, 269–89.

Gates, D. M. (1965). Energy exchange between organisms and environment. In *Biometeorology*, ed. W. P. Lowry, pp. 1–22. Corvallis: Oregon State University Press.

Gilbert, L. E. & Raven, P. H. (eds.) (1975). General introduction. In *Coevolution of Animals and Plants*, pp. ix–xiii. Austin: University of Texas Press.

Golley, F. B. (1961). Energy values of ecological materials. *Ecology*, **42**, 581–4.

Haas, H. J. & Evans, C. E. (1957). *N and C changes in Great Plains Soils as Influenced by Cropping and Soil Treatment. Technical Bulletin 1164.* United States Department of Agriculture.

Hadley, R. B. & Kieckhefer, B. J. (1963). Productivity of two prairie grasses in relation to fire frequency. *Ecology*, **44**, 389–95.

Hairston, N. O., Smith, F. E. & Slobodkin, L. B. (1960). Community structure, population control, and competition. *American Naturalist*, **94**, 421–5.

Harley, J. L. & Waid, J. S. (1955). A method for studying active mycelia on living roots and other surfaces in the soil. *Transactions of the British Mycological Society*, **38**, 104–18.

Harris, J. O. (1971). *Microbiological Studies on the Osage Site, 1970, US/IBP Grassland Biome, Technical Report 102.* Fort Collins; Colorado State University.

Harris, L. D. (1971). A précis of small mammal studies and results in the grassland biome. In *Preliminary Analysis of Structure and Function in Grasslands. Range Science Department Science Series, 10*, ed. N. R. French, pp. 213–40. Fort Collins: Colorado State University.

Harris, L. D. & Paur, L. (1972). *A Quantitative Food Web Analysis of a Shortgrass Community. US/IBP Grassland Biome, Technical Report 154.* Fort Collins: Colorado State University.

Harris, P. (1974). A possible explanation of plant yield increases following insect damage. *Agro-Ecosystems*, **1**, 219–25.

Heady, H. F. (1975). *Rangeland Management*. New York: McGraw-Hill Book Co.

Hilden, O. (1965). Habitat selection in birds. *Annales Zoologici Fennici*, **2**, 53–75.

Hitchcock, A. S. (1950). *Manual of the Grasses of the United States. USDA Miscellaneous Publication 200*. Washington, DC: US Government Printing Office.

Hobbs, J. A. & Brown, P. L. (1957). *N and organic Carbon Changes in Cultivated Western Kansas Soils. Bulletin 89*. Kansas Agricultural Experiment Station.

Horn, H. S. (1966). Measurement of 'overlap' in comparative ecological studies. *American Naturalist*, **100**, 419–24.

Houston, W. R. & Adams, R. E. (1971). Interseeding for range improvement in the northern Great Plains. *Journal of Range Management*, **24**, 457–61.

Janzen, D. H. (1971). Escape of *Cassra grandis* L. beans from predators in time and space. *Ecology*, **52**, 964–79.

Johnston, A. (1962). Effects of grazing intensity and cover on the water-intake rate of fescue grassland. *Journal of Range Management*, **15**, 79–82.

Kucera, C. L., Dahlman, R. C. & Koelling, M. R. (1967). Total net productivity and turnover on an energy basis for tallgrass prairie. *Ecology*, **48**, 536–41.

Kumar, R., Lavigne, R. J., Lloyd, J. E. & Pfadt, R. E. (1976). *Insects of the Central Plains Experiment Range, Pawnee National Grassland*. Science Monograph 32. Laramie: Agricultural Experiment Station, University of Wyoming.

Lacey, M. L. (1942). The effect of climate and different grazing and dusting intensities upon the yield of the shortgrass prairies in western Kansas. *Transactions of the Kansas Academy of Sciences*, **45**, 111–23.

Lauenroth, W. K. & Sims, P. L. (1973). *Effects of Water and Nitrogen Stresses on a Shortgrass Prairie Ecosystem. US/IBP Grassland Biome, Technical Report 232*. Fort Collins: Colorado State University.

Levin, D. A. (1971). Plant phenolics: an ecological perspective. *American Naturalist*, **105**, 157–81.

Lewis, J. K. (1971). The grassland biome: a synthesis of structure and function, 1970. In *Prelimary Analysis of Structure and Function in Grasslands. Range Science Department Science Series, 10*, ed. N. R. French, pp. 317–87. Fort Collins: Colorado State University.

Lindeman, R. L. (1942). The trophic dynamic aspect of ecology. *Ecology*, **23**, 399–418.

Lodge, R. W. & Campbell, J. B. (1971). *Management of the Western Range*. Publication 1425 Canada Department of Agriculture.

Lowe, W. E. & Paul, E. A. (1974). *Soil Microorganisms: III. Numerical Taxonomy of Aerobic, Heterotrophic Bacteria. Canadian Committee for IBP, Matador Project Technical Report 66*. Saskatoon, Canada: University of Saskatchewan.

McCalla, T. M. (1967). Effect of tillage on plant growth as influenced by soil organisms. In *Tillage for Greater Crop Production* (Conference Proceedings), pp. 19–25. American Society of Agricultural Engineers.

McGill, W. B., Paul, E. A. & Sorensen, H. L. (1974). *The Role of Microbial Metabolites in the Dynamics of Soil Nitrogen. Canadian Committee for IBP, Matador Project Technical Report 46*. Saskatoon, Canada: University of Saskatchewan.

Maher, W. J. (1974*a*). *Birds: II. Avifauna of the Matador Area. Canadian Committee for IBP, Matador Project Technical Report 58*. Saskatoon, Canada: University of Saskatchewan.

Maher, W. J. (1974*b*). *Birds: I. Populations Dynamics. Canadian Committee for IBP,*

Matador Project Technical Report 34. Saskatoon, Canada: University of Saskatchewan.

Martel, Y. A. & Paul, E. A. (1974). Effects of cultivation on the organic matter of grassland soils as determined by fractionation and radiocarbon dating. *Canadian Journal of Soil Science*, **54**, 419–26.

Mattson, W. J. & Addy, N. D. (1975). Phytophagous insects as regulators of forest primary production. *Science, Washington*, **190**, 515–22.

Mitchell, J. E. & Pfadt, R. E. (1974). A role of grasshoppers in a shortgrass prairie ecosystem. *Environmental Entomology*, **3**, 358–60.

Morrall, R. A. A. & Howard, R. J. (1974). *Leaf Spot Disease of Graminoids in Native Grassland. Canadian Committee for IBP, Matador Project Technical Report 48*. Saskatoon, Canada: University of Saskatchewan.

Muehbeier, J. (1958). Land-use problems in the Great Plains. In *Land*, 1958 Yearbook of Agriculture, ed. A. Stefferud, pp. 161–6. Washington, DC: United States Department of Agriculture.

Newton, J. D., Wyatt, F. A. & Brown, A. L. (1945). Effects of cultivation and cropping on the chemical composition of some western Canadian prairie province soils. *Scientific Agriculture*, **25**, 718–37.

Odum, E. P. (1960). Organic production and turnover in old-field succession. *Ecology*, **41**, 34–49.

Odum, E. P. (1969). The strategy of ecosystem development. *Science, Washington*, **164**, 262–70.

Parkinson, D. & Bhatt, G. C. (1974). *Soil Fungi: I. Studies on the Nature of Fungal Populations. Canadian Committee for IBP, Matador Project Technical Report 53*. Saskatoon, Canada: University of Saskatchewan.

Paul, E. A., Biederbeck, V. O., Lowe, W. E. & Willard, J. R. (1973). *Soil Microorganisms: I. Population Dynamics of Bacteria and Actinomycetes. Canadian Committee for IBP, Matador Project Technical Report 37*. Saskatoon, Canada: University of Saskatchewan.

Peden, D. G., Van Dyne, G. M., Rice, R. W. & Hansen, R. M. (1974). The trophic ecology of *Bison bison* L. on shortgrass plains. *Journal of Applied Ecology*, **11**, 489–98.

Pulliam, H. R. & Enders, F. (1971). The feeding ecology of five sympatric finch species. *Ecology*, **52**, 557–66.

Quinn, J. A. & Hervey, D. F. (1970). Trampling losses and travel by cattle on sandhill range. *Journal of Range Management*, **23**, 50–5.

Rauzi, F. (1975). Severe mechanical and chemical range renovation in northeastern Wyoming. *Journal of Range Management*, **28**, 319–26.

Reardon, P. O., Leinweber, C. L. & Merrill, L. B. (1972). The effect of bovine saliva on grasses. *Journal of Animal Science*, **34**, 877–98.

Reardon, P. O., Leinweber, C. L. & Merrill, L. B. (1974). Response of sideoats grana to animal saliva and thiamine. *Journal of Range Management*, **27**, 400–1.

Ricklefs, R. E. (1973). *Ecology*. Newton, Mass: Chiron Press Inc.

Riegert, P. W. & Varley, J. L. (1973*a*). *Above-ground Invertebrates: II. Population Dynamics and Biomass Production of Grasshoppers. Canadian Committee for IBP, Matador Project Technical Report 16*. Saskatoon, Canada: University of Saskatchewan.

Riegert, P. W. & Varley, J. L. (1973*b*). *Above-ground Invertebrates: III. Bioenergetics of Grasshoppers. Canadian Committee for IBP, Matador Project Technical Report 17*. Saskatoon, Canada: University of Saskatchewan.

Riegert, P. W., Varley, J. L. & Dunn, B. C. (1974). *Above-ground Invertebrates: IV. Populations and Energetics of Spiders. Canadian Committee for IBP, Matador Project Technical Report 57*. Saskatoon, Canada: University of Saskatchewan.

Riegert, P. W., Varley, J. L. & Willard, J. W. (1974). *Above-ground Invertebrates: V. A Summary of Populations, Biomass and Energy Flow. Canadian Committee for IBP, Matador Project Technical Report 67*. Saskatoon, Canada: University of Saskatchewan.

Sadler, D. A. R. & Maher, W. J. (1974). *Birds IV. Bioenergetics and Simulation of Energy Flow. Canadian Committee for IBP, Matador Project Technical Report 63*. Saskatoon, Canada: University of Saskatchewan.

Salter, R. M. & Green, T. C. (1933). Factors affecting the accumulation and loss of nitrogen and organic carbon in cropped soils. *Journal of the American Society of Agronomy*, **25**, 622–30.

Salter, R. M., Lewis, R. D. & Slipher, J. A. (1941). *The Soil. Extension Bulletin 175*. Ohio Agricultural Experimental Station.

Samtsevich, S. A. (1965). Active excretions of plant roots and their significance. *Soviet Plant Physiology*, **12**, 837–46.

Shepperd, D. H. (1972). *Small Mammals: Reproduction, Population Dynamics, Biomass and Energetics. Canadian Committee for IBP, Matador Project Technical Report 10*. Saskatoon, Canada: University of Saskatchewan.

Sims, P. L. & Singh, J. S. (1971). Herbage dynamics and net primary production in certain ungrazed and grazed grassland in North America. In *Preliminary Analysis of Structure and Function in Grasslands, Range Science Department Science Series*, ed. N. R. French, pp. 59–124. Fort Collins: Colorado State University.

Sims, P. L. & Singh, J. S. (1978*a*). The structure and function of ten western North American grasslands. II. Intra-seasonal dynamics in primary producer compartments. *Journal of Ecology*, **66**, 547–72.

Sims, P. L. & Singh, J. S. (1978*b*). The structure and function often western North American grasslands. III. Net primary production, turnover, and efficiencies of energy capture and water use. *Journal of Ecology*, **66**, 573–97.

Sims, P. L. & Singh, J. S. (1978*c*). The structure and function of ten western North American grasslands. IV. Compartmental transfers and energy flow within the ecosystem. *Journal of Ecology* (in press).

Sims, P. L. & Singh, J. S. & Lauenroth, W. K. (1978*d*). The structure and function of ten western North American grasslands. I. Abiotic and vegetational characteristics. *Journal of Ecology*, **66**, 251–85.

Singh, J. S., Lauenroth, W. K. & Steinhorst, R. K. (1975). Review and assessment of various techniques for estimating net aerial production in grasslands from harvest data. *Botanical Review*, **41**, 181–232.

Slobodkin, L. B., Smith, F. E. & Hairston, N. G. (1967). Regulation in terrestrial ecosystems, and the implied balance of nature. *American Naturalist*, **101**, 109–24.

Smoliak, S. (1956). Influence of climatic conditions on forage production of shortgrass rangeland. *Journal of Range Management*, **9**, 89–91.

Smolik, J. D. (1973). Nematode studies at the Cottonwood site. Ph.D. dissertation. Brookings: South Dakota State University.

Spencer, J. F. T., Babiuk, L. & Morrall, R. A. A. (1971). Yeasts of the Matador soil of Saskatchewan and their ability to use flavanoids. *Canadian Journal of Microbiology*, **17**, 1248–50.

Stewart, J. W. B., Halm, B. J. & Cole, C. V. (1973). *Nutrient Cycling: I. Phosphorus. Canadian Committee for IBP, Matador Project Technical Report 40*. Saskatoon, Canada: University of Saskatchewan.

Stoddart, L. A., Smith, A. D. & Box, T. W. (1975). *Range Management*. Third edn. New York: McGraw-Hill Book Co.

Struik, G. J. & Bray, J. R. (1970). Root–shoot ratios of native forest herbs and *Zea mays* at different soil-moisture levels. *Ecology*, **51**, 892–3.

Tannehill, I. R. (1947). *Drought, its Causes and Effects*. Princeton, NJ: Princeton University Press.

Thatcher, A. P. (1966). Range production improved by renovation and protection. *Journal of Range Management*, **19**, 382–3.

Tomanek, G. W. (1948). Pasture types of western Kansas in relation to the intensity of utilization in past years. *Transactions of the Kansas Academy of Sciences*, **51**, 171–96.

Trolldenier, G. (1973). The use of fluorescence microscopy for counting soil microorganisms. *Bulletins from the Ecological Research Committee, Stockholm*, **17**, 53–9.

Wallwork, J. A. (1976). *The Distribution and Diversity of Soil Fauna*. New York and London: Academic Press.

Whitman, W. C., Peterson, D. R. & Conlon, T. J. (1961). Grass studies at Dickinson: results of clipping trials with cool season grasses. *Farm Research*, **22**, 9–14.

Whittaker, R. H. (1970). *Communities and Ecosystems*. New York: MacMillan & Co.

Wiens, J. A. (1969). An approach to the study of ecological relationships among grassland birds. *Ornithological Monographs*, **8**, 1–93.

Wiens, J. A. (1973). Pattern and process in grassland bird communities. *Ecological Monographs*, **43**, 237–70.

Wiens, J. A. & Dyer, M. I. (1975). Rangeland avifaunas: their composition, energetics, and role in the ecosystem. In *Proceedings of a Symposium on Management of Forest and Range Habitats for Nongame Birds*, pp. 146–82. USDA Forest Service General Technical Report WO-1.

Wight, R. J. & Siddoway, F. H. (1972). Improving precipitation use efficiency on rangelands by surface modifications. *Journal of Soil and Water Conservation*, **27**, 170–4.

Willard, J. R. (1973). *Soil Invertebrates: V. Nematoda: Populations and Biomass. Canadian Committee for IBP, Matador Project Technical Report 21*. Saskatoon, Canada: University of Saskatchewan.

Willard, J. R. (1974). *Soil Invertebrates: VIII. Summary of Populations and Biomass. Canadian Committee for IBP, Matador Project Technical Report 56*. Saskatoon, Canada: University of Saskatchewan.

Williams, S. T., Davies, F. L. & Cross, T. (1968). Identification of genera of the Actinomycetales. In *Identification Methods for Microbiologists*, ed. B. M. Gibbs & D. A. Shapton, pp. 110–24. New York & London: Academic Press.

References to Part III

Andrzejewska, L. & Kajak, A. (1966). Metodyka entomologicznych badań ilościowych na łąkach. *Ekologia Polska B*, **12**, 241–61.

Andrzejewska, L. & Wójcik, Z. (1970). The influence of Acridoidea on the primary production of a meadow (field experiment). *Ekologia Polska*, **18**, 89–109.

Aristovskaya, T. V. (ed.) (1972). *Problems of Abundance, Biomass and Productivity of Microorganisms in Soil*. Leningrad: Nauka Publishing House. (In Russian.)

Balátová-Tuláčková, E. (1966). Synökologische Charakteristik der südmährischen Überschwemmungswiesen. *Rozpravy Československé akademie věd, MPV*, **76**, 1–40.

Balátová-Tuláčková, E. (1968). Grundwasserganglinien und Wiesengesellschaften. *Acta Scientiarum Naturalium Academiae Scientiarum Bohemoslovacae Brno, N.S.*, **2**, 1–37.

Balátová-Tuláčková, E. (1973). Zur Problematik des Erhaltens der hochproduktiven Überschwemmungswiesen in Trockengebieten. *Quaestiones Geobotanicae*, **11**, 41–54.

Baradziej, E. (1974). Net primary production of two marsh communities near Ispina in the Niepołomice Forest (southern Poland). *Ekologia Polska A*, **22**, 145–72.

Bazilevich, N. I. (1962). Exchange of mineral elements in various types of steppes and meadows on chernozems, chestnut-brown soils and solonetzs. In *Problems of Soil Science*, ed. S. V. Zonn, pp. 148–207. Moscow: Academy of Sciences of the USSR Publishing House. (In Russian.)

Bazilevich, N. I. (1970). The geochemistry of soda soils. In *Israel Programs of Scientific Translations*, ed. V. A. Kovda, p. 392. Jerusalem.

Bazilevich, N. I. (ed.) (1974). *Steppe, Meadow and Bog Grassland Ecosystems and Farm Crop Plantations in Steppe and Forest-steppe Zones of USSR*. Moscow: Academy of Sciences of the USSR, Soviet National Committee for the IBP.

Bazilevich, N. I. & Kobyakova, N. G. (1971). Experience from mapping some indicators of biological cycling. In *Biological Productivity and Mineral Cycling in the Terrestrial Plant Communities*, ed. N. I. Bazilevich & L. E. Rodin, pp. 197–206. Leningrad: Nauka Publishing House. (In Russian.)

Bedrogközy, Gy. (1965). Ecology of the halophilic vegetation of the Pannonicum. II. Correlation between alkali ('szik') plant communities and genetic soil classification in the Northern Hortobágy. *Acta Botanica Hungarica*, **11**, 1–51.

Boness, M. (1953). Die Fauna der Wiesen unter besonderer Berücksichtigung der Mahd. *Zeitschrift für Morphologie und Ökologie der Tiere*, **42**, 225–77.

Bouché, M. B. (1972). Contribution à l'approche méthodologique de l'étude des biocenoses. I. Vers l'analyse quantitative globale des prairies. *Annales de Zoologie. Ecologie animale*, **4**, 529–36.

Bouché, M. B. (1975). Fonctions des Lombriciens. III. Premières estimations quantitatives des stations françaises du PBI. *Revue d'Ecologie et de Biologie du Sol*, **13**, 25–44.

Bradshaw, M. E. & Jones, A. V. (1976). *Phytosociology in upper Teesdale: Guide to the vegetation maps of Widdybank Fell*. University of Durham.

Breymeyer, A. (ed.) (1971). Productivity investigation of two types of meadows in the Vistula Valley. *Ekologia Polska*, **19**, 93–261.

British Ecological Society Summer Meeting at Bangor, North Wales, 3–8 July, 1968. Y-Wyddfa (Snowdon) National Nature Reserve, Field Guide. Cyclostyle.

Clark, F. E. & Paul, E. A. (1970). The microflora of grassland. *Advances in Agronomy*, **22**, 375–435.

Cragg, J. B. (1961). Some aspects of the ecology of moorland animals. *Journal of Ecology*, **49**, 477–506.

Davey, P. M. (1954). Quantities of food eaten by the desert locust *Schistocerca gregaria* (Forsk.) in relation to growth. *Bulletin of Entomological Research*, **45**, 539–51.

Doskočil, K. & Hůrka, K. (1962). Entomofauna louky (svaz *Arrhenatherion elatioris*) a její vývoj. (Entomofauna der Wiese (Verband *Arrhenatherion elatioris*) und ihre Entwicklung.) *Rozpravy Československé akademie věd, MPV*, **72**, 1–110.

Druzina, V. D. (1972). Seasonal dynamics of ash elements of above-ground plant mass of a meadow community. *Rastitel'nye resursy*, **8**, 397–403. (In Russian.)

Dykyjová, D. (ed.) (1970). *Productivity of Terrestrial Ecosystems. Production Processes.* Czechoslovak IBP/PT–PP Report 1, Prague.

Ellenberg, H. (ed.) (1971). Integrated experimental ecology. Methods and results of ecosystem research in the German Solling Project. In *Ecological Studies*, **2**, 1–214. Berlin: Springer-Verlag.

Evdokimova, T. I. & Rudina, L. A. (1958). The role of grassland vegetation in a soil formation process under the conditions of the Moscow river flood land. *Pochvovedenie*, **9**, 90–7. (In Russian.)

Fiala, K. (1973). Underground biomass and estimation of annual rhizome increments in two polycormones of *Phragmites communis* Trin. In *Littoral of the Nesyt Fishpond*, ed. J. Květ. *Studie ČSAV*, **15**, 83–7.

Flora URSS (Flora Unionis Rerumpublicarum Socialisticarum Sovieticarum) (1964). Indices Alphabetici I–XXX. Moscow: Nauka Publishing House.

Franz, H. (1942). Untersuchungen über die Kleintierwelt ostalpiner Böden. I. Die freilebenden Nematoden. *Zoologische Jahrbücher. Abteilung für Systematik, Ökologie und Geographie der Tiere*, **75**, 369–546.

Gloser, J. (1976). Photosynthesis and respiration of some alluvial meadow grasses: response to irradiance, temperature and CO_2 concentration. *Acta Scientiarum Naturalium Academiae Scientiarum Bohemoslovacae Brno, N.S.*, **10**, 1–39.

Graff, O. (1971). Stickstoff, Phosphor und Kalium in der Regenwurmlosung auf der Versuchsfläche des Sollingprojektes. *Annales de Zoologie*, **3**, 503–11.

Grieg-Smith, P. (1964). *Quantitative Plant Ecology*. 2nd edn. London: Butterworths.

Gromadzka, J. & Trojan, P. (1967). Comparison of the usefullness of an entomological net, photoeclector and biocenometer for investigation of entomocenoses. *Ekologia Polska A*, **15**, 505–29.

Gyllenberg, G. (1969). The energy flow through a *Chorthippus porallelus* (Zett.) (Orthoptera) population on a meadow in Tvärminne, Finland. *Acta Zoologica Fennica*, **123**, 1–74.

Gyllenberg, G. (1970). Energy flow through a simple food chain of a meadow ecosystem in four years. *Annales Zoologici Fennici*, **7**, 283–9.

Gyllenberg, G. (1972). Some preliminary models for the energy flow of a *Chorthippus parallelus* population. *Acta Zoologica Fennica*, **134**, 1–25.

Haas, H. (1972). Schlüpfabundanz und Schlüpfphänologie von Insekten auf einer Wiese im Solling. Diplomarbeit Universität Göttingen.

Harmsen, G. W. (1964). Some aspects of nitrogen metabolism in soil. In *Abstracts of Papers. III. Soil Biology, VIIIth International Congress of Soil Science, Bucharest, 31 Aug.–9 Sept.*, 9–10.

Harmsen, G. W. & Schreven, D. A. (1955). Mineralization of organic matter in soil. *Advances in Agronomy*, **7**, 299–398.

Hartmann, P. (1974). Die Staphylinidenfauna verschiedener Waldbestände und einer Wiese des Solling. Diplomarbeit Universität Göttingen.

Heal, O. W. & Perkins, D. F. (1976). IBP studies on montane grasslands and moorlands. *Philosophical Transactions of the Royal Society of London*, **274B**, 295–314.

Hédin, L. (1973). Recherches sur la production primaire de la prairie de Borculo. In *Programme Biologique International, Le Haras du Pin, 2*, pp. 33–48. Rouen: Laboratoire de Zoologie INRA.

Hédin, L., Kerguelen, M. & de Montard, F. (1972). *Ecologie de la Prairie Permanente Française*. Paris: Masson et Cie.

Heikinheimo, O. & Raatikainen, M. (1962). Comparison of suction and netting methods in population investigations concerning the fauna of grass leys and cereal fields particularly in those concerning the leafhopper *Calligypona pellucida*. *Publications of the Finnish State Agricultural Research Board*, **191**, 1–31.

Hinton, J. M. (1971). Energy flow in a natural population of *Neophilaenus lineatus*. *Oikos*, **22**, 155–71.

Ierusalimskiy, N. D. (1949). *Nitrogen and Vitamin Nutrition of Microbes*. Moscow: Academy of Sciences of the USSR Publishing House. (In Russian.)

Ignatenko, I. V. & Kirillova, V. P. (1970). The change of total reserves of plant biomass by various regimes of utilization of the shallow grass-forb grass communities. In *Meadow Phytocenosis and Its Dynamics, Geobotanica 18*, pp. 205–11. Leningrad: Nauka Publishing House. (In Russian.)

Iwaki, H. & Midorikawa, B. (1968). Principles for estimating root production in herbaceous perennials. In *Methods of Productivity Studies in Root Systems and Rhizosphere Organisms*, pp. 72–8. Moscow: USSR Academy of Sciences.

Iwaki, H., Monsi, M. & Midorikawa, B. (1966). Dry matter production of some herb communities in Japan. In *The 11th Pacific Science Congress Tokyo*.

Iwanami, Y. (1973). Studies on burning temperatures of grasslands. In *Report of Institute for Agricultural Research, Tohoku University*, **24**, 59–105.

Jagnow, G. (1958). Untersuchungen über Keimzahl und biologische Aktivität von Wiesenböden. *Zeitschrft für Pflanzenernährung, Düngung und Bodenkunde*, **82**, 50–68.

Jakrlová, J. (1968). Flooded meadow communities. An analysis of their productivity in a wet year. *Folia Geobotanica et Phytotaxonomica*, **3**, 345–54.

Jakrlová, J. (1971). Flooded meadow communities. An analysis of productivity in a dry year. *Folia Geobotanica et Phytotaxonomica*, **6**, 1–27.

Jakubzcyk, H. (1969). Variations of microbiological activity in a meadow community. *Ekologia Polska A*, **17**, 856–78.

Jankowska, K. (1971). Net primary production during a three-year succession on an unmowed meadow of the Arrhenatheretum elatioris plant association. *Bulletin de l'Académie Polonaise des Sciences, Cl. II*, **19**, 789–94.

Jankowska, K. (1975). The ecology and primary production of the fresh meadow in the Ojców National Park and of the xerothermic grassland in the Skowronno steppe reserve near Pińczów. *Studia Naturae A*, **11**, 1–80.

Kajak, A. (1974). Analysis of the transfer of carbon. In *Analysis of a sheep pasture ecosystem in the Pieniny Mountains (the Carpathians)*, ed. K. Petrusewicz. *Ekologia Polska*, **22**, 711–32.

Kajak, A. & Olechwicz, E. (1970). The role of web spiders in elimination of Diptera in the meadow ecosystem. *Bulletin of the Museum of History and Nature, Series 2*, **41**, Supplement 1, 233–6.

Ketner, P. (1972). *Primary production of salt-marsh communities on the Island of Terschelling in the Netherlands. Verhandelingen 5*. Research Institute for Nature Management.

Klapp, E. (1965). *Grünlandvegetation und Standort*. Berlin: P. Parey Verlag.

Koike, K. & Yoshida, S. (1969). Seasonal changes of above-ground standing crop of *Miscanthus sinensis* community in IBP-subarea. In *Studies on the Productivity and Conservation of Grassland Ecosystems*, ed. M. Numata. Chiba: JIBP–PTG Committee.

Kontkanen, P. (1954). Studies on insect populations. I. The number of generations of some leafhoppher species in Finalnd and Germany. *Archivum Societatis Zoologicae Botanicae Fennicae*, **8**, 150–6.

Kopčanová, L., Řehořková, V. & Ševlíková, T. (1973). The influence of fertilizers on microbial activity in soils under permanent grass stand. *Agricultural University Nitra, Internal Report*, pp. 1–150. (In Slovakian.)

Kotańska, M. (1968). Dynamics of the standing crop of underground plant organs in several meadow communities in the Ojców National Park. In *Contributions from the Meeting on Primary Productivity*, ed. A. Medwecka-Kornaś, pp. 21–8. Kraków: Polish Academy of Sciences.

Kotańska, M. (1970). Morphology and biomass of the underground organs of plants in grassland communities of the Ojcóv National Park. *Studia Naturae*, **4**, 1–107.

Kotańska, M. (1973). Productivity of underground plant organs of some meadow and marsh communities of the Niepołomice Forest (southern Poland). *Bulletin de l'Académie Polonaise des Sciences, Cl. II*, **21**, 555–60.

Kotańska, M. (1975). Seasonal changes of biomass of underground plant organs in some meadow communities. *Zeszyty Naukowe Universytetu Jagiellońskiego, Prace Botaniczne*, **3**, 23–47. (In Polish.)

Kovács-Láng, E. (1970). Fractional humus investigation on soils under sward communities (Festucetum vaginatae danubiale, Festucetum Wagneri) growing on sandy sites. *Annales Universitatis Scientiarum Budapestinensis de Rolando Eötvös nominatae, Sectio biologica*, **12**, 163–70.

Kovács-Láng, E. (1974). Examination of dynamics of organic matter in a perennial open sandy steppe-meadow (Festucetum vaginatae danubiale) at the Csévharaszt IBP sample area (Hungary). *Acta Botanica Academiae Scientiarum Hungaricae*, **20**, 309–26.

Kovács-Láng, E. & Szabó, M. (1971). Changes of soil humidity and its correlation to phytomass production in sandy meadow associations. *Annales Universitatis Scientiarum Budapestinensis de Rolando Eötvös nominatae, Sectio Biologica*, **13**, 115–26.

Lieth, H. (1970). Phenology in productivity studies. In *Analysis of Temperate Forest Ecosystems*, ed. D. E. Reichle, *Ecological Studies*, **1**, 29–46. Berlin: Springer Verlag.

Loquet, M. (1973). *Étude de l'activité microbiologique d'une prairie permanente. Programme Biologique International. Le Haras du Pin, pp.* 2–26. Rouen: Laboratoire de Zoologie, INRA.

Lutman, J. (1977). The role of slugs in an *Agrostis-Festuca* grassland. In *Production Ecology of British Moors and Montane Grasslands*, ed. O. W. Heal & D. F. Perkins. Ecological Studies Series 27. Berlin: Springer Verlag.

Marchand, H. (1953). Die Bedeutung der Heuschrecken und Schnabelkerfe als Indikatoren verschiedener Graslandtypen. *Beiträge zur Entomologie*, **3**, 116–62.

Masclet, A. & Duval, Y. (1972). Étude du sol: des parcelles P.B.I. Domaine de Borculo. In *Programme Biologique International, Le Haras du Pin*, *1*, 9–24. Rouen: Laboratoire de Zoologie, INRA.

Medwecka-Kornaś, A. (ed.) (1967). Ecosystem studies in a beech forest and meadow in the Ojców National Park. *Studia Naturae A*, **1**, 1–213. (In Polish.)

Miroshnichenko, E. D. (1973). On the problem of litter decomposition in meadows. *Botanicheskii zhurnal*, **58**, 402–12. (In Russian.)

Miroshnichenko, E. D., Pavlova, T. V. & Ponyatovskaya, V. M. (1972). Decomposition of vegetative mass in dry meadows of Leningrad District (Karelian Isthmus). *Botanicheskii zhurnal*, **57**, 533–40. (In Russian.)

Monsi, M. (1968). Mathematical models of plant communities. In *Functioning of Terrestrial Ecosystems at the Primary Production Level*, ed. F. E. Eckardt, *Natural Resources Research*, **5**, 131–49. Paris: Unesco.

Monsi, M. & Saeki, T. (1953). Über den Lichtfaktor in den Pflanzengesellschaften und seine Bedeutung für die Stoffproduktion. *Japanese Journal of Botany*, **14**, 22–52.

Müller, H. J. (1957). Über die Diapause von *Stenocranus minutus*. *Beiträge zur Entomologie*, **7**, 203–6.
Mutoh, N., Yoshida, K., Yokoi, Y., Kimura, M. & Hogetsu, K. (1968). Studies on the production processes and net production of *Miscanthus sacchariflorus* community. *Japanese Journal of Botany*, **20**, 67–92.
Nakamura, Y. (1972). Ecological studies on the family Lumbricidae in Hokkaido. I. Ecological distribution. *Japanese Journal of Applied Entomology and Zoology*, **16**, 18–23.
Nielsen, C. O. (1949). Studies on the soil microfauna. II. The soil inhabiting nematodes. *Natura Jutlandica*, **2**, 1–131.
Numata, M. (1966). Some remarks on the method of measuring vegetation. *Bulletin of Choshi Marine Laboratory*, **8**, 71–7.
Numata, M. (1969). Progressive and retrogressive gradient of grassland vegetation measured by degree of succession. *Vegetatio*, **19**, 96–127.
Numata, M. (1970). Geographical distribution and ecology of *Miscanthus sinensis*. *Shin-Noyaku*, **24**, 8–16.
Numata, M. (ed.) (1973). Ecological studies of Japanese Grasslands. Draft of JIBP Synthesis Volume. Chiba: JIBP–PTG Committee.
Numata, M. (ed.) (1975). Ecological studies in Japanese grasslands with special reference to the IBP area. Productivity of terrestrial communities. *JIBP Synthesis*, **13**, 1–275.
Numata, M., Iizumi, S. & Iwaki, H. (1968). *Ecological studies of grassland in the IBP area for PT and CT at Kawatabi, Japan.* Interim Report, pp. 1–15. Chiba: JIBP–PTG Committee.
Oberdorfer, E. (1970). *Pflanzensoziologische Exkursionsflora für Süddeutschland und die angrenzenden Gebiete*. Stuttgart: E. Ulmer Verlag.
Olmi, M. (1968). Cicaline della Risaia da Vicenda Vercellese (*Homoptera Auchenorrhyncha*). *Centr. Ent. Alp Forest Con. Naz. Ric.*, **133**, 247–60.
Parkinson, D., Gray, T. R. G. & Williams, S. T. (1971). *Methods for studying the ecology of soil microorganisms*. IBP Handbook 19. Oxford: Blackwell.
Payne, W. J. (1970). Energy yield and growth of heterotrophs. *Annual Review of Microbiology*, **24**, 17–52.
Petal, J. (1971). Methods of investigating the productivity of ants. *Ekologia Polska*, **20**, 9–22.
Petal, J. (1974). Analysis of a sheep pasture ecosystem in the Pieniny Mountains (The Carpathians). XV. The effect of pasture management on ant population. *Ekologia Polska*, **22**, 679–92.
Petal, J. & Breymeyer, A. (1969). Reduction of wandering spiders by ants. *Bulletin de l'Academie Polonaise des Sciences. Cl. II Agrobiologie, Biologie*, **18**, 239–44.
Petrusewicz, K. (ed.) (1967). *Secondary Productivity of Terrestrial Ecosystems* (*Principles and Methods*). Warszawa: Państwowe Wydawnictwo Naukowe.
Précsényi, I. (1969). Analysis of the primary production (Phytobiomass) in an Artemisio-Festucetum pseudovinae. *Acta Botanica Academiae Scientiarum Hungaricae*, **15**, 309–25.
Précsényi, I. (1971). Turnover rate of phytomass in several plant communities at Újszentmargita. *Acta Botanica Academiae Scientiarum Hungaricae*, **17**, 105–13.
Précsényi, I. (1973). Relationship between structural and functional characteristics in steppe-meadows in Hungary. *Acta Botanica Academiae Scientiarum Hungaricae*, **18**, 155–62.
Řehořková, V. (1974). Biological activity of soils in meadows of Ipel and Slaná Rivers. MS Thesis, Agricultural University Nitra. (In Czech.)

Ricou, G. (1972). Programme Biologique International, Le Haıas du Pin, No. 1. Rouen: Laboratoire de Zoologie INRA.
Ricou, G. (1973). Programme Biologique International, Le Haras du Pin, No. 2. Rouen: Laboratoire de Zoologie INRA.
Ricou, G. (1974). Programme Biologique International, Le Haras du Pin, No. 3. Rouen: Laboratoire de Zoologie INRA.
Ricou, G. (1978). La prairie permanente du Nord-Ouest français. In *Problèmes d'Écologie: Ecosystèmes Terrestres*, pp. 17–74. Paris: Masson et Cie.
Ricou, G. & Douyer, C. (1976). L'action du feu contrôlé en prairies permanentes. Conséquences sur les biocoenoses animales. *Annales Zoologie, Écologie Animale*, **8**, 559–78.
Rodin, L. E. & Smirnov, N. N. (ed.) (1975). *Resources of the Biosphere: Synthesis of the Soviet Studies for the International Biological Programme I.* Leningrad: Nauka Publishing House.
Rothmaler, W. (ed.) (1970). *Exkursionsflora von Deutschland. Kritischer Ergänzungsband. Gefässpflanzen.* Berlin: Volks und Wissen Volkseigener Verlag.
Ruetz, W. F. (1973). The seasonal pattern of CO_2 exchange of *Festuca rubra* L. in a montane meadow community in northern Germany. *Oecologia*, **13**, 247–69.
Rychnovská, M. (ed.) (1972). *Ecosystem study on grassland biome in Czechoslovakia.* Brno: Czechoslovak IBP/PT-PP Report 2.
Rychnovská, M. (1976). Alluvial grassland hydrosere: Primary production and plant processes. *Polish Ecological Studies*, Warszawa, **2**, 103–12.
Rychnovská, M., Květ, J., Gloser, J. & Jakrlová, J. (1972). Plant water relations in three zones of grassland. *Acta Scientiarum Naturalium Academiae Scientiarum Bohemoslovacae Brno, N.S.*, **6**, 1–38.
Saeki, T. (1960). Interrelationship between leaf amount, light distribution and total photosynthesis in a plant community. *Botanical Magazine*, **73**, 55–63.
Schauermann, J. (1973). Zum Energieumsatz phytophager Insekten im Buchenwald. II. Die produktionsbiologische Stellung der Rüsselkäfer (Curculionidae) mit rhizophagen Larvenstadien. *Oecologie*, **13**, 313–50.
Šesták, Z., Čatský, J. & Jarvis, P. G. (ed.) (1971). *Plant Photosynthetic Production. Manual of Methods.* The Hague: Dr W. Junk N.V. Publisher.
Shimada, Y., Kawanabe, S., Kayama, R. & Ito, S. (1973). *Grassland Ecology.* Tokyo: Tsukiji-Shokan.
Shimada, Y. & Numata, M. (1971). Estimation of aboveground biomass by SDR. In *Productivity and Conservation of Grassland Ecosystems*, ed. M. Numata, pp. 20–9. Chiba: JIBP–PTG Committee.
Smalley, A. E. (1960). Energy of salt-marsh grasshopper population. *Ecology*, **41**, 672–7.
Sochava, V. B. (ed.) (1973). Topological aspects in the study of the behaviour of matter in geosystems. Irkutsk; Siberian Department of the USSR Academy of Sciences. (In Russian.)
Speidel, B. (1973). Solling Project of the Deutsche Forschungsgemeinschaft. IBP Grassland/PT Studies. Fort Collins: Colorado State University.
Speidel, B. & Weiss, A. (1972). Zur ober- und unterirdischen Stoffproduktion einer Goldhaferwiese beiverschiedener Düngung. *Angewandte Botanik*, **46**, 75–93.
Sugawara, K., Iizumi, S. & Shimada, Y. (1964). Phenological observation of grassland plant in Kawatabi. *Journal of Japanese Society of Grassland Science*, **9**. 88–96.

Tesařová, M. (1975). Litter disappearance in alluvial meadow plant communities. In *Biodegradation et Humification*, ed. G. Kilbertus *et al.*, 1ère Colloque International, pp. 255–66. Sarreguemines: Pierron.

Theron, J. J. (1951). The influence of plants on the mineralization of nitrogen and the maintenance of organic matter in the soil. *Journal of Agricultural Science*, **41**, 289–96.

Theron, J. J. & Haylett, D. G. (1953). The regeneration of soil humus under a grass ley. *Empire Journal of Experimental Agriculture*, **21**, 86–98.

Tischler, W. (1965). *Agrarökologie*. Jena: VEB Gustav Fischer Verlag.

Titlyanova, A. A. (1971). *Study of the biological cycle in biogeocenoses* (*Methods recommendations*). Novosibirsk: Siberian Branch of the Academy of Sciences of the USSR. (In Russian.)

Titlyanova, A. A. (1972). The variation of chemical composition of plants. *Izvestiya Sibirskogo Otdeleniya AN SSSR*, **4**, 21–32. (In Russian.)

Turček, F. J. (1972). Ecological studies of birds and mammals in mat-grassland. *Ekologia Polska*, **20**, 441–61.

Turner, F. B. (1970). The ecological efficiency of consumer populations. *Ecology*, **51**, 741–2.

Úlehlová, B. (1973*a*). Alluvial grassland ecosystems – habitat characteristics. *Acta Scientiarum Naturalium Academiae Scientiarum Bohemoslovacae Brno*, *N.S.*, **7** (4), 1–41.

Úlehlová, B. (1973*b*). Alluvial grassland ecosystems – microorganisms and decay processes. *Acta Scientiarum Naturalium Academiae Scientiarum Bohemoslovacae Brno*, *N.S.*, **7** (5), 1–43.

Úlehlová, B. (1974). Distribution of mineral elements in some seminatural alluvial meadow ecosystems. *Rostlinná výroba*, **20**, 533–41.

Úlehlová, B. (1976). Alluvial grassland hydrosere: Decomposition processes. *Polish Ecological Studies*, Warszawa, **2**, 113–9.

Úlehlová, B., Klimo, E. & Jakrlová, J. (1976). Mineral cycling in alluvial forest and meadow ecosystems in southern Moravia, Czechoslovakia. *International Journal of Ecology and Environmental Sciences*, *Jaipur*, **2**, 15–25.

Vagina, T. A. & Shatokhina, N. G. (1971). Distinctive features of plant biomass accumulation in various types of grassland vegetation of the Baraba forest-steppe zone. In *Geobotanical Investigations in Western and Middle Siberia*, pp. 163–90. Novosibirsk: Nauka Publishing House, Siberian Branch. (In Russian.)

Vasiliu, L. (1971). Cercetari sinecologice cantitative asupra artropodelor din pajisti (Copşa Mica şi Blajel, Judeţul Sibiu). *Studii şi Cercetari de Biologie, Seria Zoologie*, **23**, 269–75.

Waloff, N. & Solomon, M. G. (1973). Leafhoppers of acidic grassland. *Journal of Applied Ecology*, **10**, 189–212.

Watch, N. E. (1968). Relationships between assimilation efficiencies and growth efficiencies for aquatic consumers. *Ecology*, **49**, 755–9.

Wetzel, M. (1966). Pflanzenbestand, Nutzung und Futterwert auf Grünland des norddeutschen Küstenvorlandes. Kali-Briefe, Fachgebiet 4, Hannover, **3**, 1–8.

White, E. G. & Watson, R. N. (1972). A food consumption study of three New Zealand alpine grasshopper species. *New Zealand Journal of Agricultural Research*, **15**, 867–77.

Woldendorp, J. W. (1963). The influence of living plant on denitrification. *Mededelingen Landbouwhogeschool*, **63**, 1–100.

Zlotin, R. I. & Khodashova, K. S. (1973). Effect of animals on orthotrophic cycle of biological turnover. In *Problems of Biogeocenology*, pp. 105–17. Leningrad: Nauka Publishing House. (In Russian.)

References to Part IV

Agarwal, A. M. (1973). Ecological studies in the mineral circulation in the grasslands of 'Bhata' soils of Bilaspur (M.P.). Ph.D. Thesis, Ravishankar University, Raipur, India.

Ahuja, L. D. (1964). Effect of management practices on range improvement reseeding. *Annual Progress Report of Central Arid Zone Research Institute, Jodhpur, 1962–63.*

Ahuja, L. D. (1972). Range management in arid regions. *Bulletin of the Indian National Science Academy*, **44**, 95–102.

Ahuja, L. D. & Bhimaya, C. P. (1966). Development of pastures and their management in W. Rajasthan. *Gosamvardhan*, **14**, 16–21.

Ahuja, L. D., Bhimaya, C. P. & Prajapati, M. C. (1967). Preliminary studies on effect of different intensities of grazing stress on a desert rangeland. *Proceedings XIth Silvicultural Conference, Dehradun.*

Ahuja, L. D., Vishwanathan, M. K., Vyas, K. K. & Kundan Lal (1974). Growth of sheep of Chokla breed under different systems of grazing on rangelands in the arid zone of W. Rajasthan (India). *Annals of Arid Zone, Jodhpur*, **13**, 259–65.

Ambasht, R. S. (1970). Conservation of soil through plant cover on certain alluvial slopes in India. In *Proceedings of the International Union for Conservation of Nature and Natural Resources. XI. Technical Meeting*, pp. 44–8.

Ambasht, R. S., Maurya, A. N. & Singh, U. N. (1972). Primary production and turnover in certain protected grasslands of Varanasi, India. In *Papers from a symposium on tropical ecology with an emphasis on organic production*, ed. P. M. Golley & F. B. Golley, pp. 43–50. Athens, USA: University of Georgia.

Beard, J. S. (1967). Some vegetation types of tropical Australia in relation to those of Africa and America. *Journal of Ecology*, **55**, 271–90.

Bhaskaran, A. R. & Chakravarty, D. D. (1965). A preliminary study on the variation in the soil binding capacity of some grass roots. *Indian Journal of Agronomy*, **10**, 326–30.

Bhimaya, C. P. & Ahuja, L. D. (1969). Criteria for determining condition class of rangelands in W. Rajasthan (India). *Annals of Arid Zone, Jodhpur*, **8**, 73–8.

Bhimaya, C. P., Ahuja, L. D., Prakash, M., Gopinath, C. & Vangani, N. S. (1966). The economics and efficiency of different types of fencing for soil conservation in W. Rajasthan (India). *Annals of Arid Zone, Jodhpur*, **5**, 159–72.

Bhimaya, C. P., Rege, N. D. & Srinivasan, V. (1956). Preliminary studies on the role of grasses in soil conservation in the Nilgiris. *Journal of Soil and Water Conservation, India*, **4**, 113–7.

Billore, S. K. (1973). Net primary production and energetics of a grassland ecosystem at Ratlam, India, Ph.D. Thesis, Vikram University, Ujjain, India.

Bokhari, U. G. & Singh, J. S. (1975). Standing state and cycling of nitrogen in soil vegetation components of prairie ecosystems. *Annals of Botany*, **39**, 273–85.

Bor, N. L. (1960). *Grasses of Burma, Ceylon, India and Pakistan*. Oxford: Pergamon Press.

Bourlière, F. & Hadley, M. (1970). The ecology of tropical savannas. In *Annual Reviews of Ecology and Systematics*, **1**, 125–52. Palo Alto: Annual Reviews Inc.

Breymeyer, A. (1974). Structure of the tropical grassland ecosystem in Panama. In

Proceedings of IBP Symposium and Synthesis Meetings on Tropical Grassland Biome, pp. 41–4. Varanasi: Banaras Hindu University.

Brian, M. V. (ed.) (1977). *Production Ecology of Ants and Termites*. IBP Synthesis Series 13. Cambridge University Press.

Chakravarty, A. K. & Bhati, G. N. (1968). Selection of grasses and legumes for pasture of arid and semi-arid zone. I. Variation of morphological and physiological characters in the different strains for forage production. *Indian Forester*, **94**, 667–74.

Chakravarty, A. K., Debroy, R., Verma, C. M. & Das, R. B. (1966). Study on the pasture establishment technique. I. Effect of seed rate, methods of sowing and seed treatments on seedling emergence in *Cenchrus ciliaris* and *Lasiurus sindicus*. *Annals of Arid Zone, Jodhpur*, **5**, 145–58.

Chakravarty, A. K., Ram, R. & Singh, K. C. (1970). Grazing studies in the arid and semi-arid zones of Rajasthan. VII. Utilization of vegetation cover, grazing behaviour of sheep and seasonal variation of crude protein content of plants in different pastures. *Annals of Arid Zone, Jodhpur*, **9**, 10–16.

Chakravarty, A. K. & Verma, C. M. (1966). Effect of pelleting on germination of *Lasiurus sindicus* seeds. *Annals of Arid Zone, Jodhpur*, **7**, 265–9.

Chakravarty, A. K. & Verma, C. M. (1968). Germination of promising desert grass seeds under different depths of sowing in sandy soil. *Annals of Arid Zone, Jodhpur*, **7**, 75–81.

Chakravarty, A. K. & Verma, C. M. (1970). Study on the pasture establishment technique. V. Effect of reseeding on natural pasture with *Cenchrus ciliaris* by different soil working methods and fertilizer treatments on pasture production. *Annals of Arid Zone, Jodhpur*, **9**, 236–44.

Chatterjee, B. N. & Sen, M. K. (1964). Studies on the effect of grasses on red soils of Ranchi (India). *Journal of the British Grassland Society*, **19**, 340–2.

Choudhary, V. B. (1967). Seasonal variation in standing crop and energetics of *Dichanthium annulatum* grassland at Varanasi. Ph.D. Thesis. Banaras Hindu University, Varanasi, India.

Choudhary, V. B. (1972). Seasonal variation in standing crop and net above-ground production in *Dichanthium annulatum* grassland at Varanasi. In *Papers from a Symposium on Tropical Ecology with an Emphasis on Organic Production*, ed. P. M. Golley & F. B. Golley, pp. 51–3. Athens, USA: University of Georgia.

Dabadghao, P. M. (1959). Studies on determination of carrying capacity of natural grasslands at Pali. *Scientific Report, Central Arid Zone Research Institute, Jodhpur*.

Dabadghao, P. M. & Das, R. B. (1960–3). *Annual Reports of Central Arid Zone Research Institute, Jodhpur*.

Dabadghao, P. M. & Shankarnarayan, K. A. (1970). Studies of *Iseilema*, *Sehima* and *Heteropogon* communities of the *Sehima–Dichanthium* zone. *Proceedings of the XI International Grassland Congress*, pp. 36–8. Brisbane: University of Queensland Press.

Dabadghao, P. M. & Shankarnarayan, K. A. (1973). *The grass cover of India*. New Delhi: Indian Council of Agricultural Research.

Das, R. B., Dabadghao, P. M. & Debroy, R. (1964). Studies on the height/weight relationship of desert range grasses of India. *Journal of the British Grassland Society*, **19**, 429–33.

Dash, M. C. (1974). Numbers and distribution of Testacea (Protozoa) in tropical grassland soils from southern Orissa, India. *Indian Biologist*, **7**, 53–6.

Dash, M. C. & Cragg, J. B. (1972). Selection of microfungi by Enchytraeidae (Oligochaeta) and other members of the soil fauna. *Pedobiologia*, **12**, 282–6.

References

Dash, M. C. & Patra, U. C. (1977). Density, biomass and energy budget of a tropical earthworm population from a grassland site in Orissa, India. *Revue d'Ecologie et de Biologie du Sol*, **14**, 461–71.

Dash, M. C., Patra, U. C. & Thambi, A. V. (1974). Comparison of primary production of plant material and secondary production of oligochaetes in tropical grassland of southern Orissa, India. *Tropical Ecology*, **15**, 16–22.

Duthie, J. F. (1903–20). *Flora of the Upper Gangetic Plain and of the Adjacent Siwalik and Sub-Himalayan Tracts.* Calcutta: Botanical Survey of India. (Reprinted, 1960.)

Dutta, T. R. & Pandey, R. K. (1971). Brush killer for *Sehima–Dichanthium* pastures in India. *Down to Earth*, **27**, 16–8.

Dwivedi, R. Shankar (1959). Soil fungi on grasslands of Varanasi. Ph.D. Thesis. Banaras Hindu University, Varanasi, India.

Dwivedi, R. Shankar (1965). Ecology of soil fungi of some grasslands of Varanasi. I. Edaphic factors and fungi. *Proceedings of the National Academy of Sciences (India)*, **35 B**, 255–74.

Dwivedi, R. Shankar (1970). Decomposition aspects of the ecosystems in Ecology. In *The study of ecosystems*, ed. K. C. Misra *et al.* pp. 195–238. Allahabad: A. H. Wheeler & Co.

Dwivedi, R. Snehi (1970). Evaluation of the methods of measuring productivity in *Triticum aestivum* and *Dichanthium annulatum. Indian Journal of Agricultural Sciences*, **40**, 81–8.

Edroma, E. L. (1974). *Third Annual Research Report, July, 1973–June, 1974.* Uganda: Institute of Ecology.

Egunjobi, J. K. (1974). Annual burning and savanna productivity in Nigeria. In *Proceedings of IBP Symposium and Synthesis Meetings on Tropical Grassland Biome*, p. 49. Varanasi: Banaras Hindu University.

Erasmus, I. I. (1969). A review of work done on *Themeda–Arundinella* Grasslands at the Soil Conservation Research Centre, Chandigarh. *Ist Workshop on Forage Utilization*, Hissar, pp. 1–11.

Ganguly, B. N., Kaul, R. N. & Nambiar, K. T. N. (1964). Preliminary studies on a few top feed species. *Annals of Arid Zone, Jodhpur*, **3**, 33–7.

Gill, J. S. (1975). Herbage dynamics and seasonality of primary productivity at Pilani, Rajasthan. Ph.D. Thesis, Birla Institute of Technology and Science, Pilani, India.

Gillon, Y. & Gillon, D. (1973). Recherches ecologiques sur une savanna Sahelienne du Ferlo Septentroieonal, Senegal: Données quantitative Sur les Arthropodes. *Terre et Vie*, **27**, 297–323.

Gillon, Y. & Gillon, D. (1974). Recherches écologiques sur une savane Sahélienne du Ferlo Septentrional, Sénégal: Données quantitatives sur les Ténébrionides. *Terre et Vie*, **28**, 296–306.

Gupta, R. K. (1974). Grassland productivity under moisture stress in the arid zone of Rajasthan. In *Proceedings of IBP Symposium and Synthesis meetings on Tropical Grassland Biome*. Varanasi: Banaras Hindu University.

Gupta, R. K. (1975). Range ecology & development. In *Environmental Analysis of Thar Desert*, ed. R. K. Gupta & Ishwar Prakash. Dehradun: English Book Depot.

Gupta, R. K., Saxena, S. K. & Sharma, S. K. (1972). Aboveground productivity of grassland at Jodhpur, India. In *Proceedings of Symposium on Tropical Ecology with an Emphasis on Organic Production*, ed. P. M. Golley & F. B. Golley, pp. 75–93. Athens: University of Georgia.

Heal, O. W. (1964). Observations on seasonal and spatial distribution of Testacea (Protozoa: Rhizopoda) in *Sphagnum*. *Journal of Ecology*, **33**, 395–412.

Heal, O. W. (1965). Observations on testate amoebae (Protozoa: Rhizopoda) from Signy Island, South Orkney Islands. *British Antarctic Survey Bulletin*, **6**, 43–7.

Indian Grassland and Fodder Research Institute (1971). *Annual report*. India: Jhansi.

Jain, S. K. (1971). Production studies in some grasslands of Sagar. Ph.D. Thesis, Saugar University, Sagar, India.

Khanna, P. K. (1964). The succession of fungi on some decaying grasses. Ph.D. Thesis, Banaras Hindu University, Varanasi, India.

Klötzli, F. (1975). Zur Waldfähigkeit der Gebirgssteppen Hoch-Semiens (Nordäthiopien). Beiträge zur naturkundlichen Forschung in Südwestdeutschland, **34**, 131–47.

Kumar, A. (1971). Structure and net primary community production in the terrestrial herbaceous vegetation at Pilani (Rajasthan), with special reference to grasses. Ph.D. Thesis, Birla Institute of Technology and Science, Pilani, India.

Kumar, A. & Joshi, M. C. (1972). The effect of grazing on the structure and productivity of vegetation near Pilani, Rajasthan, India. *Journal of Ecology*, **60**, 665–75.

Kumarek, E. V., Sr. (ed.) (1972). Fire in Africa. *Proceedings of Annual Tall Timbers Fire Ecology Conference*, 22–3 April, 1971. Tallhassee: Tall Timbers Research Station.

Lakhani, K. H. & Satchell, J. E. (1970). Production of *Lumbricus terrestris* L. *Journal of Animal Ecology*, **39**, 473–92.

Lamotte, M., Barbault, R., Gillon, Y. & Lavalle, P. (1974). Production de quelques populations animales dans une savane tropicale de Cote d'Ivoire. In *Proceedings of IBP Symposium and Synthesis Meetings on Tropical Grassland Biome*, pp. 57–60. Varanasi: Banaras Hindu University.

Lepage, M. (1974). Recherches écologiques sur une savane Sahélienne du Ferlo septentrional Sénégal: Influence de la sécheresse sur le peuplement en termites. *Terre et Vie*, **28**, 76–94.

Malhotra, S. P. (1968). Nomads in Indian arid zone. *XXI International Geographical Congress Symposium on Indian Arid Zone*. Jodhpur, India.

Mall, L. P. & Billore, S. K. (1974). Production relations with density and basal area in a grassland community. *Geobios*, **1**, 84–6.

Mall, L. P., Billore, S. K. & Misra, C. M. (1973). A study of the community chlorophyll content with reference to height and dry weight. *Tropical Ecology*, **14**, 81–3.

Mall, L. P., Misra, C. M. & Billore, S. K. (1973). Primary productivity of grassland ecosystem at Ujjain and Ratlam districts of Madhya Pradesh. *Progress Report Indian National Science Academy Project*. Ujjain: School of Studies in Botany, Vikram University.

Mann, H. S. & Ahuja, L. D. (1974). Primary and secondary production in arid zone rangelands of western Rajasthan. In *Proceedings of IBP Symposium and Synthesis Meetings on Tropical Grassland Biome*, pp. 61–2. Varanasi: Banaras Hindu University.

Mishra, R. R. (1964). Seasonal variation in fungal flora of grasslands of Varanasi. Ph.D. Thesis, Banaras Hindu University, Varanasi.

Misra, C. M. (1973). Primary productivity of a grassland ecosystem at Ujjain. Ph.D. Thesis, Vikram University, Ujjain.

Misra, K. C. (1974). *Manual of Plant Ecology*. Oxford & New Delhi: IBH Publishing Company.

Misra, R. (1972). A comparative study of net primary productivity of dry deciduous forest and grassland of Varanasi, India. In *Papers from a Symposium on Tropical Ecology with an Emphasis on Organic Production*, ed. P. M. Golley & F. B. Golley, pp. 279–94. Athens: University of Georgia.

Mistry, P. C. & Chatterjee, N. B. (1965). Infiltration capacities of soils in Ranchi, India. *Journal of Soil and Water Conservation, India*, **13**, 43–7.

Monnier, Y. (1968). Les effets des feux de brousse sur une savane préforestière de Côte-d'Ivoire. *Études Éburnéennes*, **9**, 1–260.

Morel, G. & Morel, M. (1974). Recherches écologiques sur une savane sahélienne du Ferlo Septentrional, Sénégal: Influence de la Sécheresse de l'année 1972–3. sur l'avifaune. *Terre et Vie*, **28**, 95–123.

Naik, M. L. (1973). Ecological studies on some grasslands of Ambikapur. Ph.D. Thesis, Saugar University, Sagar.

Nielsen, C. O. (1955). Studies on Enchytraeidae. 5, Factors causing seasonal fluctuations in numbers. *Oikos*, **6**, 153–69.

Pandey, A. N. (1971). Effect of burning on *Dichanthium annulatum* grassland. Ph.D Thesis, Banaras Hindu University, Varanasi.

Pandey, H. N. & Kothari, A. (1972). Studies on the root distribution pattern and phosphate uptake rate of cropland and grassland species. In *Biology of Land Plants*, ed. V. Puri, Y. S. Murty, P. K. Gupta & D. Banerji, pp. 36–43. Sarita Prakashan, Meerut, India.

Pandeya, S. C. (1964). Ecology of grasslands of Sagar, Madhya Pradesh: IIb. Composition of the association open to grazing or occupying special habitat. *Journal of the Indian Botanical Society*, **43**, 606–39.

Pandeya, S. C. (1974). Autecology and genecology of Anjan grass (*Cenchrus ciliaris*) complex. *Second Progress Report of the Findings, PL. 480 Research Project*. Rajkot: Saurashtra University.

Peachey, J. E. (1963). Studies on Enchytraeidae (Oligochaeta) of moorland soils. *Pedobiologia*, **2**, 81–95.

Poulet, A. R. (1974). Recherches écologiques sur une savane Sahélienne du Ferlo Septentrional, Senegal: Quelques effets de la sécheresse sur le peuplement mammalien. *Terre et Vie*, **28**, 124–30.

Prajapati, M. C. (1970). Effect of different systems of grazing by cattle on *Lasiurus sindicus–Eluesine–Aristida* grassland in arid regions of Rajasthan vis-à-vis animal production. *Annals of Arid Zone, Jodhpur*, **9**, 114–24.

Prajapati, M. C., Phadke, A. B. & Agarwal, M. C. (1973). Studies on suitability of grasses for protection of field earthen structures in Agra region of Yamuna ravines. *Indian Forester*, **99**, 193–204.

Puri, G. S. (1960). *Indian Forest Ecology*. 2 vols. New Delhi: Oxford Book Co.

Rai, B. (1968). Succession of fungi on decaying leaves of *Saccharum munja* Roxb. Ph.D. Thesis, Banaras Hindu University, Varanasi.

Rao, A. (1970). The role of *Desmondium triflorum* in the production and the nitrogen economy of grasslands at Varanasi. Ph.D. Thesis, Banaras Hindu University, Varanasi.

Ray, S. N. & Nudgul, V. D. (1962). Studies on roughage utilization by cattle and buffaloes. *Indian Journal of Dairy Science*, **15**, 129.

Reynolds, J. W. (1973). Earthworm (Annelida; Oligochaeta) ecology and systematics. In *Proceedings of First Soil Micro-Communities Conference*, Springfield, Virginia, ed. D. L. Dindal, pp. 95–120. Department of Commerce: National Technology Information Service.

Satchell, J. E. (1967). Lumbricidae. In *Soil Biology*, ed. A. Burges & F. Raw, pp. 259–322. London & New York: Academic Press.

Satchell, J. E. (1970). Earthworms. In *Methods of Study in Quantitative Soil Ecology*, ed. J. Phillipson, pp. 107–27. IBP Handbook 18. Oxford: Blackwell Scientific Publications.

Shankarnarayan, K. A., Sreenath, P. R. & Dabadghao, P. M. (1969). Studies on the height–weight relationship of six important range grasses of *Sehima–Dichanthium* zone. *Annals of Arid Zone*, Jodhpur, **8**, 61–5.

Shanker, V., Shankarnarayan, K. A. & Rai, P. (1973). Primary productivity, energetics and nutrient cycling in *Sehima–Heteropogon* grassland. I. Seasonal variation in composition, standing crop and net production. *Tropical Ecology*, **14**, 238–51.

Sharma, P. D. (1967). Succession of fungi on decaying *Setaria glauca* Beauv. Ph.D. Thesis, Banaras Hindu University, Varanasi.

Sims, P. L. & Singh, J. S. (1971). Herbage dynamics and net primary production in certain ungrazed and grazed grasslands in North America. In *Preliminary Analysis of Structure and Function in Grassland*, ed. N. R. French, pp. 59–174. Range Science Department Science Series 10. Fort Collins: Colorado State University.

Singh, A. K. (1972). Structure and primary net production and mineral contents of two grassland communities of Chakia Hills, Varanasi. Ph.D. Thesis, Banaras Hindu University, Varanasi.

Singh, J. S. (1967). Seasonal variation in composition, plant biomass, and net community production in grasslands at Varanasi. Ph.D. Thesis, Banaras Hindu University, Varanasi.

Singh, J. S. (1968). Net aboveground community productivity in the grassland at Varanasi. In *Proceedings of a Symposium on Recent Advances in Tropical Ecology*, Part II, ed. R. Misra & B. Gopal, pp. 631–54. International Society for Tropical Ecology, India.

Singh, J. S. (1969*a*). Growth performance and dry matter yield of *Cassia tora* L. as influenced by population density. *Journal of the Indian Botanical Society*, **48**, 141–8.

Singh, J. S. (1969*b*). Growth of *Eleusine indica* (L) Gaertn. under reduced light intensities. *Proceedings of the National Institute of Science, India*, **35B**, 153–61.

Singh, J. S. (1969*c*). Influence of biotic disturbance on the preponderance and interspecific association of two common forbs in the grasslands at Varanasi, India. *Tropical Ecology*, **10**, 59–71.

Singh, J. S. (1973). A compartment model of herbage dynamics for Indian tropical grasslands. *Oikos*, **24**, 367–72.

Singh, J. S. & Misra, R. (1968). Efficiency of energy capture by the grassland vegetation at Varanasi. *Current Science*, India, **37**, 636–7.

Singh, J. S. & Misra, R. (1969). Diversity, dominance, stability and net production in the grasslands at Varanasi. *Canadian Journal of Botany*, **47**, 425–7.

Singh, J. S. & Yadava, P. S. (1971). *Primary productivity in the grassland ecosystem at Kurukshetra.* CSIR Technical Report 38, (73) 162 GAU II, Kurukshetra University.

Singh, J. S. & Yadava, P. S. (1973). Caloric values of important plant and insect species of a tropical grassland. *Oikos*, **24**, 186–94.

Singh, J. S. & Yadava, P. S. (1974). Seasonal variation in composition, plant biomass and net primary productivity of a tropical grassland at Kurukshetra, India. *Ecological Monographs*, **44**, 351–75.

Singh, J. S., Lauenroth, W. K. & Steinhorst, R. K. (1975). Review and assessment of various techniques for estimating net aerial primary production in grasslands from harvest data. *Botanical Review*, **41**, 181–232.

Singh, U. N. & Ambasht, R. S. (1975). Energy conserving efficiency of a forest grassland at Varanasi. *Acta Botanica Indica*, **3**, 132–5.

Soil and Water Conservation Centre (1971). *Development of grasslands in hilly region through soil and water conservation* (*in India*). Muzaffarabad, India.

Stout, J. D. (1962). An estimation of microfaunal populations in soils and forest litter. *Journal of Soil Science*, **13**, 314–20.

Tejwani, K. G., Gupta, S. K. & Mathur, H. N. (1975). *Soil and Water Conservation Research* (1956–71). New Delhi: Indian Council of Agricultural Research.

Tejwani, K. G. & Mathur, H. N. (1972). *Role of grass and legumes in soil conservation*. Soil Conservation Digest, Dehra Dun, India, 21–9.

Thambi, A. V. & Dash, M. C. (1973). Seasonal variation in numbers and biomass of Enchytraeidae (Oligochaeta) populations in tropical grassland soils from India. *Tropical Ecology*, **14**, 228–37.

Thornthwaite, C. M. (1948). An approach towards a rational classification of climate. *Geographical Review*, **38**, 55–94.

Ullah, W., Chakravarty, A. K., Mathur, C. P. & Vangani, N. S. (1972). Effect of contour furrows and contour bunds on water conservation in grasslands of W. Rajasthan. *Annals of Arid Zone, Jodhpur*, **11**, 169–82.

Varshney, C. K. (1972). Productivity of Delhi grassland. In *Papers from a Symposium on Topical Ecology with an Emphasis on Organic Production*, New Delhi, ed. P. M. Golley & F. B. Golley, pp. 27–42. Athens: Institute of Ecology, University of Georgia.

Vasudewaiah, R. D., Singh-Teotia, S. P. & Gupta, D. P. (1965). Runoff and soil loss determination studies at Deochanda Expt. Stn. II. Effect of annually cultivated grain crops and perennial grasses on 5 % slope. *Journal of Soil and Water Conservation in India, Hazaribagh*, **13**, 36–46.

Vuattoux, R. (1970). Observation on the evaluation of the woody and shrubby strata of the Savanna of Lamto (Ivory Coast). *Annales de l'Université d'Abidjan, Serie E: Ecologie*, **3**, 285–315.

Vyas, L. M., Garg, R. K. & Agrawal, S. K. (1972). Net aboveground production in the monsoon vegetation at Udaipur, India. In *Papers from a Symposium on Tropical Ecology with an Emphasis on Organic Production*, New Delhi, ed. P. M. Golley & F. B. Golley, pp. 75–100. Athens: Institute of Ecology, University of Georgia.

Whyte, R. O. (1964). *Grassland and fodder resources of India*. Scientific Monograph 22. New Delhi: Indian Council of Agricultural Research.

Whyte, R. O. (1974). Grasses and grasslands. In *Natural Resources of Humid Tropical Asia*, pp. 239–62. Paris: Unesco.

Wood, T. C. (1974). The effects of clearing and grazing on the termite fauna in tropical savannas and woodlands. In *Proceedings of IBP Symposium and Synthesis Meetings on Tropical Grassland Biome*, pp. 59–60. Varanasi: Banaras Hindu University.

Yadava, P. S. (1972). Primary productivity in a grassland ecosystem at Kurukshetra. Ph.D. Thesis, Kurukshetra University, Kurukshetra.

References to Part V

Agricultural Research Council (1965). *The Nutrient Requirements of Farm Livestock* 2. London.

Alexander, M. (1971). Biochemical ecology of micro-organisms. *Annual Reviews of Microbiology*, **25**, 361–92.

Alexander, M. (1975). Environmental consequences of rapidly rising food output. *Agro-Ecosystems*, **1**, 249–64.

Arnold, G. W. (1964). Factors within plant associations affecting the behaviour and performance of grazing animals. In *Grazing in Terrestrial and Marine Environments*. British Ecological Society No. 4, ed. D. J. Crisp, pp. 133–54. Oxford: Blackwell Scientific Publications.

Bailey, C. G. & Riegert, P. W. (1971). Food preferences of the dusky grasshopper, *Encoptolophus sordidus costalis* (Scudder) (Orthoptera: Acrididae). *Canadian Journal of Zoology*, **49**, 1271–4.

Barley, K. P. (1959). The influence of earthworms on soil fertility. II. Consumption of soil and organic matter by the earthworm *Allolobophora caliginosa* (Savigny). *Australian Journal of Agricultural Research*, **10**, 179–85.

Bartha, R., Lanzilotta, R. P. & Pramer, D. (1967). Stability and effects of some pesticides in soil. *Applied Microbiology*, **15**, 67–75.

Begg, J. E. (1959). Annual pattern of soil moisture stress under sown and native pastures. *Australian Journal of Agricultural Research*, **10**, 518–29.

Bell, M. K. (1974). Decomposition of herbaceous litter. In *Biology of Plant Litter Decomposition*, ed. C. H. Dickinson & G. J. F. Pugh, **1**, 37–67. London: Academic Press.

Black, J. N. (1964). An analysis of the potential production of swards of subterranean clover (*Trifolium subterraneum* L.) at Adelaide, South Australia. *Journal of Applied Ecology*, **1**, 3–18.

Breymeyer, A. & Van Dyne, G. M. (eds.) (1978). *Grasslands, Systems Analysis and Man*. IBP Volume 19. Cambridge University Press.

Carter, E. D. (1965). Some relationships between superphosphate use and consequent animal production from pasture in South Australia. *Proceedings IX International Grassland Congress, Sao Paulo*, pp. 1027–32. Sao Paulo State Department Animal Production.

Chase, F. E. (1958). Manometric studies in agricultural and forest soils. *North American Forest Soils Conference*, pp. 122–9. Guelph: Ontario Agricultural College.

Chase, F. E. & Gray, P. H. H. (1957). Application of the Warburg respirometer in studying respiratory activity in soil. *Canadian Journal of Microbiology*, **3**, 336–49.

Cook, M. G. & Malecky, J. M. (1974). Economic aspects of pasture improvement. A case study in the New England Region of New South Wales. *Wool Economic Research Report 25*. Canberra: Bureau of Agricultural Economics.

Corbett, J. L. & Furnival, E. P. (1976). Early weaning of grazing sheep. *Australian Journal of Experimental Agriculture and Animal Husbandry*. **16**, 149–66.

Dann, P. R. (1972). Fodder crops for sheep in Australia. In *Plants for Sheep in Australia*, ed. J. H. Leigh & J. C. Noble, pp. 175–81. Sydney: Angus & Robertson.

Davidson, R. L. (1964*a*). Natural regeneration of grassland on fallows with and without fertilizers in South Africa. *Empire Journal of Experimental Agriculture*, **32**, 161–5.

Davidson, R. L. (1964*b*). Theoretical aspects of nitrogen economy in grazing experiments. *Journal of the British Grassland Society*, **16**, 273–80.

Davidson, R. L. (1965). Management of sown and natural lovegrass. *Journal of Range Management*, **18**, 214–8.
Davidson, R. L. (1969). Effect of root/leaf temperature differentials on root/shoot ratios in some pasture grasses and clover. *Annals of Botany*, **33**, 561–9.
Donald, C. M. (1970). Temperate pasture species. In *Australian Grasslands*, ed. R. M. Moore, pp. 303–19. Canberra: Australian National University Press.
Dyksterhuis, E. J. (1949). Condition and management of rangeland based on quantitative ecology. *Journal of Range Management*, **2**, 104–15.
Edmund, D. B. & Hoveland, C. S. (1972). A study of the position of grass growing points in two species under different systems of management. *New Zealand Journal of Agricultural Research*, **15**, 7–18.
Engelmann, M. D. (1961). The role of soil arthropods in the energetics of an old field community. *Ecological Monographs*, **31**, 221–38.
Engelmann, M. D. (1966). Energetics, terrestrial field studies, and animal productivity. *Advances in Ecological Research*, **3**, 73–115.
Forbes, R. S. (1974). Decomposition of agricultural crop debris. In *Biology of Plant Litter Decomposition*, ed. C. H. Dickinson & G. J. F. Pugh, **2**, 723–42. London: Academic Press.
Freney, J. R. & Spencer, K. (1960). Soil sulphate changes in the presence and absence of growing plants. *Australian Journal of Agricultural Research*, **11**, 339.
George, J. M., Hutchinson, K. J. & Mottershead, B. E. (1970). Spear thistle (*Cirsium vulgare*) invasion of grazed pastures. *Proceedings XI International Grassland Congress*, Surfers Paradise, pp. 685–8. Brisbane: University of Queensland Press.
Hall, D. M. & Grossbard, Erna (1972). The effects of grazing or cutting a perennial ryegrass and white clover sward on the microflora of the soil. *Soil Biology Biochemistry*, **4**, 199.
Hamilton, B. A., Hutchinson, K. J., Annis, P. C. & Donnelly, J. B. (1973). Relationships between the diet selected by grazing sheep and the herbage on offer. *Australian Journal of Agricultural Research*, **24**, 271–7.
Hamilton, B. A., Hutchinson, K. J. & Swain, F. G. (1972). The effects of fodder conservation on the production of sheep grazing four temperate perennial grasses. *Proceedings Australian Society Animal Production*, **9**, 214–20.
Hartley, W. & Williams, R. J. (1956). Centres of distribution of cultivated pasture grasses and their significance for plant introduction. *Proceedings Seventh International Grassland Congress, Palmerston North, New Zealand*, pp. 190–201.
Healey, I. N. (1967). The energy flow through a population of soil Collembola. In *Secondary Productivity of Terrestrial Ecosystems*, ed. K. Petrusewicz, **2**, 695–708. Warszawa–Krakow: Panstwowe Wydawnictwo Naukowe.
Hemmingsen, A. M. (1960). Energy metabolism as related to body size and respiratory surfaces and its evolution. *Reports of the Steno Memorial Hospital and the Nordisk Insulinlaboratorium*, Volume IX, Copenhagen.
Henzell, E. F. & Ross, P. J. (1973). The nitrogen cycle of pasture ecosystems. In *Chemistry and Biochemistry of Herbage*, ed. G. W. Butler & R. W. Bailey, pp. 222–46. London: Academic Press.
Hilder, E. J. (1963). Performance of several grasses in swards. *Field Station Record, Division of Plant Industry. Commonwealth Scientific and Industrial Research Organization (Australia)*, **2**, 9–24.
Hilder, E. J. & Mottershead, B. E. (1963). The redistribution of plant nutrients through free-grazing sheep. *Australian Journal of Science*, **26**, 88–9.

Hinton, J. M. (1971). Energy flow in a natural population of *Neophilaenus lineatus* (Homoptera). *Oikos*, **22**, 155–71.
Hounam, E. C. (1969). Revised regression equations for estimation of solar radiation over Australia. *Australian Meteorological Magazine*, **17**, 91–4.
Hutchinson, K. J. (1970). The persistence of perennial species under intensive grazing in a cool temperate environment. *Proceedings XI International Grassland Congress*, Surfers Paradise, pp. 611–14. Brisbane: University of Queensland Press.
Hutchinson, K. J. (1971). Productivity and energy flow in grazing/fodder conservation systems. *Herbage Abstracts*, **41**, 1–10.
Jones, R. J. & Sandland, R. L. (1974). The relation between animal gain and stocking rate. Derivation of the relation from the results of grazing trials. *Journal of Agricultural Science*, **83**, 335–42.
Ketellapper, H. J. (1960). The effect of soil temperature on the growth of *Phalaris tuberosa* L. *Physiologia Plantarum*, **13**, 641–7.
King, K. L. & Hutchinson, K. J. (1976). The effects of sheep stocking intensity on the abundance and distribution of mesofauna in pastures. *Journal of Applied Ecology*, **13**, 41–55.
Kitazawa, Y. (1967). Community metabolism of soil invertebrates in forest ecosystems of Japan. In *Secondary Productivity of Terrestrial Ecosystems*, ed. K. Petrusewicz, **2**, 649–62. Warszawa–Krakow: Panstwowe Wydawnictwo Naukowe.
Klekowski, R. Z., Prus, T. & Zyromska-Rudzka, H. (1967). Elements of energy budget of *Tribolium castaneum* (Hbst) in its developmental cycle. In *Secondary Productivity of Terrestrial Ecosystems*, ed. K. Petrusewicz, **2**, pp. 859–79. Warszawa–Krakow: Panstwowe Wydawnictwo Naukowe.
Langlands, J. P. & Bennett, I. L. (1973*a*). Stocking intensity and pastoral production. I. Changes in the soil and vegetation of a sown pasture grazed by sheep at different stocking rates. *Journal of Agricultural Science*, **81**, 193–204.
Langlands, J. P. & Bennett, I. L. (1973*b*). Stocking intensity and pastoral production. III. Wool production, fleece characteristics, and the utilization for maintenance and wool growth by Merino sheep grazed at different stocking rates. *Journal of Agricultural Science*, **8**, 211–8.
Langlands, J. P. & Bowles, J. E. (1974). Herbage intake and production of Merino sheep grazing native and improved pastures at different stocking rates. *Australian Journal of Experimental Agriculture and Animal Husbandry*, **14**, 307–15.
Lawton, J. H. (1970). Feeding and food energy assimilation in larvae of the damselfly *Pyrrhosoma nymphula* (Sulz.) (Odonata: Zygoptera). *Journal of Animal Ecology*, **39**, 669–89.
Luxton, M. (1972). Studies on the oribated mites of a Danish beechwood soil. *Pedobiologia*, **12**, 434–63.
Lynch, J. J. & Marshall, K. J. (1969). Shelter: A factor increasing pasture and sheep production. *Australian Journal of Science*, **32**, 22–3.
MacLean, S. F. Jr. (1973). Life cycle and growth energetics of the arctic crane fly *Pedicia hannai antenatta*. *Oikos*, **24**, 436–43.
McMillan, J. H. & Healey, I. N. (1971). A quantitative technique for the analysis of the gut contents of Collembola. *Revue d'Ecologie et du Biologie du Sol*, **8**, 295–300.
McNeill, S. (1971). The energetics of a population of *Leptopterna dolabrata* (Heteroptera: Miridae). *Journal of Animal Ecology*, **40**, 127–40.
McNeill, S. & Lawton, J. H. (1970). Annual production and respitation in animal populations. *Nature, London*, **225**, 472–4.

Marchant, R. & Nicholas, W. L. (1974). An energy budget for the free-living nematode *Pelodera* (Rhabeditidae). *Oecologia*, **16**, 237–52.

Martin, J. P. (1966). Influence of pesticides on soil microbes and soil properties. In *Pesticides and their Effects on Soils and Water*, ed. S. A. Breth & M. Stelly, pp. 95–108. A.S.A. Special publication 8. Madison: Soil Science Society of America.

Matsumura, F. & Boush, G. M. (1971). Metabolism of insecticides by microorganisms. In *Soil Biochemistry*, ed. A. D. McLaren & J. Skujins, **2**, 320–36. New York: Marcel Dekker.

May, P. F., Till, A. R. & Cumming, M. J. (1972). Systems analysis of ^{35}sulphur kinetics in pasture grazed by sheep. *Journal of Applied Ecology*, **9**, 25–49.

May, P. F., Till, A. R. & Cumming, M. J. (1973). Systems analysis of the effect of application methods on the entry of sulphur into pastures grazed by sheep. *Journal of Applied Ecology*, **10**, 607–26.

Mead, K. J. & Blunt, R. A. (1964). Crop rotations for the New England Tableland. *Agricultural Gazette of New South Wales*, **75**, 871–5, 917.

Menhinick, E. F. (1967). Structure, stability and energy flow in plants and arthropods in a *Sericea lespedeza* stand. *Ecological Monographs*, **37**, 255–72.

Moore, N. W. (1967). A synopsis of the pesticide problem. In *Advances in Ecological Research*, ed. J. B. Cragg, **4**, pp. 75–129. London: Academic Press.

Moore, R. M. (1970). South-eastern temperate woodlands and grasslands. In *Australian Grasslands*, ed. R. M. Moore, pp. 169–90. Canberra: Australian National University Press.

Moulder, B. C. & Reichle, D. E. (1972). Significance of spider predation in the energy dynamics of forest-floor arthropod communities. *Ecological Monographs*, **42**, 473–98.

Nakamura, M. (1965). Bio-economics of some larval populations of pleurostict Scarabaeidae on the flood plain of the River Tamagawa. *Japanese Journal of Ecology*, **15**, 1–18.

Nielsen, C. O. (1949). Studies on the soil microfauna. II. The soil inhabiting nematodes. *Natura Jutlandica*, **2**, 1–131.

Odum, E. P., Connell, C. E. & Davenport, L. B. (1962). Population energy flow of three primary consumer components of old-field ecosystems. *Ecology*, **43**, 88–96.

Paton, D. F. & Hosking, W. J. (1970). Wet temperate forests and heaths. In *Australian Grasslands*, ed. R. M. Moore, pp. 141–58. Canberra: Australian National University Press.

Petal, J. (1967). Productivity and the consumption of food in the *Myrmica laevinodis* Nyl. population. In *Secondary Productivity of Terrestrial Ecosystems*, ed. K. Petrusewicz, **2**, 841–58. Warsawa–Krakow: Panstwowe Wydawnictwo Naukowe.

Petrusewicz, K. & MacFadyen, A. (1970). *Productivity of Terrestrial Animals. IBP Handbook 13*. Oxford: Blackwell Scientific Publications.

Reed, K. F. M. (1972). The performance of sheep grazing different pasture types. In *Plants for Sheep in Australia*, ed. J. H. Leigh & J. C. Noble, pp. 193–204. Sydney: Angus & Robertson.

Reichle, D. E. (1969). Energy and nutrient metabolism of soil and litter invertebrates. In *Productivity of Forest Ecosystems*, ed. P. Duvigneand. *Proceedings Brussels Symposium, 1969*. Unesco, 1971.

Reuss, J. O., Cole, C. V. & Innis, G. S. (1973). US/IBP Grassland Biome technical report 220. Fort Collins: Colorado State University.

Richards, B. N. (1974). *Introduction to the Soil Ecosystem*. Harlow: Longman Group.

Rocks, R. L., Lutton, J. J. & McCabe, T. P. (1973). Estimation of total sulphur^{35} in soil and biological materials using automatic equipment. *Proceedings of the Conference on Science and Technology*. Abstracts pp. 13–4. Adelaide: Australia and New Zealand Association for the Advancement of Science, South Australia, Inc.

Roe, R. (1947). *Preliminary survey of the natural pastures of the New England district of New South Wales and a general discussion on their problems.* Bulletin 210. Commonwealth Scientific and Industrial Research Organisation, Australia.

Roe, R., Southcott, W. H. & Turner, H. N. (1959). Grazing management of native pastures in the New England region of New South Wales. I. Pasture and sheep production with special reference to systems of grazing and internal parasites. *Australian Journal of Agricultural Research*, **10**, 530–54.

Rogers, L., Lavigne, R. & Miller, J. L. (1972). Bioenergetics of the western harvester ant in the shortgrass plains ecosystem. *Environmental Entomology*, **1**, 763–8.

Satchell, J. E. (1974). Litter – interface of animate/inanimate matter. In *Biology of Plant Litter Decomposition*, ed. C. H. Dickinson & G. J. F. Pugh, **1**, xiii–xliv. London: Academic Press.

Shaw, W. M. & Robinson, B. (1960). Pesticide effects in soils on nitrification and plant growth. *Soil Science*, **90**, 320–3.

Slobodkin, L. B. (1962). Energy in animal ecology. *Advances in Ecological Research*, **1**, 69–101.

Smalley, A. E. (1960). Energy flow of a salt marsh grasshopper population. *Ecology*, **41**, 672–7.

Smith, P. H. (1972). The energy relations of defoliating insects in a hazel coppice. *Journal of Animal Ecology*, **41**, 567–87.

Southcott, W. H., Wheeler, J. H., Hill, M. K. & Hedges, D. A. (1972). Effect of subdivision, stocking rate, anthelmintic and selenium on the productivity of Hereford heifers. *Proceedings of the Australian Society of Animal Production*, **9**, 408–11.

Stiven, G. (1952). Production of antibiotic substances by the roots of a grass, *Trachypogon plumosus* (H.B.K.) Nees, and of *Pentanisia variabilis* (E. Mey) Harv. (Rubiaceae). *Nature, Lond.*, **170**, 712.

Stiven, G. (1957). A study of edaphic factors in secondary succession on the Transvaal Highveld. Ph.D. Thesis, University of the Witwatersrand, Johannesburg.

Striganova, B. R. & Rachmanov, R. R. (1972). Comparative study of the feeding activity of Diplopods in Lenkoran Province of Azerbaijan. *Pedobiologia*, **12**, 430–3.

Tadmor, N. H., Eyal, E. & Benjamin, R. W. (1974). Plant and sheep production on semiarid annual grassland in Israel. *Journal of Range Management*, **27**, 427–32.

Till, A. R. (1976). The efficiency of sulphur utilization in grazed pastures. *Proceedings of the Australian Society of Animal Production*, **11**, 313–6.

Till, A. R. & May, P. F. (1970). Nutrient cycling in grazed pastures. II. Studies on labelling of the grazed pasture system by solid [^{35}S] gypsum and aqueous $Mg^{35}SO_4$. *Australian Journal of Agricultural Research*, **21**, 465–63.

Till, A. R. & May, P. F. (1971). Nutrient cycling in grazed pastures. IV. The fate of sulphur^{35} following its application to a small area in a grazed pasture. *Australian Journal of Agricultural Research*, **22**, 391–400.

Tu, C. M. (1970). Effects of four organophosphate insecticides on microbial activities in soil. *Applied Microbiology*, **19**, 479–84.

Turnbull, A. L. & Nicholls, C. F. (1966). A 'quick-trap' for area sampling of arthropods in grassland communities. *Journal of Economic Entomology*, **59**, 1100–4.

Umbreit, W. W., Burris, R. H. & Stauffer, J. F. (1964). *Manometric Techniques*. 4th edn. Minneapolis: Burgess.

Van Hook, R. I., Jr. (1971). Energy and nutrient dynamics of spider and orthopteran populations in a grassland ecosystem. *Ecological Monographs*, **41**, 1–26.

Van Hook, R. I., Jr. & Dodson, G. J. (1974). Food energy budget for the Yellow-Poplar weevil, *Odontopus calceatus* (Say). *Ecology*, **55**, 205–7.

Vickery, P. J. (1972). Grazing and net primary production of a temperate grassland. *Journal of Applied Ecology*, **9**, 307–14.

Vickery, P. J. & Hedges, D. A. (1972). *Mathematical relationships and computer routines for a productivity model of improved pasture grazed by Merino sheep*. Technical Paper 4. Australia: CSIRO Animal Research Laboratories.

Webb, N. R. & Elmes, G. W. (1972). Energy budget for adult *Steganacarus magnus* (Acari). *Oikos*, **23**, 359–65.

Wheeler, J. L. (1965). The improvement of winter feed in year-long grazing programmes. *Proceedings IX international Grassland Congress*. Sao Paulo, **2**, 975–80. Sao Paulo: State Department of Animal Production.

Whittaker, R. H. (1970). *Communities and Ecosystems*, pp. 82–3. London: Macmillan.

Wiegert, R. G. (1964). Population energetics of meadow spittlebugs (*Philaenus spumarium* L.) as affected by migration and habitat. *Ecological Monographs*, **34**, 217–41.

Wiegert, R. G. (1965). Energy dynamics of the grasshopper populations in old field and alfalfa field ecosystems. *Oikos*, **16**, 161–76.

Wiegert, R. G. & Evans, F. C. (1964). Primary production and the disappearance of dead vegetation on an oldfield in south eastern Michigan. *Ecology*, **45**, 49–62.

Wilkinson, S. R. & Lowrey, R. W. (1973). Cycling of mineral nutrients in pasture ecosystems. In *Chemistry and Biochemistry of Herbage*, ed. G. W. Butler & R. W. Bailey, pp. 247–315. London: Academic Press.

Willoughby, W. M. (1966). Methods of determining optimum stocking rates. *Wool Technology and Sheep Breeding*, **13**, 94–8.

Willoughby, W. M. (1970). Feeding value and utilization of pasture. *Proceedings of the Australian Society of Animal Production*, **8**, 415–21.

Young, B. A. & Corbett, J. L. (1968). Energy requirements for maintenance of grazing sheep measured by calorimetric techniques. *Proceedings of the Australian Society Animal Production*, **7**, 327–34.

References to Part VI

Aleynikova, M. M. & Utrobina, N. M. (1969). Zhivotnoye naselynie pochv v agrobiogeocenozah severhogo Povolzhya. In *Zhivotnoye naselyenie pochv agrobiocenozov i ego izmienenye pod vlyaniem selskohozyastvennogo proizvodstwa*, ed. A. Aleynikova, pp. 3–61. Kazan: Izdatielstvo Kazanskogo Universiteta.

Aristovskaya, T. V. (1972). *Voprosy czislennosti, biomasy i produktivnosti pochvennyh mikroorganizmov*. Leningrad: Nauka Publishing House.

Azzi, G. (1956). *Agricultural Ecology*. London: Constable & Company.

Babiuk, L. A. & Paul, E. A. (1970). The use of fluorescein isothiocyanate in the determination of the bacterial biomass of grassland soil. *Canadian Journal of Microbiology*, **16**, 57–62.

Bray, J. R. (1963). Root production and the estimation of net productivity. *Canadian Journal of Botany*, **41**, 65–72.

Bray, J. R., Lawrence, D. B. & Pearson, L. C. (1959). Primary production in some Minnesota terrestrial communities for 1957. *Oikos*, **10**, 38–49.

Bucherer, H. (1966). Zur Frage der Differentialfärbung lebender und toter Microorganismen mit Acridinorange nach Struger. *Zeitshrift für Biologie*, **15**, 175–84.

Casida, L. E. (1971). Microorganisms in unamended soil as observed by various forms of microscopy and staining. *Applied Microbiology*, **21**, 1040–5.

Dabrowska-Prot, E. & Karg, J. (1975). An ecological analysis of Diptera in agrocenoses. *Polish Ecological Studies*, **1**, 123–37.

Dabrowska-Prot, E., Karg, J. & Ryszkowski, L. (1974). An attempt to estimate the role of invertebrates in agrocenotic economies. In *Ecological Effects of Intensive Agriculture*, ed. L. Ryszkowski, pp. 41–62. Warsaw: Polish Scientific Publishers.

Dębosz, K. (1975). Dynamika węgla i azotu w kwasach huminowych w hodowlach z inoculum glebowym. In *Procesy mikrobiologiczne w glebie*, ed. J. Gołębiowska, pp. 80–2. Poznań: Akademia Rolnicza w Poznaniu.

Freytag, H. R. (1971). To some natural phenomenons in the course of mineralization and humification. *Transactions of International Symposium 'Humus et Planta'*, pp. 95–102. Prague: Academia.

Gołębiowska, J., Margowski, Z. & Ryszkowski, L. (1974). An attempt to estimate the energy and matter economy in the agrocenoses. In *Ecological Effects of Intensive Agriculture*, ed. L. Ryszkowksi, pp. 19–40. Warsaw: Polish Scientific Publishers.

Gołębiowska, J., Sawicka, A. & Skalski, B. (1972). Nitrogen nutrition of some soil bacteria. *Polish Journal of Soil Science*, **5**, 45–51.

Greaves, M. P., Wheatly, R. E., Shepherd, H. & Knight, A. H. (1973). Relationship between microbial populations and adenosine triphosphate in a basin peat. *Soil Biology and Biochemistry*, **5**, 685–7.

Herbich, M. (1969). Primary production of a rye field. *Ekologia Polska*, **17**, 343–50.

Holm-Hansen, O. (1970). ATP levels in algal cells as influenced by environmental conditions. *Plant and Cell Physiology*, **11**, 689–700.

Holm-Hansen, O. & Bootch, Ch. R. (1966). The measurement of adenosine triphosphate in the ocean and its ecological significance. *Limnology and Oceanography*, **11**, 510–19.

Iwaki, H. (1974). Comparative productivity of terrestrial ecosystems in Japan, with emphasis on the comparison between natural and agricultural systems. In *Proceedings of the First International Congress of Ecology*, pp. 40–5. Wageningen: Pudoc.

Jopkiewicz, K. (1972). Zagęszczenie i przepływ energii przez populację dżdżownic. *Zeszyty Naukowe Instytutu Ekologii PAN*, **5**, 227–36.

Kabacik-Wasylik, D. (1975). Research into the number, biomass and energy flow of Carabidae (Coleoptera) communities in rye and potato fields. *Polish Ecological Studies*, **1**, 111–21.

Kaczmarek, W., Kaszubiak, H. & Pędziwilk, Z. (1975). Badania nad zawartoscią ATP w glebie. In *Procesy mikrobiologiczne w glebie*, ed. J. Gołębiowska, pp. 26–8. Poznań: Akademia Rolnicza w Poznaniu.

Karg, J. (1975). Heteroptera of rye and potato cultures. *Bulletin Academie Polonaise des Sciences, CL II*, **23**, 379–82.

Kaszubiak, H., Kaczmarek, W. & Durska, G. (1975). Badania nad rozkładem i syntezą komórek drobnoustrojów glebowych. In *Procesy mickrobiologiczne w glebie*, ed. J. Gołębiowska, pp. 83–5. Poznań: Akademia Rolnicza w Poznaniu.

Kleinhempel, D., Freytag, H. E. & Steinbrenner, K. (1971). Grundlagen und Aspekte der Steuerung des Umsatzes organischer Stoffe in Boden. *Archiw für Bodenfrüchtbarkeit und Pflanzenproduktion*, **15**, 155–76.

Kozlowskaya, L. S. (1965). *Vliyanye bezpozvonochnih zhivotnyh na aktivizacyu azota, fosfora i kalya v torfianyh pochvah. Osobiennosti bolotoobrazovanya v nektoryh lesnyh i predgornyh rayonah Sibirii i Dalnovo Vostoka.* Moscow: Nauka Publishing House.

Kukielska, C. (1972). Produkcja pierwotna pól uprawnych. *Zeszyty Naukowe Instytutu Ekologii PAN*, **5**, 165–8.

Kukielska, C. (1973*a*). Primary productivity of crop fields. *Bulletin Academie Polonaise des Sciences, Cl II*, **19**, 109–15.

Kukielska, C. (1973*b*). Studies on the primary production of the potato field. *Ekologia Polska*, **21**, 813–26.

Lee, C. C., Harris, R. F., Williams, J. D. H., Armstrong, D. E. & Syers, J. K. (1971*a*). Adenosine triphosphate in lake sediments. I. Determination. *Proceedings of the Soil Science Society of America*, **35**, 82–6.

Lee, C. C., Harris, R. F., Williams, J. D. H., Syers, J. K. & Armstrong, D. E. (1971*b*). Adenosine triphosphate in lake sediments. II. Origin and significance. *Proceedings of the Soil Science Society of America*, **35**, 86–91.

Lieth, H. (1962). *Die Stoffproduction der Pflanzendecke*. Stuttgart: Fischer.

Łucrak, J. (1975). Spider communities of the crop fields. *Polish Ecological Studies*, **1**, 93–110.

Macfadyen, A. (1971). Soil metabolism in relation to ecosystem energy flow and to primary and secondary production. In *Methods of Study in Soil Ecology*, ed. J. Phillipson, pp. 167–71. Paris: UNESCO.

McGill, W. B., Paul, E. A., Shields, J. A. & Lowe, W. E. (1973). Turnover of microbial populations and their metabolites in soil. In *Modern Methods in the Study of Microbial Ecology*, ed. T. Rosswall, pp. 293–302. Bulletin 17. Stockholm: Ecological Research Committee.

Martin, J. P. & Haider, K. (1971). Microbial activity in relation to soil humus formation. *Soil Science*, **111**, 54–63.

Mihaylova, E. N. & Nikitina, Z. I. (1972). Chislennost, biomasa mikroorganizmov pochv Onon-agrunskovo stiepiennovo landshafta i vlianye faktorow sriedy na nikh. In *Voprosy chislennosti, biomasy i produktivnosti pochvennyh mikroorganizmov*, ed. T. V. Aristovskaya, pp. 120–5. Leningrad: Nauka Publishing House.

Minderman, G. & Vulto, J. C. (1973). Comparison of techniques for the measurement of carbon dioxide evolution from soil. *Pedobiologia*, **13**, 73–80.

Muszynska, M. (1974). Microbiological turnover of nitrogen in soil. Part I. The influence of some chemical substances on the dynamics of microorganisms active in nitrogen turnover. *Polish Journal of Soil Sciences*, **7**, 159–68.

Muszynska, M. & Andrzejewski, M. (1974). Microbiological turnover of nitrogen in soil. Part II. The influence of plant material on the dynamics of microorganisms active in nitrogen turnover. *Polish Journal of Soil Sciences*, **7**, 159–68.

Parkinson, D., Gray, T. R. G. & Williams, S. T. (1971). *Methods for studying the ecology of soil microorganisms. IBP Handbook 19.* Oxford: Blackwell Scientific Publications.

Pasternak, D. (1974). Primary production of field with winter wheat. *Ekologia Polska*, **22**, 369–78.

Paul, E. A., Biederbeck, V. O. & Rosha, N. S. (1970). Investigation of the metabolism of soil micro-organisms by the use of C^{14}. In *Methods of Study in Soil Ecology*, ed. J. Phillipson, pp. 111–6. Paris: UNESCO.

Pielowski, Z., Pinowski, J., Ryszkowski, L. & Strawiński, S. (1974). Tentative appraisal of the role of vertebrates economically important in an agricultural landscape. In *Ecological Effects of Intensive Agriculture*, ed. L. Ryszkowski, pp. 63–83. Warsaw: Polish Scientific Publishers.

Radomski, Cz., Łykowski, B. & Madany, R. (1974). Preliminary climatological and meteorological characteristics of cultivated fields in the neighbourhood of Turew. In *Ecological Effects of Intensive Agriculture*, ed. L. Ryszkowski, pp. 7–17. Warsaw: Polish Scientific Publishers.

Rauner, Y. L. (1972). *Teplovi bzlans rastitelnovo pokrova*. Leningrad: Gidrometeoizdat.

Ryszkowski, L. (ed.). (1974). *Ecological Effects of Intensive Agriculture*. Warsaw: Polish Scientific Publishers.

Ryszkowski, L. (1975). Energy and matter economy of ecosystems. In *Unifying Concepts in Ecology*, ed. W. H. van Dobben & R. Lowe-McConnell, pp. 109–26. The Hague: Junk.

Shields, J. A., Paul, E. A. & Lowe, W. E. (1974). Factors influencing the stability of labelled microbial materials in soils. *Soil Biology and Biochemistry*, **6**, 31–7.

Tischler, W. (1965). *Agrarökologie*. Jena: Fischer Verlag.

Vömel, A. (1974). *Der Nahstoffumsatz in Boden und Pflanze aufgrund von Lisimeterversuchen*. Berlin & Hamburg: Verlag Prey.

Wasilewska, L. (1974). Number, biomass and metabolic activity of nematodes of two cultivated fields in Turew. *Zeszyty Problemowe Postępów Nauk Rolniczych*, **154**, 419–42.

Wasylik, A. (1974). The mites (Acarina) of potato and rye fields in the environs of Choryn. *Polish Ecological Studies*, **1**, 83–91.

Wójcik, Z. (1973). Productivity of a sandy ryefield. *Ekologia Polska*, **21**, 340–57.

Index

Page numbers in italic type indicate reference to a table, figure etc.

Index

Index